THE UNIVERSE
& HOW IT WORKS

THE UNIVERSE
& HOW IT WORKS

ANDROMEDA

THE UNIVERSE & HOW IT WORKS

Consultant Editor: Robin Kerrod
Art Director: John Ridgeway
Designer: David West
Text Editors: Steve Luck,
    Caroline Sheldrick
Production: Steve Elliott

Media conversion and typesetting:
Peter MacDonald, Una Macnamara and
    Vanessa Hersey

Planned and produced by:
Andromeda Oxford Ltd
11–15 The Vineyard, Abingdon,
Oxfordshire OX14 3PX

Copyright © Andromeda Oxford Ltd 1993

All rights reserved. No part of this publication may be reproduced or utilized in any form or by any means, electronic or mechanical, including photocopying, recording, or by an information storage and retrieval system, without permission in writing from the publisher.

ISBN 1 871869 12 9

Published in Great Britain by
Andromeda Oxford Ltd
This edition specially produced for
Selectabook Ltd

Origination by Hong Kong Reprohouse Co Ltd, Hong Kong

Printed in Hong Kong by Dai Nippon Ltd

Authors:
Robin Kerrod
Peter Lafferty

Contents

INTRODUCTION 7

SOLIDS, LIQUIDS AND GASES 10
Moving molecules • Solids • Liquids • Gases • Changes of state

ATOMS AND MOLECULES 20
Inside the atom • Atomic structure • Particles galore • Radioactivity • Nuclear energy

CHEMICAL ELEMENTS 30
Elements and compounds • The Periodic Table • Forming crystals • Forming molecules • Chemical reactions

MOLECULES IN MOTION 40
Heat and temperature • Travelling heat • Sound

ELECTRICITY 46
Electric charges and fields • Static electricity • Electrons on the move • Using electricity • Semiconductors

MAGNETISM 56
Magnets • Magnetic field • Electricity and magnetism • Electromagnetism • Generators and motors

LIGHT AND RADIATION 64
Reflection and refraction • Curved mirrors and lenses • Waves of light • Colour • Lasers • The electromagnetic spectrum

FORCES, ENERGY AND MOTION 76
Force and movement • Laws of motion • Energy and work • Gravity

MEASURING 84
Standards • Weight • Length and distance • Time • Electricity

SEEING NEAR 94
Early microscopes • Optical microscopes • Electron microscopes

SEEING FAR 100
Early telescopes • Refractors and reflectors • Astronomical observatories • Radio telescopes • Space telescopes

ANALYSING AND PROBING 110
Research • Chemical analysis • Radiation methods • Sound methods • Laser methods • Atom-smashing

THE SOLAR SYSTEM 122
Our star, the Sun • Spots and flares •
The planets

PLANETS NEARBY 128
Mercury • Venus • Mars •
The Martian landscape

FAR-DISTANT WORLDS 136
Jupiter • Jupiter's weather • Saturn •
Rings and ringlets • Outer planets

DEBRIS OF THE SOLAR SYSTEM 146
Asteroids and comets • Sizes and orbits •
Meteors and meteorites

MYRIADS OF MOONS 152
The Moon • The lunar surface •
Moons of the planets • New discoveries

MEASURING THE HEAVENS 160
Scale of the Universe •
The celestial sphere • Constellations

THE STARRY HEAVENS 166
Varieties of stars •
Brightness, size and speed •
The H-R diagram • Variable stars •
Bright nebulae • Between the stars

BIRTH AND DEATH OF STARS 176
A star is born • Stellar life cycles •
Violent death

GALAXIES GALORE 182
The Milky Way •
Spirals, ellipticals and irregulars •
Active galaxies • Quasars •
Clusters of galaxies

BIG BANG, BIG CRUNCH 192
An expanding Universe •
After the Big Bang • Open or closed?

INTO SPACE 198
Beating gravity •
Comsats and weather satellites •
Earth-survey satellites •
Probes to other worlds • *Voyager 2*

SPACE TRANSPORTATION 208
Early days • Apollo • Soyuz •
The Shuttle system • Shuttle hardware •
Shuttle operations

HUMANS IN SPACE 220
Surviving the hazards • Living in orbit •
Spacesuits • Spacewalking

SPACE STATIONS 228
Salyut • *Skylab* • Spacelab • *Mir*

INDEX 236

PICTURE CREDITS 240

Introduction

Studying the Universe and how it works is an infinitely challenging and complex task. Some of the greatest minds of each generation have wrestled with explanations which might bring some order to the variety of different phenomena that surround us. Yet it is only during the latter part of this century that advances in scientific knowledge have allowed us to come a little nearer to understanding the Universe. This encyclopedia brings together in a single volume a wealth of information written by scientists who are also experienced authors. In so doing, it provides a clear introduction for readers of any age to this all encompassing subject. Each topic is copiously illustrated with colour artworks and diagrams as well as photographs thus making the presentation as lively and as meaningful as possible.

The first section of the encyclopedia examines the basic constituents of the Universe: the nature of matter. Matter appears in an infinite variety of different guises: pebbles on the beach, dew on the grass, clouds in the sky, wildlife on the prairies. However, the many millions of different substances that exist are in fact made up of only about 90 simple "building blocks" called chemical elements. These elements exist as tiny particles called atoms, usually linked together in groups in the form of molecules. Atoms contain even tinier particles such as protons and electrons. It is the number and arrangement of these particles inside atoms and molecules that determine the nature of matter, such as whether it occurs as a solid, liquid or gas.

Atoms and molecules store energy. Our understanding of the structure of these particles has allowed us to produce energy in the form of nuclear power. Recent developments in the release of energy from nuclear fusion – a process that occurs in the Sun and other stars – could be the answer to the energy needs of the 21st century. In this first section of the encyclopedia, nuclear energy is explained.

In the second section of *The Universe & How It Works*, you can find out about the forces that govern our Universe. One of these is electricity. Not only is this one of the most useful forms of energy that exists, but it also holds the key to the nature of matter. The atoms of matter consist of electrically charged particles, and are largely held together by electrical attraction. Electricity is thus a great universal force, dictating the nature of matter everywhere. Closely linked to electricity is magnetism. The two go hand in hand, considered within the science of electromagnetism. They also travel together through space as a wave motion, forming a family of electromagnetic waves. Most familiar are the waves by which we see – light. This section reveals why different colours exist and, among other topics, how a rainbow is formed or how a laser works. Finally examined is the force that acts throughout the whole Universe, effectively holding it together: gravity. This attracts objects and makes them move.

Throughout history, we have devised many kinds of instruments that might enable us to discover still more about the Universe. Such instruments enable us, for example, to see things normally invisible to the eye, such as minute viruses, and to hear sounds normally inaudible to the ear, such as ultrasonic waves. This next section examines instruments used by scientists and engineers. They vary from the kind of scales used in Ancient Egypt to the modern world's orbiting space telescopes. The improvement in man's measuring devices over even the last fifty years can be seen in improvements in the accuracy of clocks. In the 1950s, quartz crystal clocks were accurate to within one second every three to thirty years but now improved cesium atomic clocks are accurate to within one second every 30,000 years and scientists in the United States are developing a clock that should be accurate to within one second every 10 billion years! This increasing technology and sophistication has allowed man to develop an understanding of the Universe which would have been unthinkable a hundred years ago. The most obvious example of this can be seen in the exploration of space. The birth of modern astronomy can be dated back to the pioneering studies of Galileo in 1609 when he invented the first telescope. Today, scientists using radio telescopes can detect such things as the faint traces of radiation which were produced when the Universe began more than 15,000 million years ago.

Our home the Earth seems big to us, but it is a mere speck in a boundless Universe. Very much bigger is the Sun. This holds the Earth captive in space by its enormous gravity. It also holds in check eight other large bodies, which we call the planets. The Earth and the planets are the main members of the Sun's family, or Solar System. The family also includes many moons, rock lumps we call asteroids, and comets.

From man landing on the Moon and walking in space to the extraordinary photographs of far distant galaxies, the study of the Universe has produced some of the most fantastic images of this century. This last section of the encyclopedia is illustrated with such images and, supported by diagrams and artworks, vividly brings to life the wonder of space. Each planet is described in detail, from those near the Sun, such as Venus which is so hot and inhospitable that if you were set down on its surface you would at the same time be suffocated

and roasted, to the very extremities of the Solar System where Pluto, actually smaller than our Moon, travels its lonely 248-year orbit.

Deep space has produced an endless series of discoveries from intriguing black holes and variable stars to clues on how the Universe began. The science that studies the Universe is astronomy. It began in the Middle East at least 5,000 years ago, and so is one of the oldest of the sciences. Using the latest telescopes and Space-Age techniques, astronomers are continually coming up with astonishing new discoveries.

Space and what is in it has intrigued human beings since the dawn of civilization. But only since the late 1950s have we had the technology to send objects into space to investigate. A few years later, in 1961, human beings began travelling in space. Since then the space frontier has well and truly been pushed back. Unmanned spacecrafts have travelled billions of kilometres to explore other planets and their moons orbiting in the Solar System. Astronauts have walked on the moon and have spent over a year in space in Earth-orbiting space stations.

The encyclopedia ends with a detailed look at various space-station projects which will allow man to study the Universe from permanently manned satellites above Earth.

Spot facts and fact boxes
Each section of *The Universe & How it Works* is prefaced by a short introduction and a series of memorable "spot facts" which highlight some of the subjects covered in the pages that follow. Scattered throughout the encyclopedia are fact boxes that focus on particular subjects to help reinforce the general text – for example, on how x-rays work or how thermos flasks maintain liquids at the same temperature.

The Universe & How it Works has been designed to explain as clearly and as vividly as possible the marvels of the Universe in which we live, and, at the same time, to encourage readers of all ages to delve still deeper into the mysteries of science.

Solids, liquids and gases

Spot facts

- The densest material in the Universe occurs inside the smallest stars, called neutron stars. A pin's head of this material weighs more than two supertankers.

- Solids are not always denser than liquids. Ice is less dense than water. This is why icebergs float in the sea. Pumice, a kind of lava, can float on water too – the only rock to do so.

- Smoke and fog are mixtures of small particles and gases. These mixtures are called aerosols.

- The molecules in air at ordinary temperatures are moving at about 10 times the speed of the winds in a hurricane.

The matter of the everyday world comes in three familiar states called solids, liquids and gases. Solids, such as a piece of steel, have a fixed shape which stays the same at ordinary temperatures. Liquids like water and milk have no fixed shape. They take on the shape of the container that holds them. Gases, such as the air in a toy balloon, are also shapeless and they fill the whole of their container. These properties of solids, liquids and gases come about because their atoms or molecules – the tiny particles that make them up – are held together with differing strengths. Changes from one state to another occur when the atomic or molecular arrangement changes usually because of a change in temperature.

▶ The water in the lake is a liquid. Ice and snow – water in the solid state – can be seen on the mountain slopes. The gaseous form of water – water vapour – is invisible but it is present in the air. The clouds on the mountain top are formed when water vapour condenses into water droplets.

Moving molecules

A solid is dense and rigid because its molecules are bound tightly in place. The attractive force between the molecules is strong. Because of this, the molecules are held in a more or less fixed position. However, they can vibrate, or move back and forth slightly, in all three directions. Their vibrations increase in speed as the solid is heated. If a solid such as a metal is heated sufficiently, the molecules vibrate so much that the attractive forces cannot hold the metal atoms in their rigid structure.

In a liquid, the attraction between molecules is weaker. They are able to move about like a person can move about in a crowd. This leaves empty spaces into which other molecules can move. This movement of molecules enables the liquid to flow easily.

In gases, the molecules move about completely freely. Their speed increases as the gas is heated. This way of looking at solids, liquids and gases is called the kinetic theory.

▼ A volcanic eruption involves matter in all three states. The volcano and the surrounding land are made of solid rock. Molten rock bubbling out is a liquid. And the fumes blown high into the air are gases.

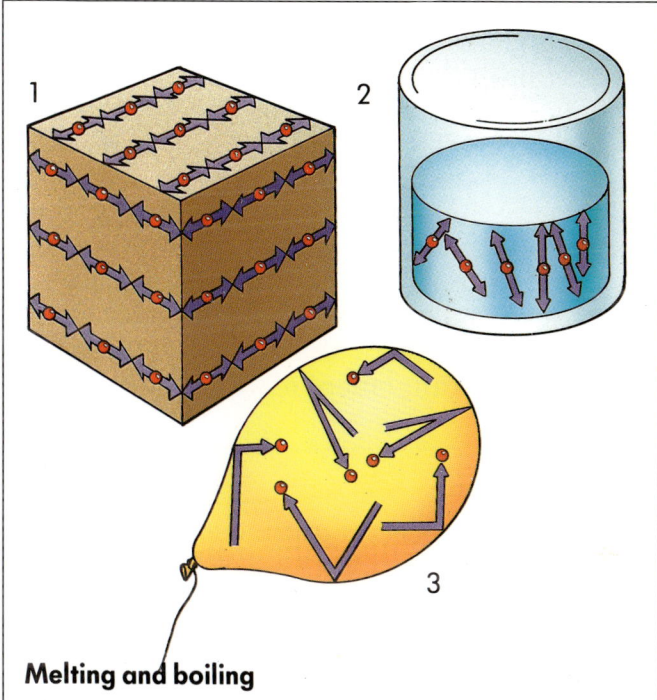

Melting and boiling

Heating a solid (1) makes its molecules vibrate more vigorously. Eventually, when the temperature is high enough, the molecules give up their fixed positions. The solid melts and turns into a liquid (2). Further heating makes the molecules vibrate even more. Eventually the liquid boils and turns to a gas (3).

Solids

Most solids are made up of crystals, pieces of material that have flat surfaces with straight edges. Salt and sugar are familiar examples of crystals. However, a microscope reveals that other solids, such as steel and copper, are also made up of small crystals. X-rays show that crystals are composed of a regular arrangement of atoms. They are spaced very close together – only a few tenths of a nanometre apart (a nanometre is a thousandth of a millionth of a metre). The geometric shape of a crystal reflects the regular arrangement of its atoms.

Different arrangements

Many properties of a solid depend upon its atomic or molecular arrangement. An interesting example is the element carbon, which occurs in two very different forms. One form, diamond, is a very hard material. Its hardness results from the very strong chemical bonds that form between the atoms in a diamond crystal. The other form of carbon, graphite, is one of the softest substances. It has a structure in which the carbon atoms lie in layers. There are only weak links, or bonds, between the

Crystal shapes

A crystal has flat faces angled to each other. The shape of the faces and the angles between them depend on the arrangement of the atoms in its molecule. In different crystals of the same mineral, the angles between faces are always the same (right). The colour of a crystal also depends upon the molecular make-up of the crystal. Quartz (below) is clear, whereas feldspar (below right) is pearly.

layers. As a result, graphite is a relatively weak material, used as a lubricant.

Many other qualities of a solid also depend upon the strength of the forces between its atoms or molecules. The melting temperature is an example. A solid with strong bonds between its atoms needs lots of heat energy to melt it. The amount a solid expands when it is heated also depends upon the strength of the forces between its atoms or molecules. As a solid is heated, the atoms or molecules vibrate more rapidly. They are able to move apart slightly if the forces holding them together are not too strong. So the average distance between the molecules increases, and the material expands.

Elasticity also depends upon the forces between atoms or molecules. If a weight is hung from a wire, the wire stretches slightly. As long as the weight is not too heavy, the wire returns to its original length when the weight is removed. This is called elastic stretching. Materials that stretch least are those with strong forces between their atoms. As long as the weight is not too heavy, the force between the atoms pulls them together again when the weight is removed. But if too great a weight is used, the wire does not return to its original length when the weight is removed. It remains stretched. In this case, the bonds between the atoms in the wire are permanently lengthened.

▼ Large structures such as bridges have to built of strong materials. They have to be strong enough to support the structure itself as well as any extra weight it has to carry. They also have to be designed to withstand the effects of expansion in hot weather.

An expansion joint in a railway line

When railway tracks are laid, small gaps are left between the ends of the rails. When the tracks become hot in summer, the metal rails expand and fill the gaps. Without expansion joints, the expanding track would buckle.

Liquids

The forces between molecules in a liquid are weaker than those in a solid. Nevertheless, these forces produce effects that can easily be seen in everyday life. One effect is called surface tension. It causes the surface of a liquid to behave as if it is covered with a thin rubber skin. This tension causes the roundness of small drops of liquid, and lets small insects walk on water. The tension arises because the molecules at the liquid surface are pulled towards the centre of the liquid by the attracting force of the other molecules.

Surface tension makes water climb up a fine glass tube that is dipped into water. This effect is called capillarity. It happens because the attractive forces between the glass molecules and the water molecules are greater than those between the water molecules themselves. The surface of the water is pulled upwards at the edges of the tube, creating a curved surface where the water wets the glass. Some liquids, such as mercury, drop down small tubes. This happens because the mercury does not "wet" the glass. Its molecules attract one another more than they attract the glass molecules.

▼ Sugary syrup running off a spoon shows all the features of a viscous liquid. It is thick and sticky, and it pours only with difficulty.

▼ A pond skater walks on the surface of a pond. At the surface of a liquid, forces pull the molecules inwards. This surface tension is enough to support the insect.

Liquids flow easily because there are only weak forces between the molecules of a liquid. However, there is some resistance to flow caused by these forces. This resistance is called viscosity. A liquid with a high viscosity – a viscous liquid – flows only slowly, like syrup.

Some liquids are good solvents – that is, other substances can dissolve in them to form a solution. Water is a good solvent for many substances, such as salt and sugar, which are said to be soluble in water. Gases like oxygen and carbon dioxide can also dissolve in water. A third type of solution is formed when two liquids mix together. For example, water and alcohol mix to form such a liquid solution.

The amount of a substance that dissolves in a given amount of a solution is called its solubility. For a solid substance, solubility depends on the temperature. Sugar is much more soluble in hot water than in cold water. For a gas, solubility also depends on pressure. The higher the pressure, the greater is the solubility of the gas.

▼ A dam stores energy in a vast lake of water. The pressure of the water increases with depth, and is greatest near the base of the dam. For this reason, the dam is much thicker at the base than at the top. The energy is carried in flowing water, which spins turbines to generate electricity.

▼ The weight of the ice in a glacier causes it to flow downhill, like a very viscous liquid. A thin film of water helps the glacier to slip over the rocks.

15

Gases

A gas consists of molecules travelling at high speed. Each cubic centimetre of air we breathe contains about 20 million million million molecules at ordinary temperature and pressure. The molecules dart about at 450 metres per second. Each molecule travels only a short distance before it collides with another – about one ten-millionth of a metre. The gradual mixing of one gas with another called diffusion, involves many millions of collisions.

Minute molecules

Molecules of gas are too small to be seen directly. However, when smoke particles suspended in air are examined with a microscope, they are seen to make small irregular movements. This motion is called Brownian motion after the British scientist Robert Brown who first saw them in 1827. The movements of the smoke particles are caused by air molecules constantly hitting them.

▼ A hot-air balloon rises because the heated air inside it is lighter than the unheated and denser gas surrounding it. Heated gas becomes less dense because its molecules move more rapidly, so that the average distance between them increases.

Gas can be compressed very easily. This is because there is a relatively large amount of space between each gas molecule, so it easy to squeeze them closer together. One of the first scientists to study the connection between the volume of a gas and the pressure on it was the Irish scientist Robert Boyle. He showed in 1662 that if the pressure on a gas is doubled, the volume is halved, as long as the temperature does not change. In general, the product of the pressure and the volume is constant for a given mass of gas. This is called Boyle's law.

Another important law describes the way the volume of a gas changes when it is heated or cooled. This is called Charles's law. It states that if the volume of a gas is known at 0°C, the volume increases or decreases by 1/273 of this value for every degree rise or fall in temperature, as long as the pressure does not change. Thus at a temperature of −273°C, a gas will have zero volume. This temperature, called absolute zero, must therefore be the lowest possible temperature.

Inside the Sun

The matter at the centre of the Sun can reach temperatures of about 14 million degrees. At these temperatures, the atoms of the material inside a star such as the Sun are ripped apart. An extremely hot gas made up of sub-atomic particles called protons and electrons is formed. The gas is called a plasma.

▼ A car being tested in a wind tunnel to see how well it slips through the air. Streamlining the car helps to reduce air resistance.

▼ The winds that push a sailing ship along are caused by pressure differences in the atmosphere. Over hot regions the air rises, creating an area of low pressure below it. Air from cooler regions flows into this low-pressure zone as the wind.

17

Changes of state

The temperature of a substance is a measure of the average energy of its molecules. The higher the temperature, the greater is the average energy. Not every molecule has the same energy, however. As the temperature of a crystalline solid rises, the number of molecules with enough energy to move freely increases and the crystal starts to melt. As more heat is added, even more molecules move away from their fixed positions. While this takes place the temperature of the substance remains constant until it is completely melted. The solid has changed state into a liquid.

The heat energy required to melt a solid completely is called the latent heat of fusion. This heat may be supplied by a candle flame or an electric cooker, for example, or it may come from the solid's surroundings. This explains why a block of ice cools its surroundings. Heat is extracted from the surroundings in order to melt the ice.

A similar effect occurs when a liquid evaporates, or turns into a gas. Energy is needed for the molecules of the liquid to escape from their neighbours and form a gas. When water evaporates from your skin, for example, the skin feels cooler because it has lost heat to the evaporating water.

◀ The energy needed to melt steel during laser welding is supplied by a powerful beam of light. The laser beam is concentrated on a very small area at a time, allowing precise cutting and welding.

Much more energy is needed for molecules in a liquid to form a gas than for molecules of a solid to form a liquid. For example, at room temperature, only a tiny fraction of the water molecules in a bowl have enough energy to evaporate. However, within a few days all the water evaporates away because the water molecules diffuse away from the bowl. Eventually all the molecules acquire sufficient energy to escape from the liquid.

The process of evaporation can be speeded up by heating a liquid. When the liquid is hot enough, it boils. But even at the boiling point, more heat energy is needed to convert all of the liquid into a gas, or vapour. This energy is called the latent heat of vaporization. The bubbles in a pan of boiling water, for example, are bubbles of water vapour that form at the bottom of the pan and then rise to the surface.

▼ Mist is caused when a rise in temperature evaporates water to form a gas, or vapour. This is evaporation. When the vapour meets a layer of colder air, it forms tiny droplets of mist. This is condensation.

▲ A skater moves on a thin film of water formed when pressure melts the ice under the skates.

▼ Clouds of mist form when dry ice — solid carbon dioxide — is thrown onto a stage. The change of a solid into a gas is called sublimation.

Atoms and molecules

Spot facts

- A gold wedding ring contains about 10,000 million million million atoms. About 100,000 million atoms would fit on this full stop.

- Atoms are mostly empty space. If the nucleus were the size of a tennis ball, the nearest electron would be a kilometre away.

- A proton in an atom is about 2,000 times heavier than an electron.

- Most molecules are made up of small numbers of atoms, but many contain more. Aspirin molecules contain 21 atoms. Rubber molecules may have up to 65,000 atoms. Sugar may have up to 150,000 atoms in its molecules.

Atoms and molecules are very small, yet scientists have discovered many things about them. They have found that there are even smaller particles inside atoms. Most of the mass of an atom is concentrated at its centre, in the nucleus. The nucleus is made up of particles called protons and neutrons. Even smaller particles called electrons circle around the nucleus. The number of electrons and the way they are arranged around the nucleus are among the most important features of an atom. They determine how atoms join together to form different chemical substances. Molecules are formed when atoms join together. Simple molecules may have only two atoms. Complex molecules, such as those found in living things, may have many thousands of atoms.

▶ Particles much smaller than an atom can be detected in a bubble chamber. As they travel through the chamber, they leave behind trails of bubbles which are then photographed. By studying these tracks, scientists are able to identify the particles.

Inside the atom

The idea that matter is made up of atoms is a very old one. The ancient Greek thinker Democritos, who lived in about 400 BC, argued that all matter was made up of very small particles. These particles were, he thought, indestructible. In 1802 the English scientist, John Dalton, provided the first evidence that atoms were real. He agreed with Democritos that atoms were indestructible and indivisible. They were like small billiard balls, Dalton thought. He was able to measure the weights of some atoms and explain how atoms joined together to form molecules. His atomic theory was accepted for many years.

Then in 1897 another English scientist, J. J. (Joseph John) Thomson, proposed that atoms were not indivisible. They contained even smaller particles, called electrons. Thomson thought electrons were scattered throughout the atom, like currants in a Christmas cake.

The centre of an atom

Additional insight into the atom was provided in 1911 by Ernest Rutherford, a New Zealander who did most of his scientific work in England. Rutherford bombarded a thin piece of gold with alpha particles. Alpha particles are rays given off by certain radioactive materials, such as radium. He was amazed to find that a few alpha particles bounced off the gold rather than going straight through.

This could happen only if most of the mass in the atom were concentrated into a very small region at its centre. We now call this very dense region the nucleus. The electrons circled around the nucleus in a sort of "electron cloud", Rutherford said. In 1913 the Danish scientist Niels Bohr suggested that the electrons circled around the nucleus in a series of orbits at different distances from the nucleus, in much the same way as the planets orbit the Sun.

◀ The English scientist J. J. Thomson sitting in front of a cathode-ray tube. He used the cathode-ray tube to study electrons. From his experiments he was able to demonstrate that electrons are present in all atoms.

▼ In 1802 John Dalton pictured atoms as solid objects like billiard balls. When J. J. Thomson discovered the electron, he imagined electrons scattered through the atoms. In 1911 Ernest Rutherford discovered the nucleus and pictured electrons in orbit around it. In 1913 Niels Bohr showed these orbits are arranged in layers, or "shells".

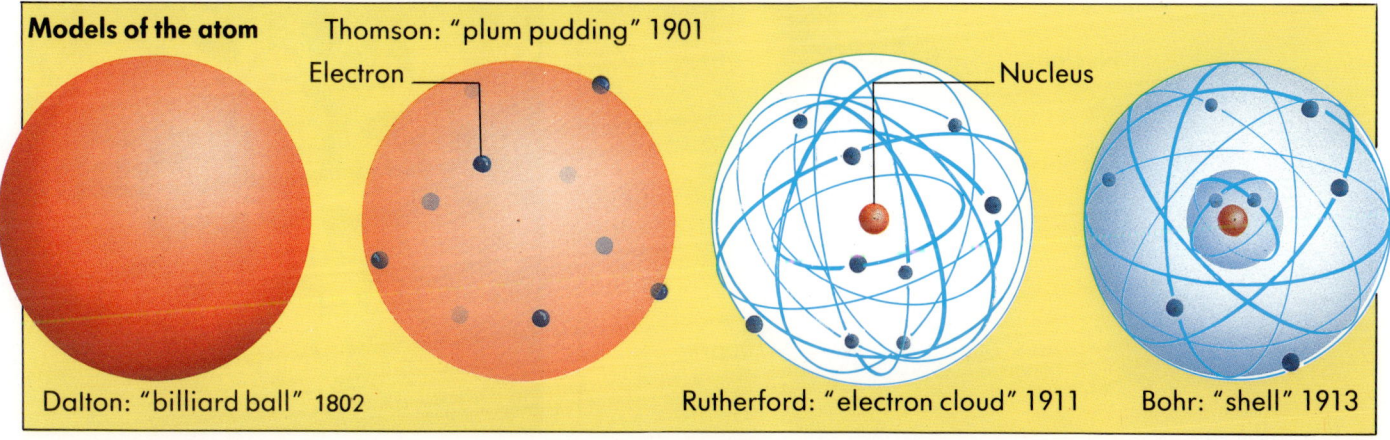

Atomic structure

We now know that the atomic nucleus contains particles called protons and neutrons. Like electrons, protons carry a tiny amount of electricity – they are said to be charged. Protons, however, carry a different kind of electricity from electrons. Scientists say that the proton has a positive charge, and the electron has a negative charge. The neutron, the other particle found in the nucleus, has no charge. It is electrically neutral.

Niels Bohr thought that electrons move in clearly defined paths, or orbits, around the nucleus. In each of these orbits, they require a certain amount of energy to keep them from the nucleus. The electrons in orbits close to the nucleus require more energy than those farther out. When an electron moves from one orbit to another orbit closer to the nucleus, the electron gives off energy, which may be in the form of a small flash of light.

Wave patterns

Later workers took Bohr's ideas into account but added the idea that electrons, and other small particles, can sometimes behave like waves. When electrons pass through a very narrow slit, for example, they appear to spread out like a wave does when going through a slit. Inside the atom, the wave-like behaviour of electrons means that it is impossible to say exactly where an electron is at any time. So, instead of saying that the electrons are in precise orbits, scientists draw maps of where electrons are likely to be.

The region of space occupied by an electron is called an orbital. The simplest orbital is like a ball. It can hold two electrons. Other orbitals have more complicated shapes, such as a dumb-bell and an hourglass. By discovering how electrons fit into an atom's orbitals, scientists can explain its chemical properties.

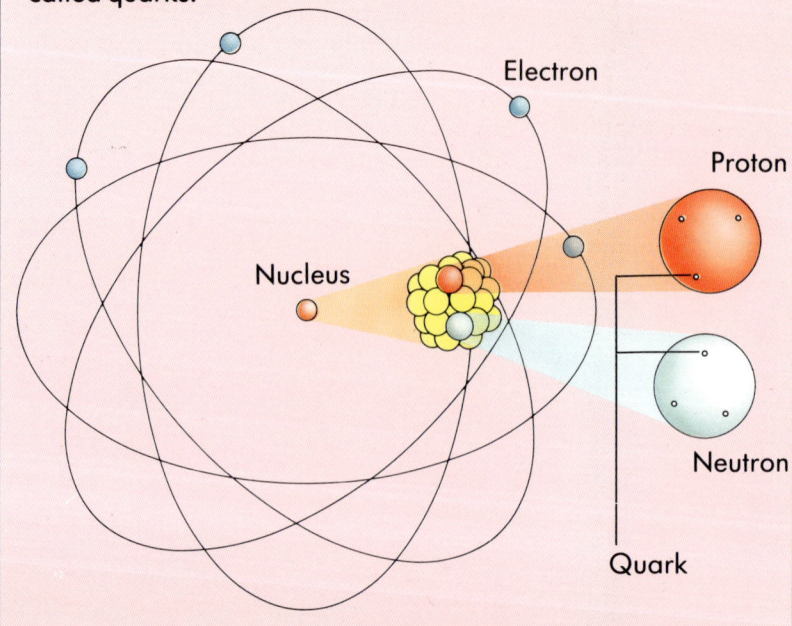

Inside the nucleus

Within the nucleus of an atom there are positively charged protons and neutrons, which have no electrical charge. The number of protons equals the number of negatively charged electrons circling the nucleus, so that the atom has no overall electric charge. Modern experiments use beams of high-energy electrons to probe inside the protons and neutrons themselves. These experiments show that they contain point-like particles called quarks.

▼ A group of uranium atoms. The picture was taken using a scanning electron microscope. In this instrument a fine beam of electrons is swept, or scanned, across the object being examined. Some of the electrons are reflected off the object and used to form the picture. The latest electronic imaging techniques can be used to photograph individual atoms which are only 30 billionths of a centimetre across.

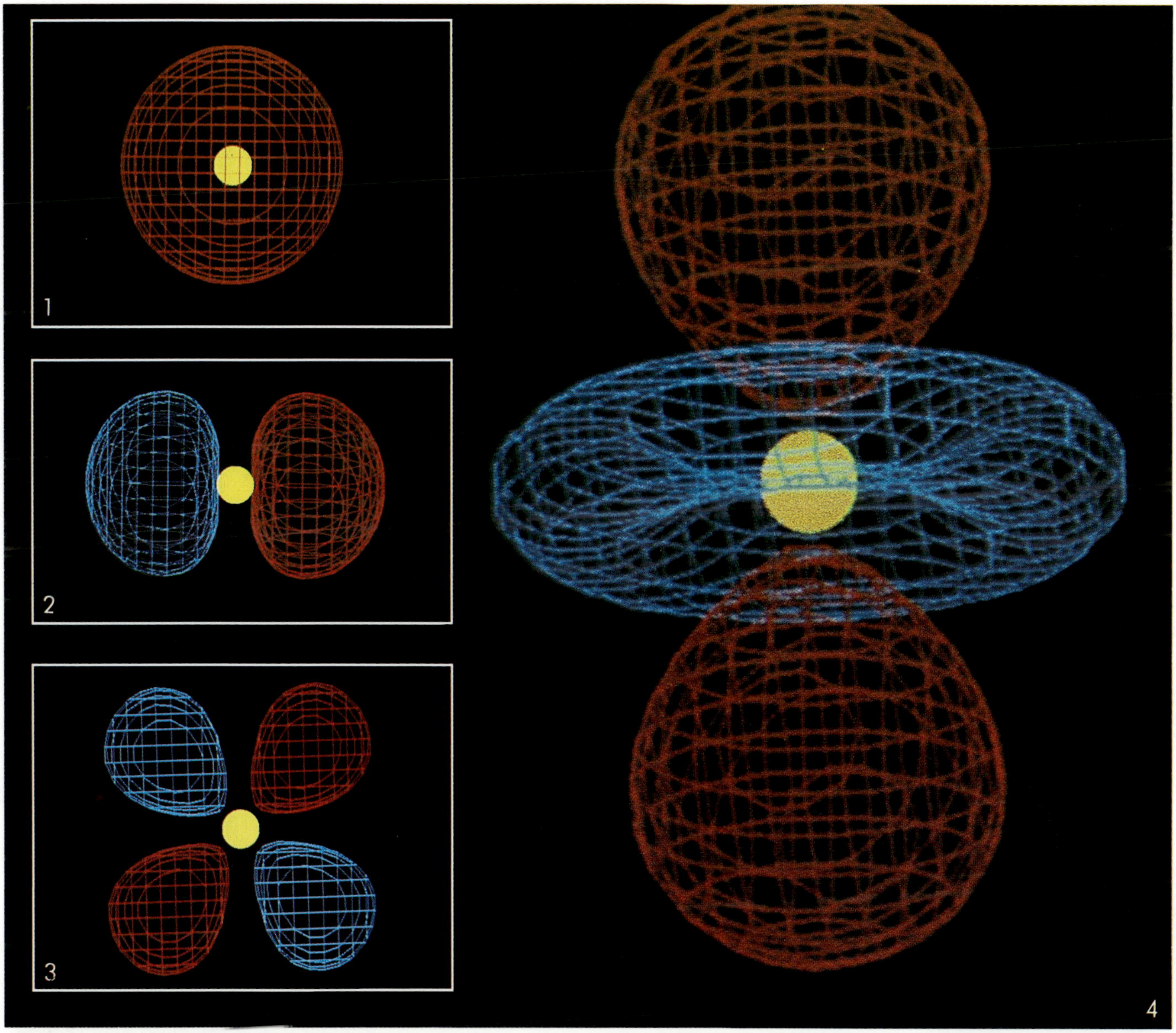

▲ In the wave picture of the atom, scientists draw "orbitals". Orbitals are regions of space where the electron is likely to be, although its exact location cannot be stated. There are four simple shapes of orbital: (1) spherical, or ball-like, (2) dumb-bell, (3) four-leaf clover, (4) hourglass and ring. Other orbitals are much too complicated to draw.

▶ The particle picture of the atom has electrons circling the nucleus, like planets around the Sun, as in the atoms shown here. The first orbital out from the nucleus can hold two electrons, the second orbital can hold eight electrons, and so on. The simplest atom, hydrogen, has one electron in the first orbital. Oxygen, with a total of eight electrons, has two in the first orbital and six in the second. Magnesium has two electrons in its third orbital.

The build-up of electrons

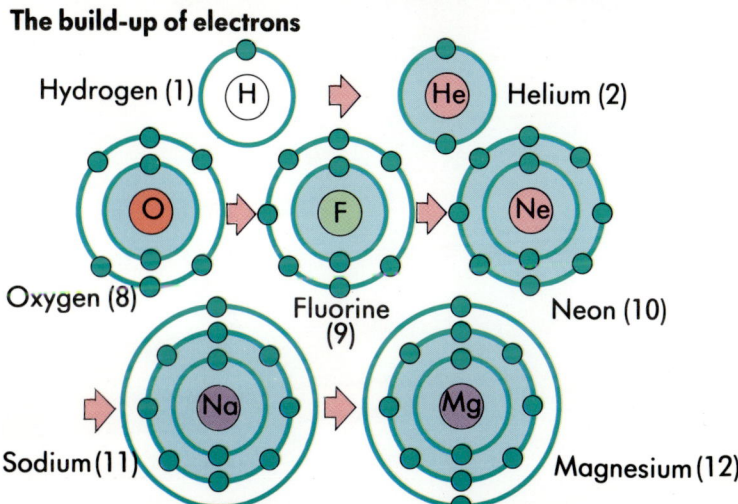

Particles galore

To study the particles that make up the nucleus of an atom, scientists use particle accelerators, popularly known as atom-smashers. These are large machines in which particles are boosted, or accelerated, to high speeds. The accelerated particles collide with the nuclei in a target, producing a shower of other particles, which have very short lifetimes. The tracks of these particles are recorded using many types of detectors, such as a bubble chamber. As the particles travel through the liquid in the chamber, they leave tracks of bubbles behind. Magnets around the chamber are used to bend the tracks of electrically charged particles. By seeing how the particles are affected by the magnets, scientists are able to identify them.

These experiments reveal that there are many different subatomic particles. There are heavy particles, such as the pion and the kaon. Like the neutron and the proton, these particles are made up of quarks. The proton and neutron each have three quarks. The pion and kaon are made up of two quarks. There are also light particles, such as the neutrino. This particle appears to have no mass at all. Neutrinos are not made up of quarks. Neither are electrons and other slightly heavier particles called the tau and the muon. The story is not yet over, because scientists still have more things to discover about particles.

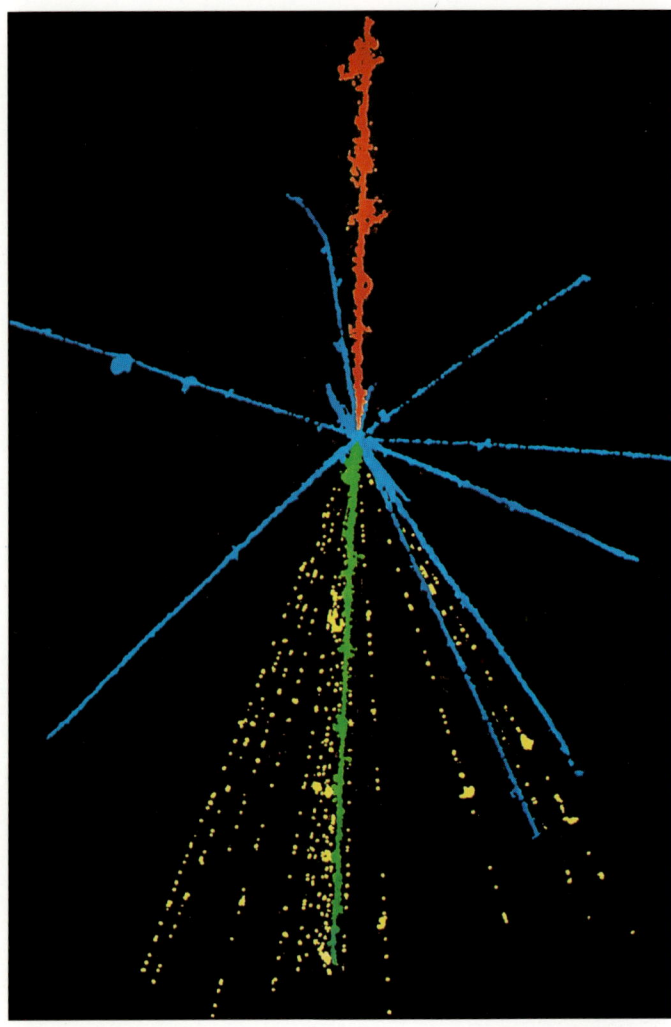

▲ Tracks of cosmic rays. They are high-energy particles that reach Earth from outer space. They have been captured in a photographic emulsion. The resulting picture was later coloured to identify the tracks. A sulphur nucleus (red) has collided with a nucleus in the film to produce a spray of particles including a fluorine nucleus (green), 16 pions (yellow) and several other nuclear fragments (blue).

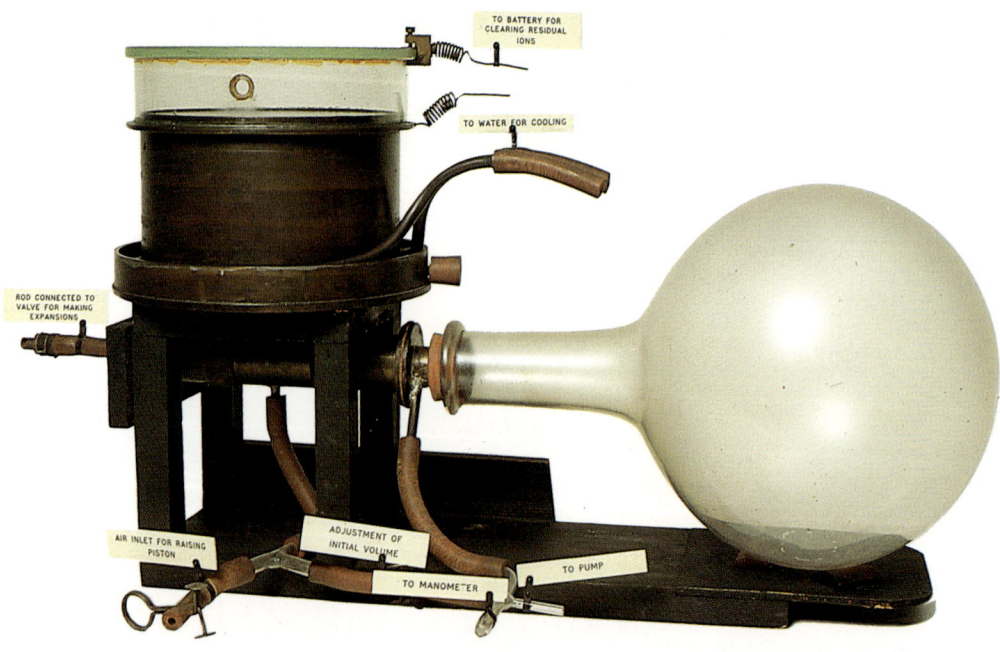

◀ An early type of particle detector was the cloud chamber, invented by Scottish physicist Charles Wilson in 1895. If the globe is filled with moist air and the pressure reduced suddenly, water droplets form and make particle tracks visible.

Tracking particles

In this photograph of particle tracks in a bubble chamber, colours have been added. Negative particles, called kaons, enter the picture from below. Their tracks curve to the right slightly, showing that the chamber's magnetic field deflects negative particles (shown in purple, pale blue and green) to the right. Positive particles (orange and red) are deflected to the left.

The two spirals are caused by electrons, the only particles light enough to curl so tightly in the magnetic field. The tiny spiral is from an electron knocked from an atom in the bubble-chamber liquid. The large spiral comes from the break-up of the particle that made the pale blue track. This must have been a muon.

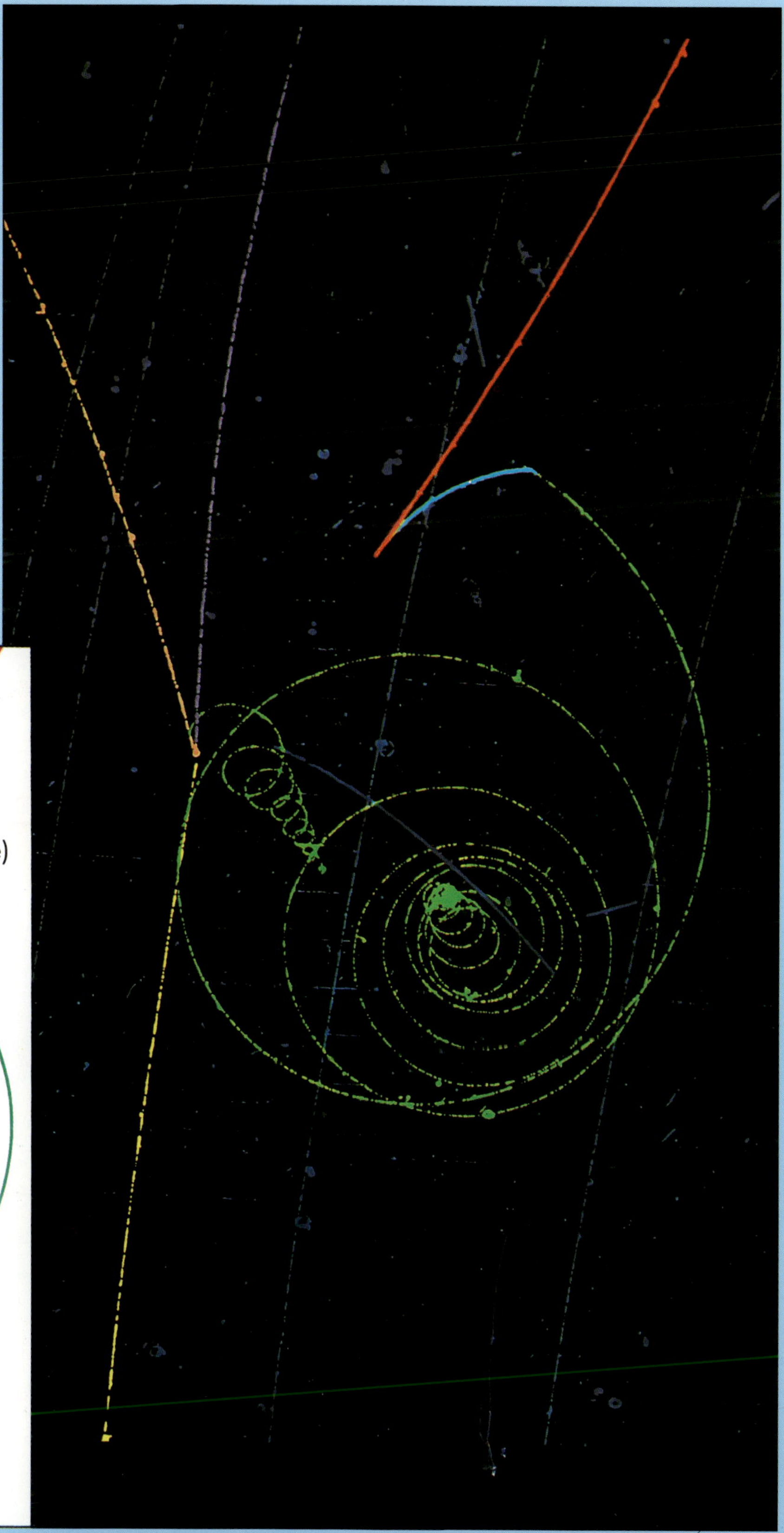

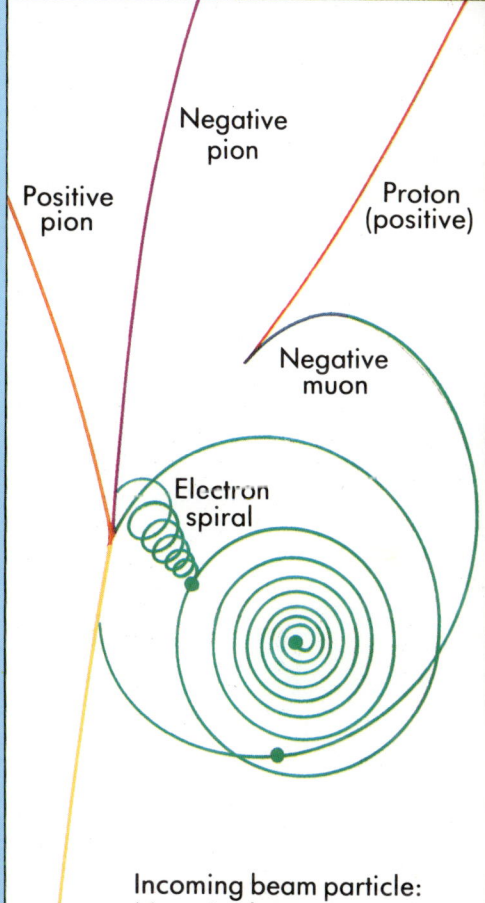

Radioactivity

All the atoms of a particular element have the same number of protons in their nuclei. But the number of neutrons may vary. Atoms that have the same number of protons, but a different number of neutrons, are called isotopes.

Nearly all chemical elements have several isotopes. There are three isotopes of hydrogen, for example. The most common isotope of hydrogen has a single proton in its nucleus. The other isotopes, called deuterium and tritium, have two and three particles in their nuclei, respectively.

Unstable isotopes
The nuclei of many isotopes are stable and unchanging. However, some isotopes are unstable. The nuclei of these isotopes break up, or decay. These isotopes are said to be radioactive. They give out energy in the form of radiation in order to become more stable. Three different types of radiation are given out by radioactive isotopes. One type is called alpha radiation. This consists of a stream of tiny particles, called alpha particles, each made up of two protons and two neutrons. The second type is called beta radiation. It consists of high-energy electrons, or beta particles. The third type of radiation is gamma radiation, which resembles high-energy X-rays.

Half-life
Not all the nuclei of a radioactive isotope decay at the same time. One nucleus may decay quickly, another may take a long time. But the decay process always proceeds at the same rate, on average. The time taken for half the nuclei in a sample to decay is always the same. This time is called the half-life of the isotope.

The half-lives of some isotopes are as short as a fraction of a second, whereas other isotopes decay much more slowly, with half-lives of millions of years. This provides a clue to which isotopes occur naturally. Those with short half-lives, compared to the age of the Earth, have long since decayed and disappeared. Some, such as carbon, take longer to decay. By measuring the amount of radioactive isotopes remaining in samples of old materials, it is possible to estimate how old they are. This process is called radiocarbon dating.

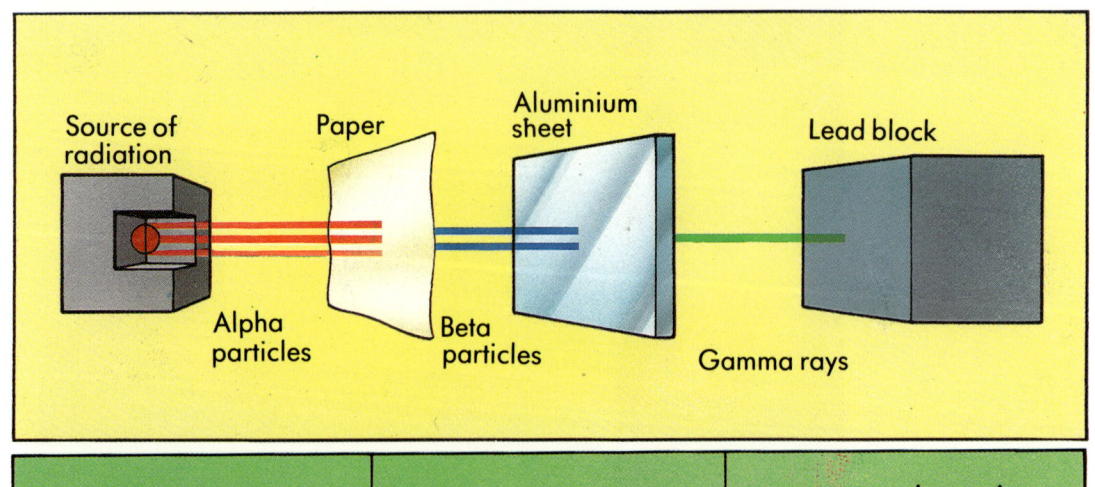

◄ The three kinds of radiation given out by a radioactive substance have different penetrating power. Alpha particles are stopped by a sheet of paper; beta particles by a thin sheet of metal; gamma rays by a thick lead block.

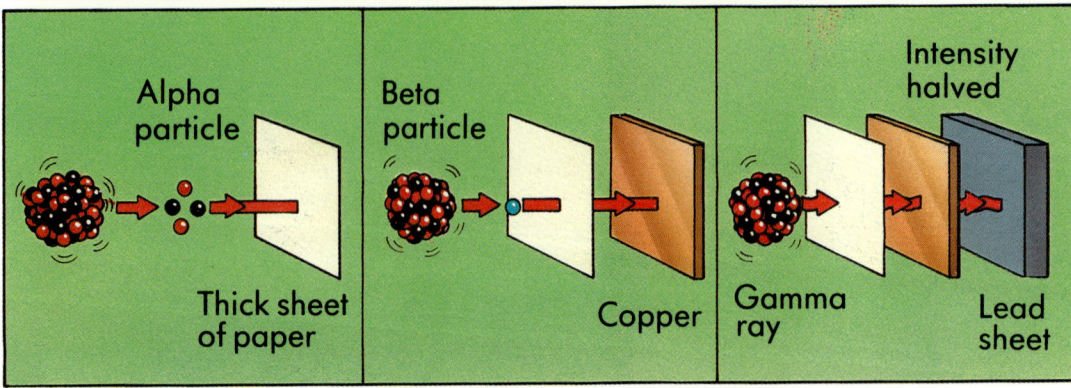

◄ Alpha particles have the lowest energy. Beta particles are more penetrating than alpha particles because they are smaller and move at higher speeds. More penetrating still are gamma rays. They are similar to X-rays but have a shorter wavelength.

▼ A radioactive substance decays in a regular way. The graph shows how the amount of a sample of the radioactive gas radon decreases as it decays into polonium.

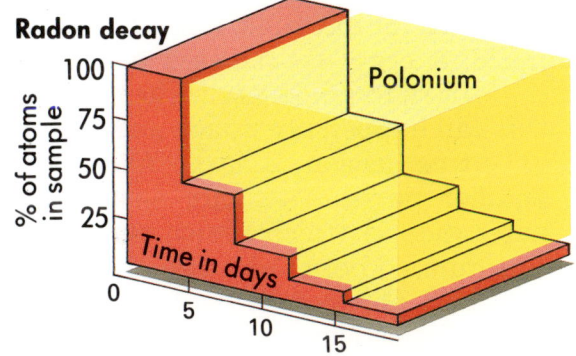

▲ A Geiger counter measures levels of radioactivity. Here one is being used to check produce after the explosion of a nuclear reactor at Chernobyl in Russia in 1986.

▶ This body, well preserved in the bogs of central England, is about 1,800 years old, according to radiocarbon dating measurements.

▼ Radiocarbon dating is used to measure the age of once-living material. The age of the sample is calculated by measuring the proportion of radioactive carbon, carbon-14, present.

27

Nuclear energy

Atoms and molecules are storehouses of energy. Chemical reactions can release some of this energy by rearranging the atoms and making them combine into different molecules. This happens, for example, when gas burns. But far more energy is released when an atomic nucleus rearranges itself. The best-known ways this can happen are fission and fusion.

During fission, a large nucleus splits into smaller parts. The only naturally occurring substance in which fission can occur easily is uranium. There are two main isotopes of uranium. One is called uranium-235, because it has 235 particles in its nucleus. It can be split easily. The uranium-235 nucleus absorbs an extra neutron. This makes the nucleus unstable. It splits into two roughly equal parts, releasing energy and a few extra neutrons. The extra neutrons are important because they can go on to split yet more nuclei.

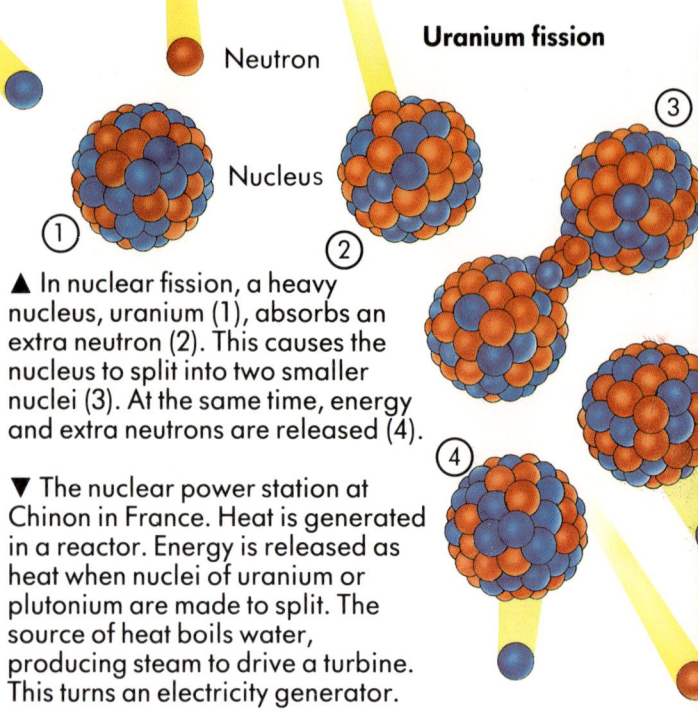

Uranium fission

▲ In nuclear fission, a heavy nucleus, uranium (1), absorbs an extra neutron (2). This causes the nucleus to split into two smaller nuclei (3). At the same time, energy and extra neutrons are released (4).

▼ The nuclear power station at Chinon in France. Heat is generated in a reactor. Energy is released as heat when nuclei of uranium or plutonium are made to split. The source of heat boils water, producing steam to drive a turbine. This turns an electricity generator.

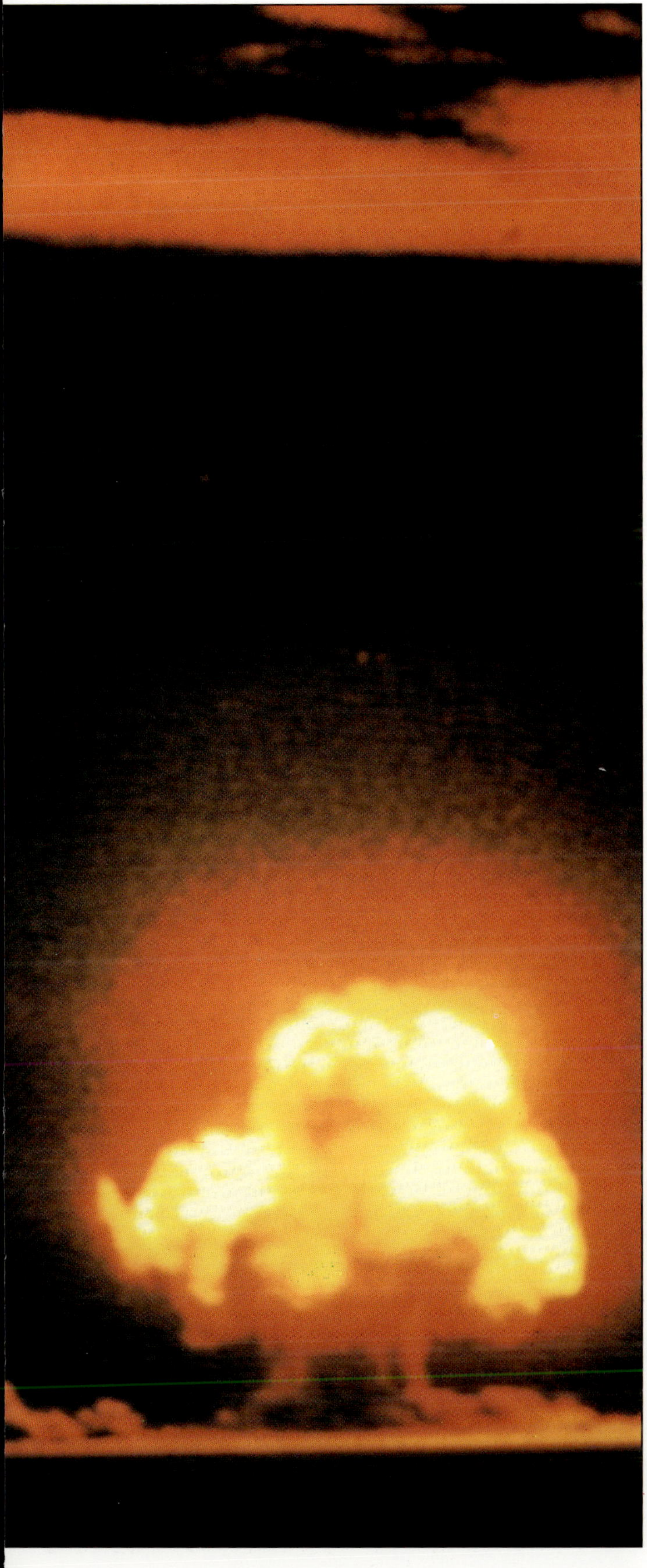

The second way to release nuclear energy is fusion. This is the joining of light nuclei, such as an isotope of hydrogen, to make heavier nuclei. Fusion is the process that produces energy inside the Sun and other stars. For many years, scientists have been trying to produce energy from hydrogen fusion here on Earth. They have been successful in producing hydrogen bombs, which release fusion energy in a sudden, violent manner. But they still have not produced fusion energy in a controlled manner.

The main problem is that a very high temperature, about 100 million degrees, is needed before hydrogen nuclei can be made to combine. At lower temperatures, their electric charges cause them to push apart, or repel each other. Fusion would be a cheap source of energy because the hydrogen fuel needed could be extracted from water. However, it will be many years before fusion power is a reality.

◀ The first atomic bomb explosion in New Mexico on 16 July 1945 worked on the principle of fission. In the more powerful hydrogen bomb, an atomic bomb produces the high temperature needed for fusion.

▼ The aim in nuclear fusion is to bring together two heavy isotopes of hydrogen, tritium and deuterium, so that they fuse, or join. The result is a helium nucleus, a neutron and the release of energy.

Hydrogen fusion

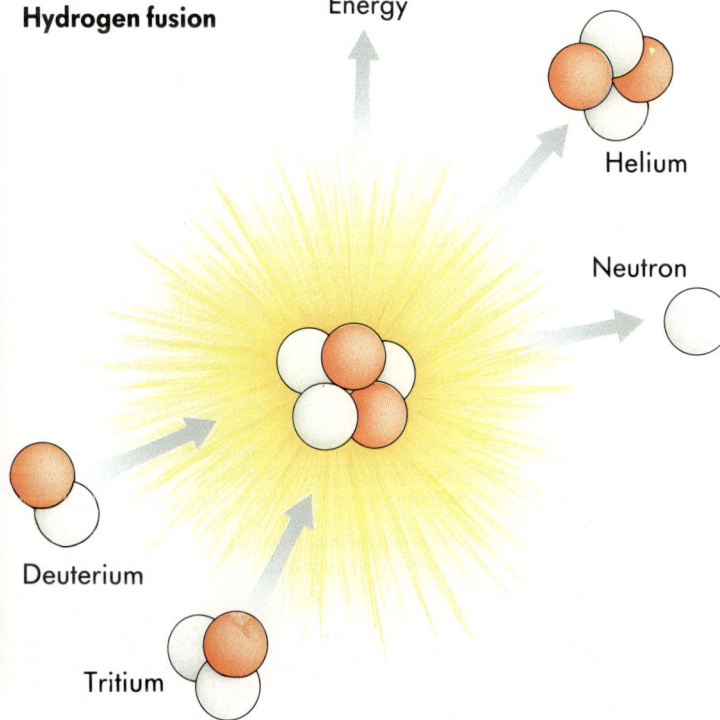

Chemical elements

Spot facts

- Of the 92 naturally occurring chemical elements, only two are liquids at ordinary temperatures: bromine and mercury. Only 11 elements are gases. All the other elements are solids, mostly metals.

- The most common element in the Universe is hydrogen. Most stars are made up of nine-tenths hydrogen and one-tenth helium.

- The rarest naturally occurring element is astatine. It is believed that there is only a third of a gram of astatine in the whole of the Earth's crust.

▶ With their primitive apparatus, seen in this 18th century painting, early chemists, or alchemists, could perform only simple experiments. They searched for a "philosopher's stone", which could turn common metals into gold. They also sought medicines that could prolong life for ever. Despite failing in these aims, the alchemists made many chemical discoveries and laid the foundations of the science of chemistry.

All the countless different materials that exist in the Universe consist of combinations of simple substances, called elements. The elements are the basic building blocks from which all other substances are made up. When elements combine chemically, they form compounds. Chemists explain the properties of the elements, how they combine to form compounds, and how chemical reactions occur, by looking at the way the electrons are arranged in the atoms.

Elements and compounds

Substances can be elements, compounds or mixtures. A mixture is a substance that can be separated into different materials by means, such as filtering, that can be easily reversed. These changes are called physical changes. Seawater is a mixture. The salt and water can be separated by heating the seawater until the water has evaporated, leaving the salt behind. The change can be reversed easily by pouring the salt back into water and stirring. A mixture consists of atoms or molecules that are not connected together and so are easy to separate.

Compounds can be separated into different substances too. But the changes involved, called chemical changes, are difficult to reverse. When wood burns, it undergoes a chemical change. It produces new substances, smoke and ash. However, it is very difficult to reverse the change. Compounds are substances that are made up of two or more elements whose atoms combine to form molecules. The molecules are broken down into atoms or changed into other molecules during a chemical change.

An element is a substance that is made up of a single kind of atom. These atoms cannot be broken down by chemical means. So elements are substances that cannot be separated into simpler substances by chemical changes.

◀ This figure of gold and copper alloy is about 20 cm high. It was made by the Indians of Colombia. Gold was one of the earliest known elements because it occurs naturally as nuggets. It is easily worked and has been used in jewellery for hundreds of years.

▼ The mineral galena is a compound of the elements lead and sulphur. Its molecules consist of an atom of lead and an atom of sulphur chemically bonded.

The Periodic Table

In the Periodic Table, elements are arranged in order of increasing atomic number. The atomic number is the number of protons in an element's nucleus. An element resembles those above and below it in the table – that is, in the same "group". It also resembles the elements on each side of it – that is, in the same "period".

For example, helium, neon, argon, krypton, xenon and radon are found in the group on the right-hand side of the table. All these gases are very similar and very unreactive. Moving to the left across the table, the elements gradually become more reactive. The elements in the left-hand column are the most reactive of all.

Electron configurations

Scientists have found that the position of an element in the Periodic Table is related to the way the electrons are arranged.

The noble gases on the right of the table have eight electrons in their outer layer. This is a very stable arrangement, and it is difficult to remove electrons to allow chemical bonds, or links with other elements, to form.

The atoms of the elements on the left of the table, such as sodium, have only a single electron in their outer layer. This electron can easily be removed, so the elements are very reactive. The atoms of the elements in the last-but-one period, such as chlorine, have seven electrons in their outer layer. They can easily take up another electron, and are also very reactive. For this reason, sodium reacts violently with chlorine to form the chemical compound sodium chloride (common salt).

▲ Magnesium, when clean and freshly prepared, is a silvery metal found in seawater and several minerals. It burns with an intense white flame. Magnesium is used in fireworks, camera flashbulbs and lightweight alloys.

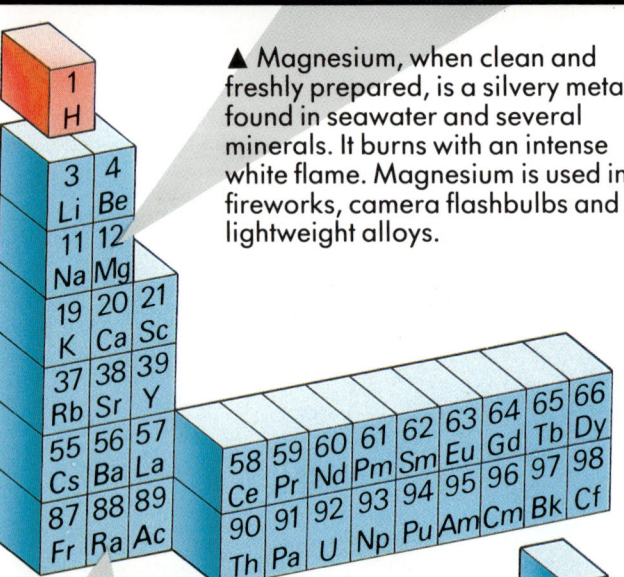

Non-metal / Metal / Atomic number / Chemical symbol

Atomic structure

The basic structure of an atom consists of a central nucleus surrounded by electrons. The electrons occupy "shells" which surround each other like the layers of an onion. The innermost shell holds only 2 electrons; the second holds 8; and the third holds 18.

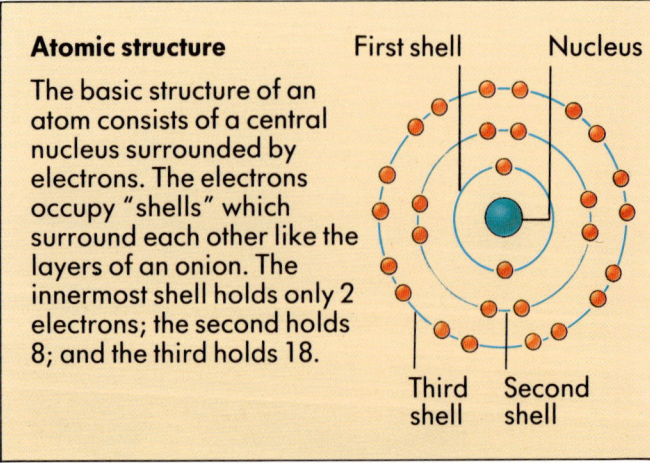

First shell / Nucleus / Third shell / Second shell

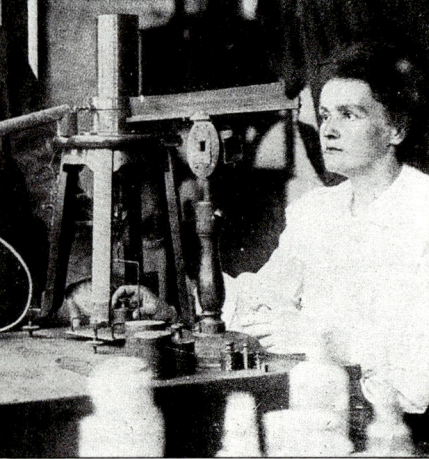

▲ Marie Curie and her husband Pierre discovered radium in 1899.

▶ Drilling tool, the tips of which are made of tungsten carbide. Tungsten is a very strong silvery white metal.

32

◀ Parts of Concorde are made of titanium alloy. Titanium is a strong, light metal. It is found in many minerals.

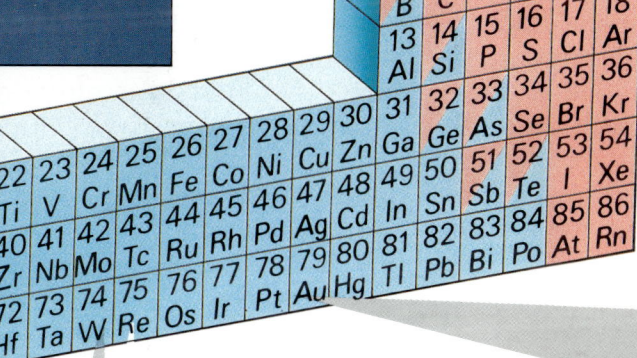

▲ Gold is a very stable element, which does not corrode. For this reason, and because of its beauty and value, gold is used in jewellery.

▲ The Periodic Table contains three distinct types of elements. The representative elements include both metals, such as magnesium (Mg), and non-metals, such as chlorine (Cl). The transition elements are a series of metals, including iron (Fe) and gold (Au). The inner transition elements, including lawrencium (Lr), all have similar properties.

The elements

	Symbol	Atomic number		Symbol	Atomic number		Symbol	Atomic number
Actinium	Ac	89	Hafnium	Hf	72	Praseodymium	Pr	59
Aluminium	Al	13	Helium	He	2	Promethium	Pm	61
Americium	Am	95	Holmium	Ho	67	Protactinium	Pa	91
Antimony	Sb	51	Hydrogen	H	1	Radium	Ra	88
Argon	Ar	18	Indium	In	49	Radon	Rn	86
Arsenic	As	33	Iodine	I	53	Rhenium	Re	75
Astatine	At	85	Iridium	Ir	77	Rhodium	Rh	45
Barium	Ba	56	Iron	Fe	26	Rubidium	Rb	37
Berkelium	Bk	97	Krypton	Kr	36	Ruthenium	Ru	44
Beryllium	Be	4	Lanthanum	La	57	Samarium	Sm	62
Bismuth	Bi	83	Lawrencium	Lr	103	Scandium	Sc	21
Boron	B	5	Lead	Pb	82	Selenium	Se	34
Bromine	Br	35	Lithium	Li	3	Silicon	Si	14
Cadmium	Cd	48	Lutecium	Lu	71	Silver	Ag	47
Caesium	Cs	55	Magnesium	Mg	12	Sodium	Na	11
Calcium	Ca	20	Manganese	Mn	25	Strontium	Sr	38
Californium	Cf	98	Mendelevium	Md	101	Sulphur	S	16
Carbon	C	6	Mercury	Hg	80	Tantalum	Ta	73
Cerium	Ce	58	Molybdenum	Mo	42	Technetium	Tc	43
Chlorine	Cl	17	Neodymium	Nd	60	Tellurium	Te	52
Chromium	Cr	24	Neon	Ne	10	Terbium	Tb	65
Cobalt	Co	27	Neptunium	Np	93	Thallium	Tl	81
Copper	Cu	29	Nickel	Ni	28	Thorium	Th	90
Curium	Cm	96	Niobium	Nb	41	Thulium	Tm	69
Dysprosium	Dy	66	Nitrogen	N	7	Tin	Sn	50
Einsteinium	Es	99	Nobelium	No	102	Titanium	Ti	22
Erbium	Er	68	Osmium	Os	76	Tungsten	W	74
Europium	Eu	63	Oxygen	O	8	Uranium	U	92
Fermium	Fm	100	Palladium	Pd	46	Vanadium	V	23
Fluorine	F	9	Phosphorus	P	15	Xenon	Xe	54
Francium	Fr	87	Platinum	Pt	78	Ytterbium	Yb	70
Gadolinium	Gd	64	Plutonium	Pu	94	Yttrium	Y	39
Gallium	Ga	31	Polonium	Po	84	Zinc	Zn	30
Germanium	Ge	32	Potassium	K	19	Zirconium	Zr	40
Gold	Au	79						

Forming crystals

Many substances are found in crystal form, with a regular arrangement of atoms. The metallic elements are examples. In metals, the atoms are packed closely together like stacks of identical balls. There are three ways that the atoms (balls) can be stacked together. For this reason, there are three different atomic arrangements found in metals. The first is called a body-centred cube. In this arrangement, the atoms are stacked in cube shapes, with an atom in the centre of each cube. Sodium and vanadium crystals have this arrangement.

The second arrangement is called a face-centred cube. It is cube-shaped, with an additional atom in the centre of each face. Aluminium and gold have crystals with this arrangement. The third is called hexagonal close-packed. It has the atoms arranged in hexagonal, or six-sided groups. Magnesium has this arrangement.

The bonds, or links, between the atoms in a metal crystal are different from the bonds found in other crystals or molecules. The metal atoms are held together by a "sea" of electrons, which have broken free from the metal atoms. The electrons move freely among the atoms, acting as a kind of glue. The presence of free-moving electrons explains why metals are such good conductors of heat and electricity. The electrons carry the heat and electricity as they move about. Also, because the atoms are not strongly bonded to their neighbours, metals can be bent, hammered into sheets and drawn into thin wire without breaking.

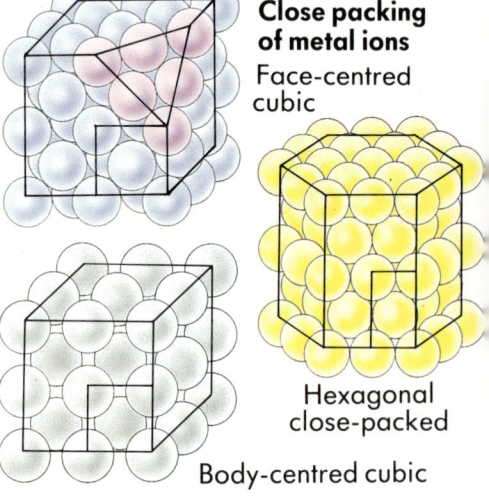

Close packing of metal ions
Face-centred cubic

Hexagonal close-packed

Body-centred cubic

◀ Under a microscope, we can see the small crystals that make up a piece of aluminium.

▼ In an electron microscope, it is possible to locate the atoms (black dots) in the crystals.

34

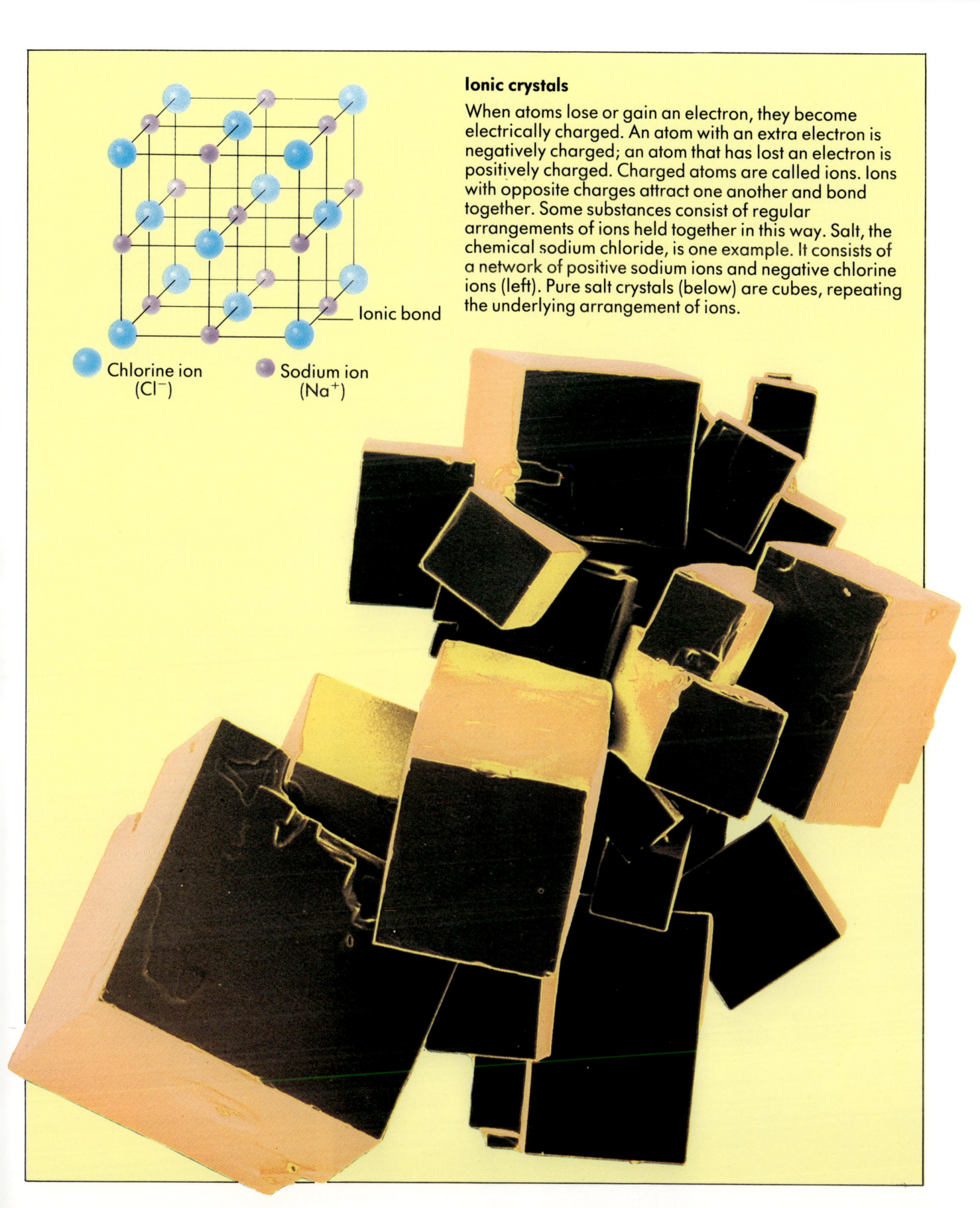

Ionic crystals

When atoms lose or gain an electron, they become electrically charged. An atom with an extra electron is negatively charged; an atom that has lost an electron is positively charged. Charged atoms are called ions. Ions with opposite charges attract one another and bond together. Some substances consist of regular arrangements of ions held together in this way. Salt, the chemical sodium chloride, is one example. It consists of a network of positive sodium ions and negative chlorine ions (left). Pure salt crystals (below) are cubes, repeating the underlying arrangement of ions.

Ionic bond

Chlorine ion (Cl^-)

Sodium ion (Na^+)

Forming molecules

Atoms that readily gain or lose electrons to form ions do so in order to form a stable outer layer of electrons. An outer electron layer that has eight electrons is especially stable. For this reason, an atom with seven electrons in its outer layer, such as chlorine, can achieve a stable outer layer by gaining an electron. An atom with nine electrons in its outer layers, such as sodium, readily loses an electron in order to form a stable outer layer. The ions formed in this way readily group together to form ionic crystals. There are ionic bonds between the charged atoms in such crystals.

Covalent bonds
Chlorine atoms can also form another kind of bond in which electrons are shared between atoms. If two chlorine atoms come together, a pair of electrons, one from each atom, can be shared. This arrangement completes the outer electron layer of each atom and forms a stable molecule of two chlorine atoms. This type of bond, in which electrons are shared, is called a covalent bond.

Water molecules form because covalent bonds link two hydrogen atoms to an oxygen atom. The oxygen atom normally has six electrons in its outer layer, so it needs to share two electrons. One shared electron comes from each of the two hydrogen atoms. The hydrogen atoms form a stable electron layer with two electrons in it. It is unusual for atoms to be stable with two electrons in the outer layer. Only hydrogen and a few other elements are able to do this.

Scientists have a simple way of representing a water molecule. They write H_2O. This shows that the molecule is made up of two hydrogen atoms and one oxygen atom. But it does not show how the electrons are shared between the atoms. Other diagrams, using dots or short lines, are used to show how the electrons are shared and where the covalent bonds are. Even these diagrams do not show how the molecule really looks. Computer drawings are sometimes used to picture molecules more accurately.

Sometimes scientists make models of molecules, using balls to represent the atoms and sticks to show the bonds holding the molecule together. The most realistic models are called space-filling models. These use balls of different sizes to represent atoms of different sizes.

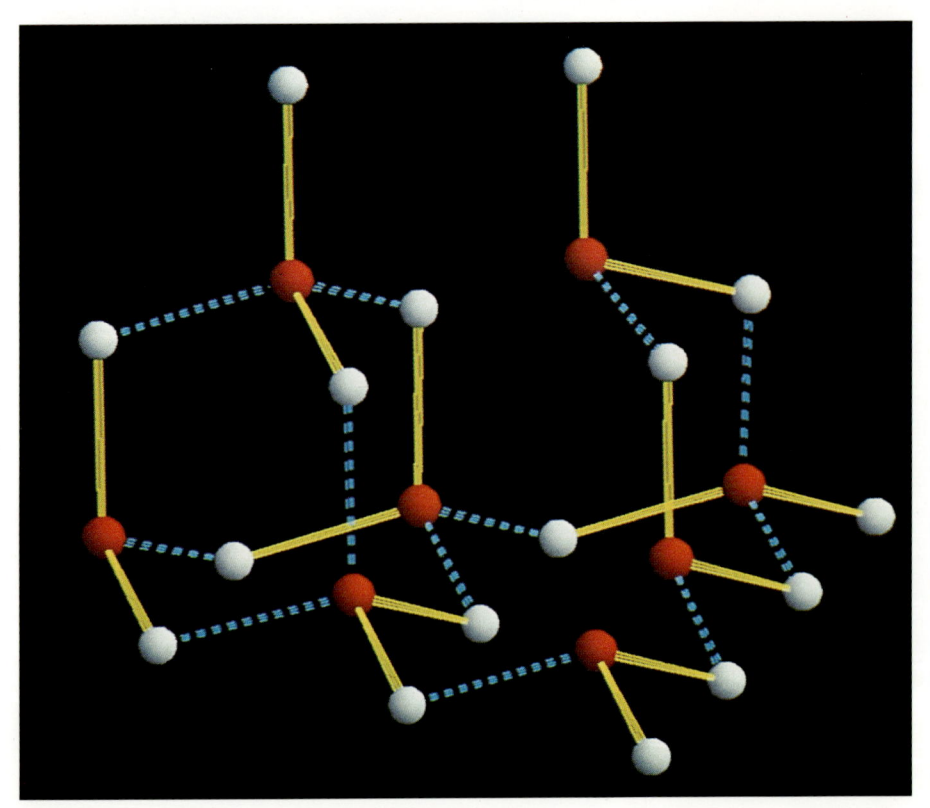

◀▼ In molecules of water, the oxygen atoms and hydrogen atoms are joined to each other by covalent bonds (shown in yellow). But separate molecules are also weakly bonded together. The hydrogen atoms in one molecule are weakly bonded to the oxygen atoms in other molecules by hydrogen bonds (shown in blue).

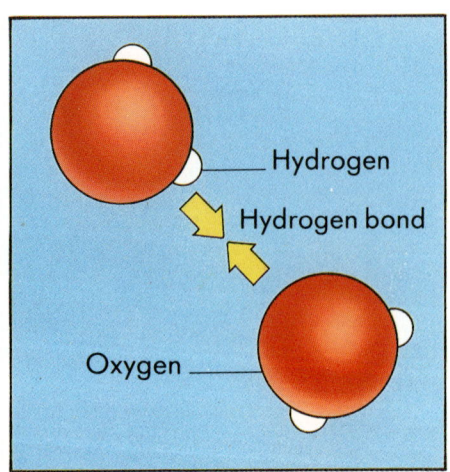

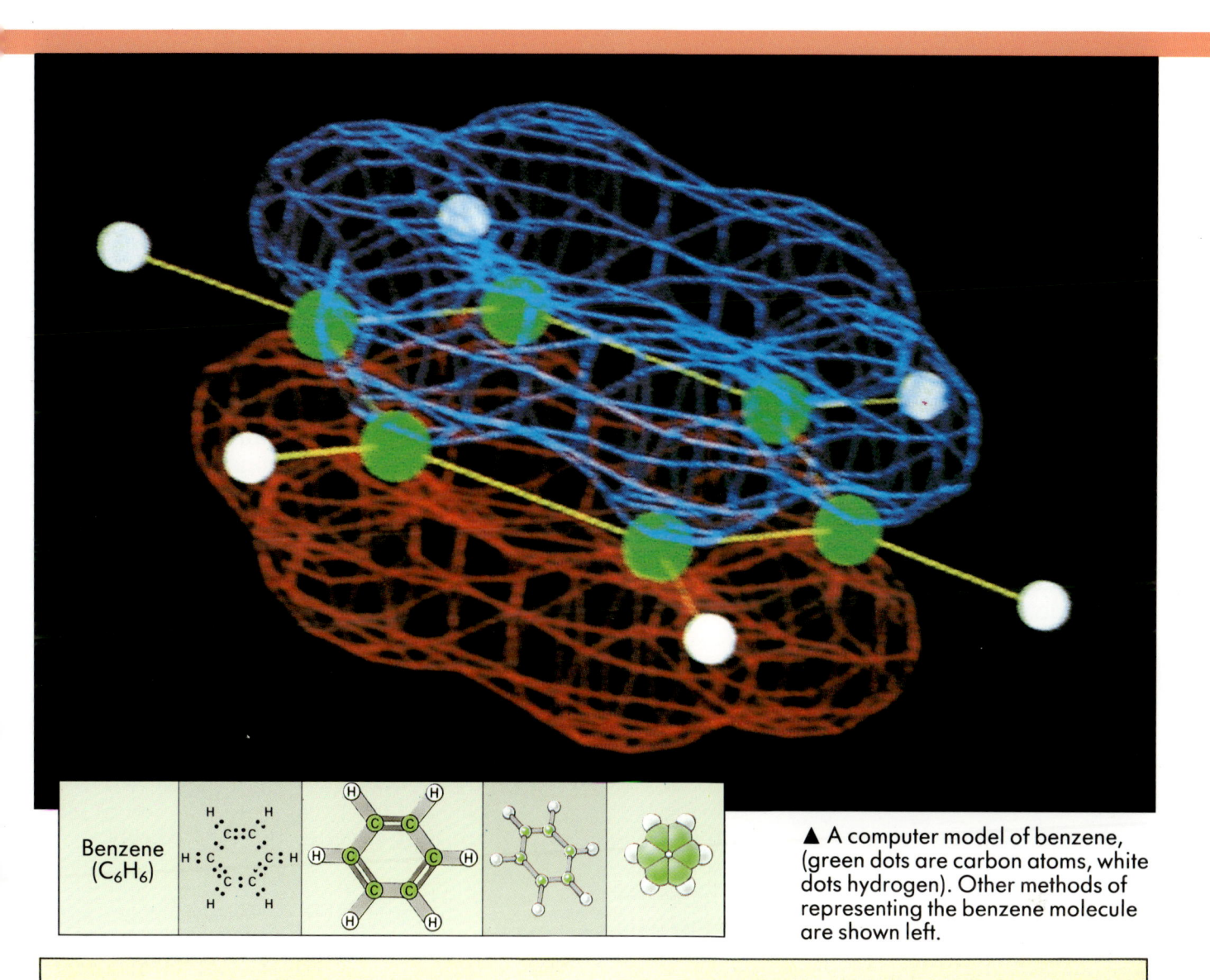

▲ A computer model of benzene, (green dots are carbon atoms, white dots hydrogen). Other methods of representing the benzene molecule are shown left.

Benzene (C₆H₆)

Types of chemical bond

There are three common types of chemical bonding, ionic, covalent and metallic. Ionic bonding relies on the attraction of the opposite electrical charges of the ions. Covalent bonding uses shared electrons to link atoms in molecules, such as water. The linking of atoms in a metal crystal – metallic bonding – is brought about by a "sea" of free electrons moving around the atoms.

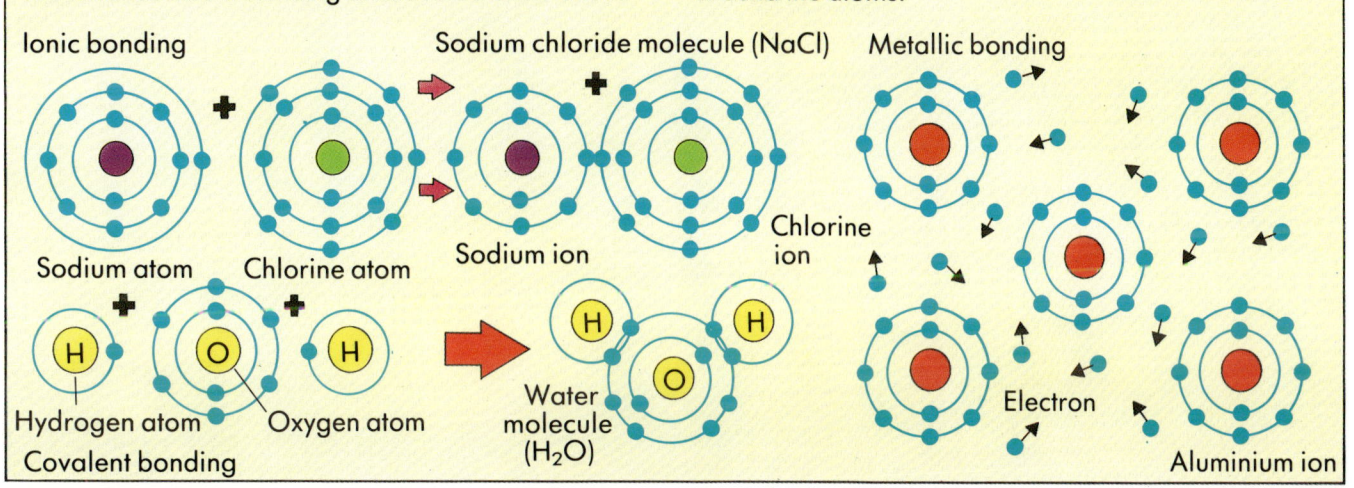

Ionic bonding

Sodium atom + Chlorine atom → Sodium chloride molecule (NaCl)

Sodium ion + Chlorine ion

Metallic bonding

Hydrogen atom + Oxygen atom + Hydrogen atom → Water molecule (H₂O)

Covalent bonding

Electron

Aluminium ion

37

Chemical reactions

When coal burns, or dynamite explodes, or a metal ship rusts, a chemical reaction, or change takes place. These changes involve the formation or breaking up of molecules. When coal burns, for example, carbon atoms in the coal link up with oxygen atoms from the air to form molecules of carbon dioxide. Scientists use a kind of chemical shorthand to write an equation to show the process:

$$C + O_2 \rightarrow CO_2$$

Coal is made up of carbon. The chemical symbol for an atom of carbon is C. Air contains oxygen molecules made up of two oxygen atoms. These molecules are written as O_2. The carbon dioxide molecule that is formed is represented as CO_2. It is made up of one carbon atom joined to two oxygen atoms.

When the gas methane burns in air, it produces carbon dioxide and water. The reaction describing this process is:

$$CH_4 + 2O_2 \rightarrow CO_2 + 2H_2O$$

This equation is more complicated. The methane molecule is written as CH_4. It is made up of a carbon atom joined to four hydrogen atoms. The equation shows how the methane molecule reacts with two molecules of oxygen to produce a carbon dioxide molecule and two molecules of water represented by $2H_2O$.

Interacting molecules

All chemical changes involve energy changes. Many reactions need a supply of heat energy before they can start. Coal must be heated before it begins to burn, for example. In general, once a reaction starts, it may produce heat, or it may continue to take in heat and require heat to make any change take place.

In a molecule, the electrons are arranged in layers, or orbitals, as in an atom. But the shapes of the orbitals are more complicated because there is more than one atom in a molecule. In some molecules, the electrons are arranged so that they have lots of energy. These are high-energy molecules, such as methane.

If the atoms in a high-energy molecule are rearranged, the energy can be released. One way of doing this is to break the molecule into parts and reassemble the parts into low-energy molecules. This is what happens when methane, for example, is burned. In plants, a series of chemical reactions converts low-energy carbon dioxide gas from the air into high-energy sugar molecules, which the plant uses for food. This process, which takes place only in sunlight, is called photosynthesis.

▼ All chemical reactions involve energy changes. Some reactions, such as the burning of waste methane gas in the desert near an oil field, give out energy. Other reactions extract energy from their surroundings. Photosynthesis extracts energy in the form of sunlight.

Speeding up chemical reactions

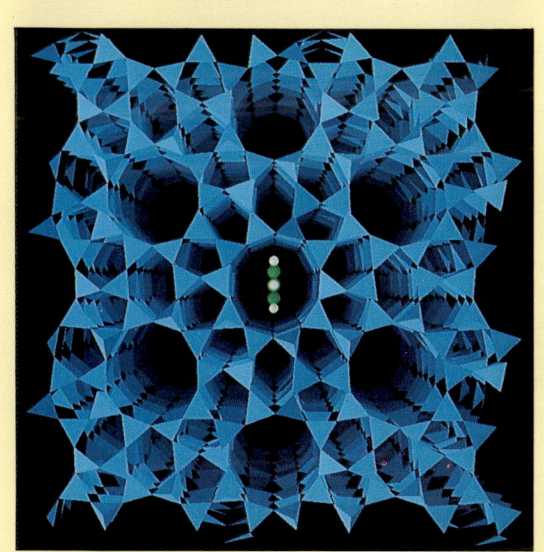

A catalyst is a substance that speeds up a chemical reaction without itself being used up. The hardener in a two-part epoxy adhesive (top) is a catalyst that speeds up the setting process. Minerals known as zeolites are used as catalysts in the manufacture of petrol from the alcohol methanol. Inside the zeolite (above) are minute channels and cavities. The methanol molecules enter these channels, where they give up their oxygen atoms, combine and form petrol. Zeolites are also often used in water softening. They change hard water to soft by taking out certain chemicals.

◀ The demolition of a hotel. Explosives release energy much faster than it can spread into the surroundings. This raises the temperature, increasing the speed of the reaction even further. The reaction creates a large volume of expanding gas, and an explosion takes place.

Molecules in motion

Spot facts

- There is only one temperature at which molecules stop moving: −273.15°C. This is the lowest temperature of all, called absolute zero.

- At very low temperatures, around −270°C, the gas helium becomes a liquid. A cupful of liquid helium empties as the liquid flows up the side of the cup. It is like water flowing uphill.

- There is more heat energy locked up in an iceberg than in a cup of boiling water. This is because the iceberg, although it is colder, is much larger.

- Sounds travel through air at a speed of about 1,224 km/h, depending on the temperature. Sound travels 15 times faster through steel than it does through air.

The molecules in objects around us are always moving. This movement is a form of energy that we call heat. The molecular movements are random vibrations – small back-and-forth movements. The higher the temperature, the faster are the molecular vibrations. But molecules can also move in more regular ways. Sounds are caused by molecules of the air vibrating in a regular, wave-like manner. We can use sound waves to carry information, as in speech. We also get pleasure from music because of the regular vibrations produced by musical instruments. Sound is important to us as a way of communicating. Animals, too, use sound to communicate. In addition, some animals use very high-pitched sounds to find their way around. Such ultrasounds are also finding many uses in science and medicine.

▶ Drums and cymbals are percussion instruments. They produce sounds when tightly stretched skin or metal is struck. The vibrating metal or skin makes molecules of air vibrate. The vibrating air molecules create the sounds.

Heat and temperature

There is a difference between the temperature of an object and the amount of heat it contains. The temperature measures how fast the molecules of the object are moving. In a hot object, the molecules vibrate back and forth at great speed. In a cold object, they vibrate more slowly. But the amount of heat energy in an object is the total energy of all its molecules. This depends upon both the temperature and the amount of material in the object. Even though its temperature is higher, a very hot small object does not contain as much heat as a larger object that is cooler. A large iceberg contains enough heat to boil water if only it could be collected and used.

▼ In a mercury or alcohol thermometer, the liquid in the bulb expands and moves up the thin tube. The scale indicates the temperature. In a bimetallic thermometer, a strip made of steel and brass fixed together is used. The brass expands more than the steel as the strip gets warmer. This makes the strip bend, moving the needle

There are three common scales for measuring temperature. On the Celsius, or centigrade, scale, the temperature at which water freezes is 0°C, and water boils at 100°C. The scale is named after the Swedish physicist Anders Celsius. On the Fahrenheit scale, named after the German scientist Gabriel Fahrenheit, the freezing point of water is 32°F, and the temperature of boiling water is 212°F. The third scale is called the Kelvin scale, after a Scottish scientist, Sir William Thomson, who became Lord Kelvin. On this scale, the lowest possible temperature, called absolute zero, is 0K. One kelvin is equal to 1°C. On the Kelvin scale, 0°C becomes 273K, and 100°C is 373K.

over the scale. The thermocouple thermometer is made from wires of two different metals. If the welded ends of the wires are at different temperatures, a small voltage is produced. This moves the needle over the meter scale. The digital thermometer has a thermocouple in its probe, and an integrated circuit creates a digital read-out.

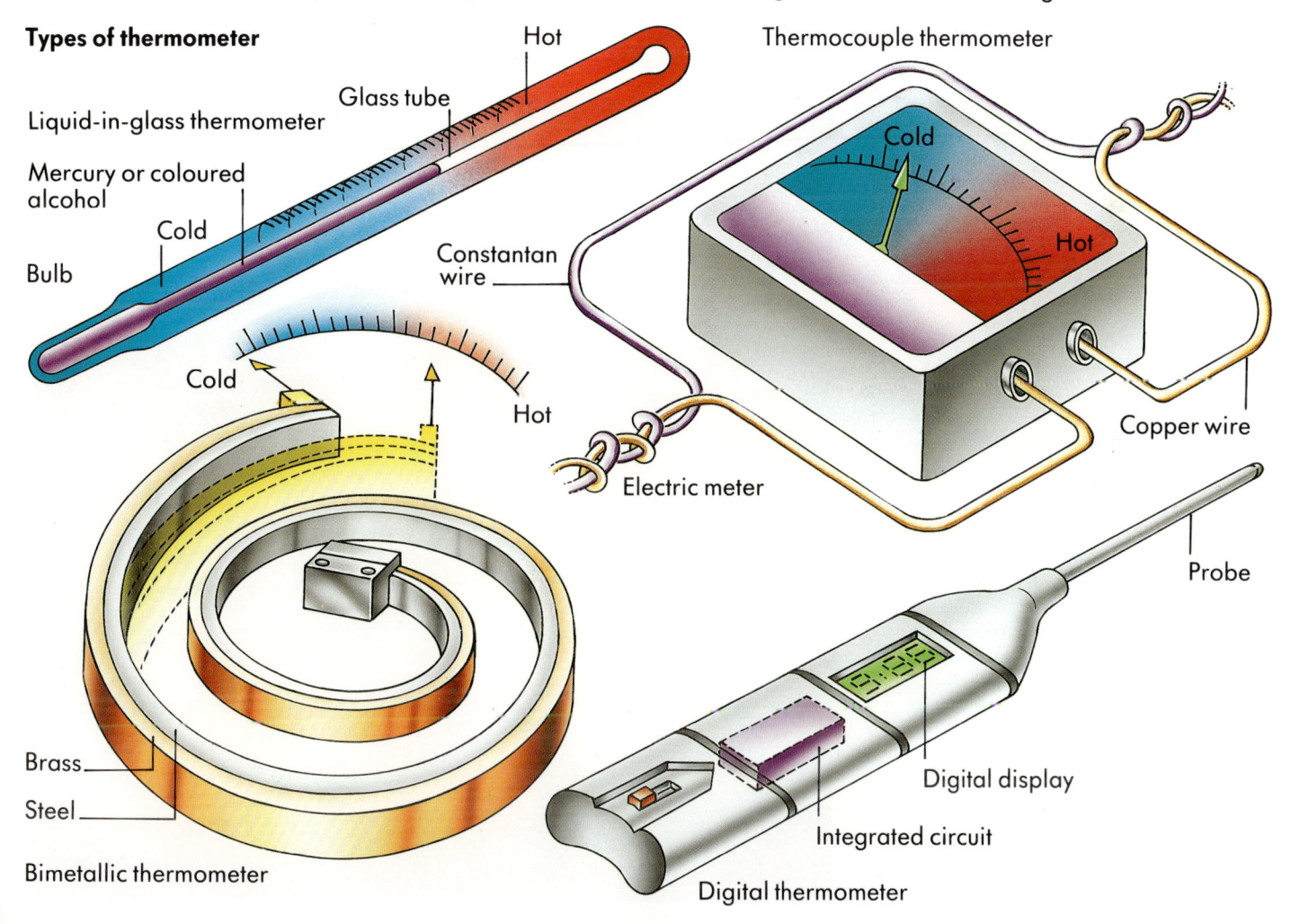

Types of thermometer

Travelling heat

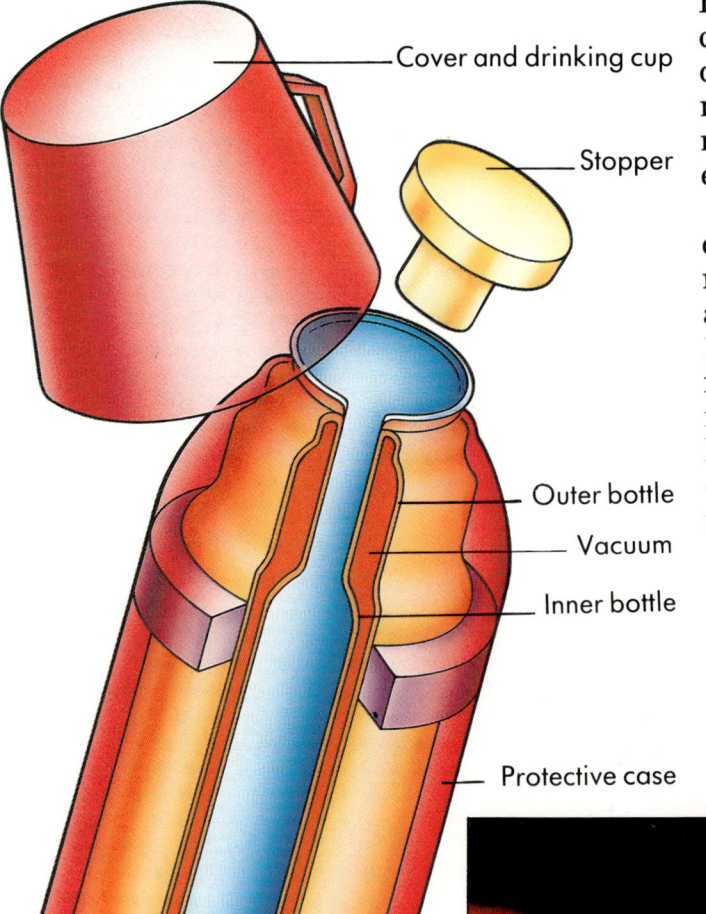

Heat can travel in three ways: by convection, conduction and radiation. Convection and conduction can take place only where there is matter. They are processes that involve moving molecules. But heat can also travel through empty space, where there is no matter.

Heat travels through empty space as energy-carrying waves, called radiation or infrared rays. William Herschel, a famous English astronomer, discovered infrared rays in about 1800. Using a prism, he split the light coming from the Sun into the colours of the rainbow. He noticed that heat was coming from the red light. He realized that there must be invisible rays carrying heat in the sunlight. He called them infrared rays.

Infrared rays are very similar to light rays and travel at the same speed, 300,000 km/s. Like light rays, they can be reflected and absorbed by matter. Infrared rays are given out by all objects. They are given out by our bodies, for example, and doctors can use the rays to detect some diseases.

▲ The Thermos, or vacuum, flask keeps liquids hot or cold because its walls are airless. Heat that would normally pass from or to the contents of the flask by convection or conduction is prevented from doing so by the vacuum.

▶ The NASA Space Shuttle is covered with tiles made of silica, a poor conductor of heat. The tiles protect the craft from the high temperatures produced when it re-enters the Earth's atmosphere.

Convection carries heat in liquids and gases. When liquids and gases are heated, they expand and become less dense. The warmer material rises. This movement of heated material sets up a current, called a convection current. Such currents are seen on a large scale in the atmosphere, where they cause winds and breezes. Convection currents in the oceans carry heat from the tropics to colder parts of the world. On a smaller scale, convection currents spread heat through liquid in a saucepan.

When a solid object is heated, molecules near the source of heat vibrate more rapidly than those farther from it. The rapidly vibrating molecules bump into the molecules next to them, passing on some of their energy. These molecules, in turn, bump against their neighbours. In this way the heat flows to all parts of the object. This heat-spreading process is called conduction. Some materials, such as metals, are good conductors of heat. Materials which do not conduct heat easily, such as rubber, are called insulators.

▼ A glider pilot must seek out rising currents of warm air, called thermals, to carry the glider upwards. Gliding birds also use these convection currents to gain height.

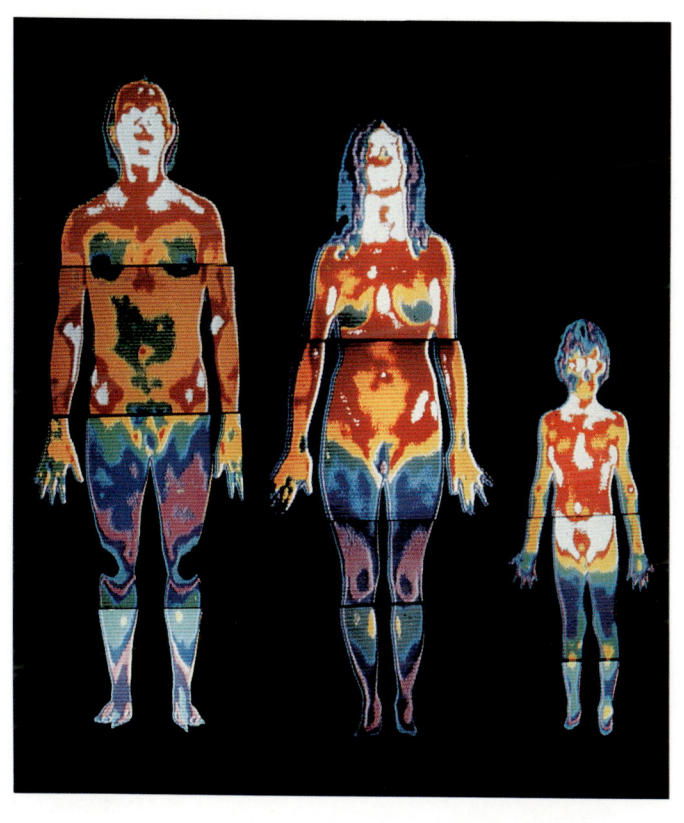

▲ A thermal image of a family. Hot objects give off infrared radiation, which can be photographed. The hottest parts of the people's bodies are brightest.

43

Sound

All sounds are made by vibrating objects. For example, when you speak, your vocal chords vibrate. As a vibrating object moves, it pushes the air molecules in front of it. This creates a region of high pressure, where the air molecules are pressed together. When the object moves backwards, it creates a region of low pressure, where the molecules are farther apart than normal. These high- and low-pressure regions travel out from the vibrating object as sound waves, like ripples on a pond.

When the sound waves enter a person's ear, an intricate system of tiny bones and a very thin membrane causes the eardrum to vibrate, and the sound is heard. The size of a vibration or wave is called its amplitude. The greater the amplitude of a sound wave, the louder it is. The loudness of a sound is measured in units that are called decibels.

Frequency

The number of complete vibrations or waves per second is called its frequency. The greater the frequency of a sound wave, the higher is the pitch of the sound we hear. Frequency is measured in units called hertz, after Heinrich Hertz who worked on radio waves between 1885 and 1889. One hertz equals one complete vibration per second.

Most people can hear sounds with frequencies as low as 20 hertz. Young people can hear sounds with frequencies up to 20,000 hertz, but most older people can hear sounds only up to 16,000 hertz. Cats can hear sounds up to 25,000 hertz and dogs do even better – they can hear sounds up to 35,000 hertz.

Sounds travel faster through liquids and solids than through gases. In seawater, for example, the speed of sound is nearly 1,500 m/s,

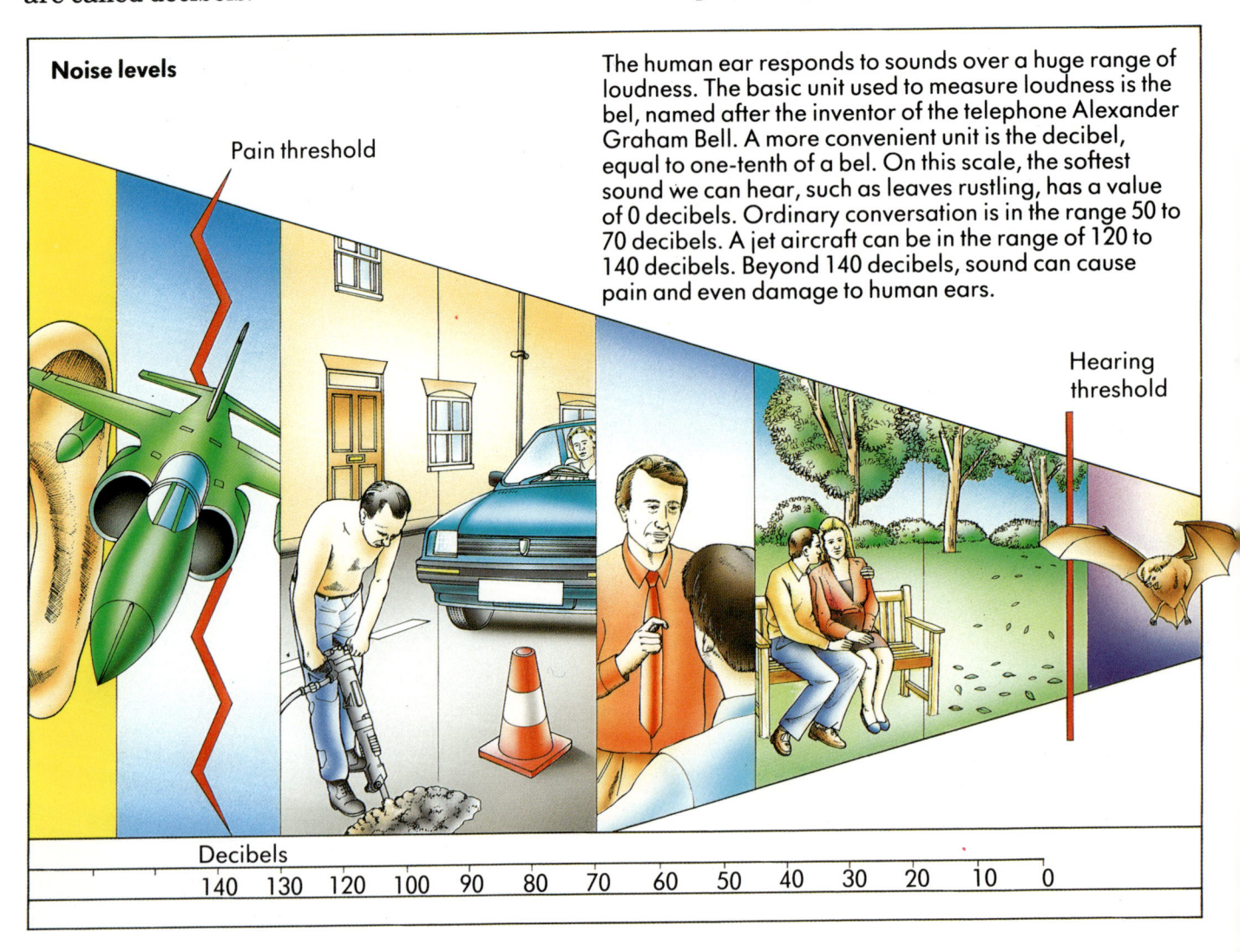

Noise levels

The human ear responds to sounds over a huge range of loudness. The basic unit used to measure loudness is the bel, named after the inventor of the telephone Alexander Graham Bell. A more convenient unit is the decibel, equal to one-tenth of a bel. On this scale, the softest sound we can hear, such as leaves rustling, has a value of 0 decibels. Ordinary conversation is in the range 50 to 70 decibels. A jet aircraft can be in the range of 120 to 140 decibels. Beyond 140 decibels, sound can cause pain and even damage to human ears.

four times the speed in air. The speed of sound in air is a little less than 350 m/s. Speed also depends upon temperature. The higher the temperature, the greater is the speed. This is why sounds seem louder and travel farther at night than during the day. At night, the air near the ground is cooler than the air above it. Sound waves travelling upwards into the warmer air are bent back towards the ground, carrying sound a greater distance. The bending of waves in this way is called refraction.

Sound waves can also be reflected. They can bounce off surfaces they hit, causing echoes. Sounds, like all waves, can also bend around corners. This is called diffraction. The amount a wave is bent depends upon its frequency. Lower frequencies are bent more than higher ones. This is why words overheard round the corner of an open door sound mumbled.

▼ When the string of a piano is struck, the profile of the string assumes a wave pattern. This is called a standing wave. The simplest standing wave is the fundamental. The higher-pitched harmonics correspond to standing waves of higher frequency.

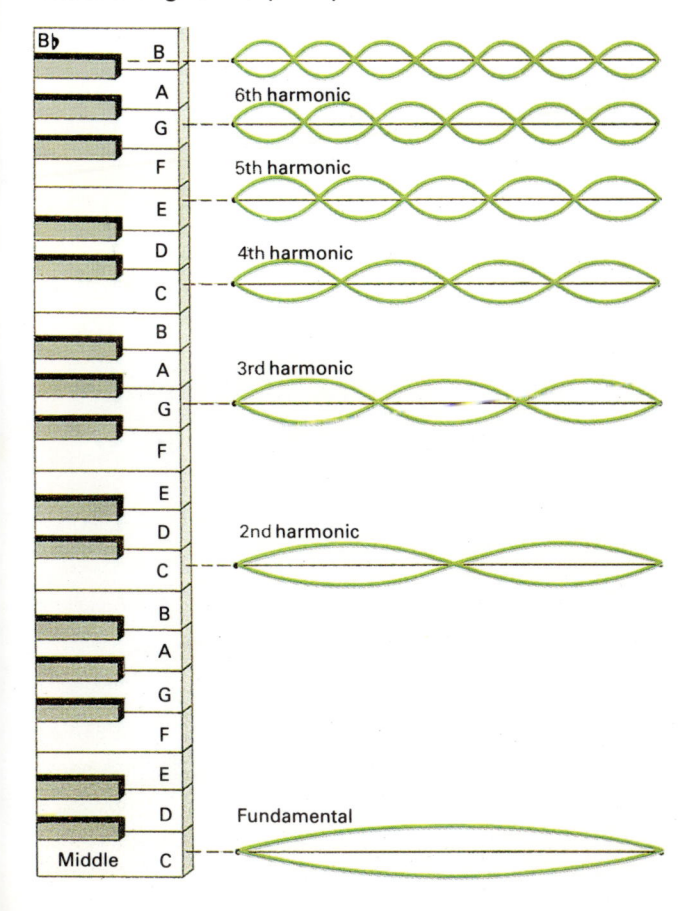

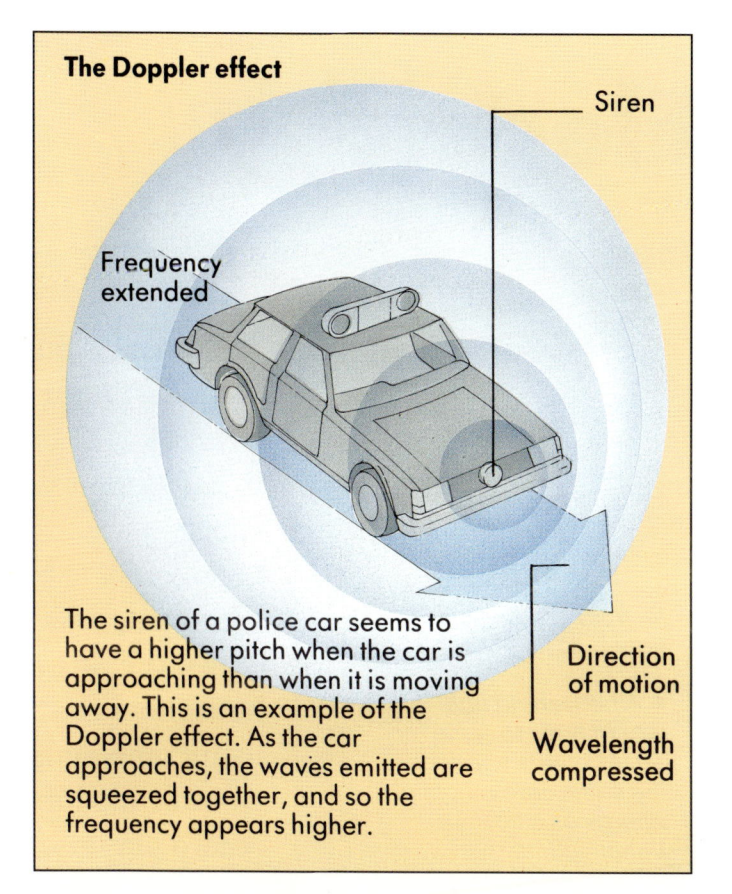

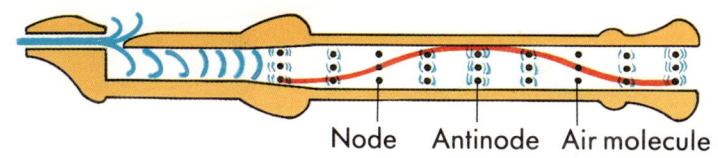

The Doppler effect

The siren of a police car seems to have a higher pitch when the car is approaching than when it is moving away. This is an example of the Doppler effect. As the car approaches, the waves emitted are squeezed together, and so the frequency appears higher.

The recorder

▲ When somebody plays a wind instrument, such as a recorder, a standing wave is produced inside the instrument. It consists of a stationary pattern of vibrating air. A point where there is no vibration is called a node; an antinode is where vibration is greatest.

Electricity

Spot facts

- About 3 million million million electrons pass through a burning light bulb every second.

- Electric eels store enough electricity in their tails to light up 12 light bulbs.

- A household light bulb would have to shine for 10,000 years to release the same amount of light energy as a flash of lightning.

- Electrons are not only particles, they are also waves! Electron waves are used instead of light waves in electron microscopes.

The world is built from atoms made from a small number of different particles. Most of the particles carry electric charges. It is the force between these charges that helps to hold atoms together. Taming the energy held in these charges – seen uncontrolled in a lightning flash – has been one of the most important successes of science. Electricity has become our obedient servant, lighting our cities, turning the wheels of industry and powering our household gadgets. New ways of controlling electricity, using materials called semiconductors, have made possible modern electronics, such as those used in radios, record players, televisions and computers.

▶ The bright lights of Los Angeles. Electricity is the servant and messenger of the modern world. It is the most convenient source of power ever discovered. It supplies light and heat as well as mechanical power. It is also easily and cheaply carried along wires or cables.

Electric charges and fields

The origin of electricity lies inside the atoms that make up matter. Electrons and protons carry tiny amounts of electricity. They are said to have an electric "charge". There are two kinds of electric charge. The electron has one kind, called a negative charge. Protons have the other kind, called a positive charge. If two negative charges or two positive charges are brought close together, they repel each other and push apart. If a positive charge is brought near a negative charge, the charges attract each other and pull together. In other words, like charges repel each other, and unlike charges attract each other.

Normal atoms have no overall electric charge. The same number of negatively charged electrons orbit the nucleus as there are positively charged protons in it: the negative and positive charges are balanced. However, atoms and the objects that they make up can become electrically charged by losing or gaining electrons. If an object gains electrons, it becomes negatively charged. If the object loses electrons, it becomes positively charged.

Electric fields

The region around a charged object or particle, where the electric force can be determined, is called an electric field. The strength of the field at a particular point depends on the size of the charge on the object and the distance between the point and the object. The field becomes stronger the closer the point is to the object.

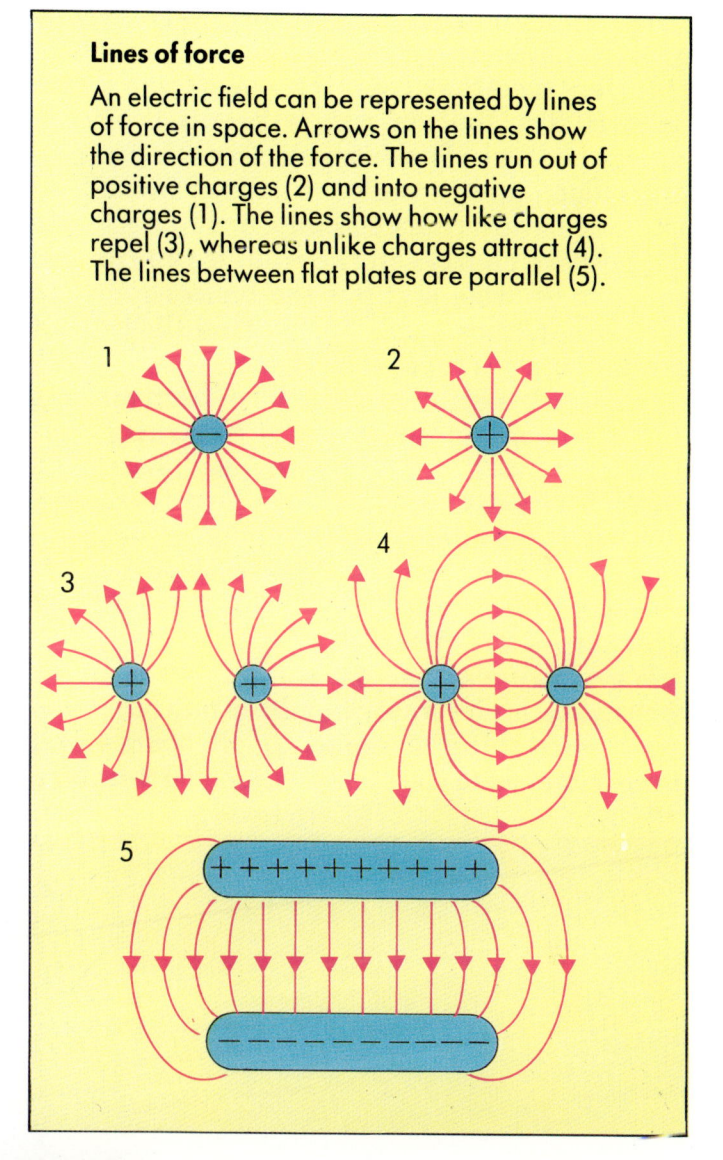

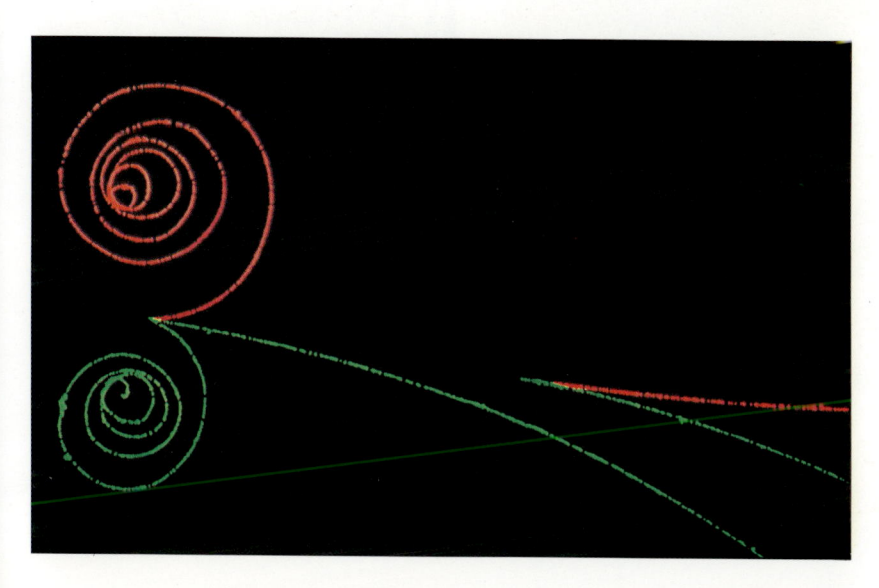

Lines of force
An electric field can be represented by lines of force in space. Arrows on the lines show the direction of the force. The lines run out of positive charges (2) and into negative charges (1). The lines show how like charges repel (3), whereas unlike charges attract (4). The lines between flat plates are parallel (5).

◀ Here, particles with opposite charges move in a bubble chamber. A magnetic field makes the particle tracks spiral in different directions. The green tracks are made by electrons with a negative charge. The red tracks are made by positrons, particles with the same mass as electrons but carrying a positive charge. Electrons are normally found orbiting the central nucleus of an atom. Electrons are also the origin of electricity.

Static electricity

If you rub a balloon on a woollen sweater, you give the balloon an electric charge. The rubbing transfers electrons from the sweater on to the balloon. The balloon gets a negative electric charge because of its extra electrons. If you hold the balloon up to a wall, the balloon sticks to it. The negatively charged balloon is attracted to the positive charges in the wall. If you rub two balloons on wool and put the balloons next to each other, they push apart. This shows that similar charges repel each other.

In these experiments, you have been making and using static electricity – electricity that does not move but stays in one place. The ancient Greeks, about 2,500 years ago, did similar experiments by rubbing a piece of amber – a fossilized resin material – with fur. The word electricity comes from elektron, the Greek word for amber.

When you walk across some types of nylon carpet, static electricity builds up on you. If you touch something metallic, small sparks will jump from you to the metal. When you take off a nylon shirt or blouse, you can sometimes hear a crackling sound, and see small sparks. These sparks are like miniature lightning flashes.

Lightning conductor

In 1752 a famous American scientist and inventor called Benjamin Franklin did a dangerous experiment. He flew a kite during a lightning storm. Electricity flowed down the string of the kite, making a small spark on a metal key near his hand. This showed that lightning was just a large electric spark. Later, Franklin used his discovery to invent the lightning conductor. This is a metal strip that runs from the top of a tall building to the ground. It carries the electricity safely away if lightning strikes the building.

▶ Lightning is caused when a large electric charge builds up on a cloud as the result of ice and water particles in the cloud rubbing together. Positive charges build up at the top of the cloud and negative electrons at the bottom. The electrons suddenly leap from the cloud to the ground, or to another cloud.

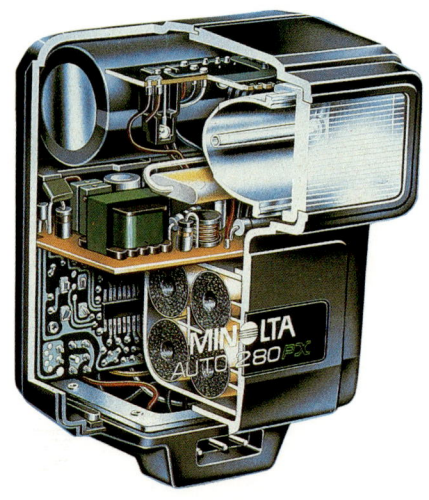

▲ An electronic flashgun uses devices called capacitors to store electric charges. Capacitors are also used in computer memories and radio circuits.

◀ This child's hair is standing on end because it is electrically charged. Some electrons have rubbed off her hair on to the comb, giving her hair a positive charge. Because each of her hairs has the same charge, they repel each other and stick out.

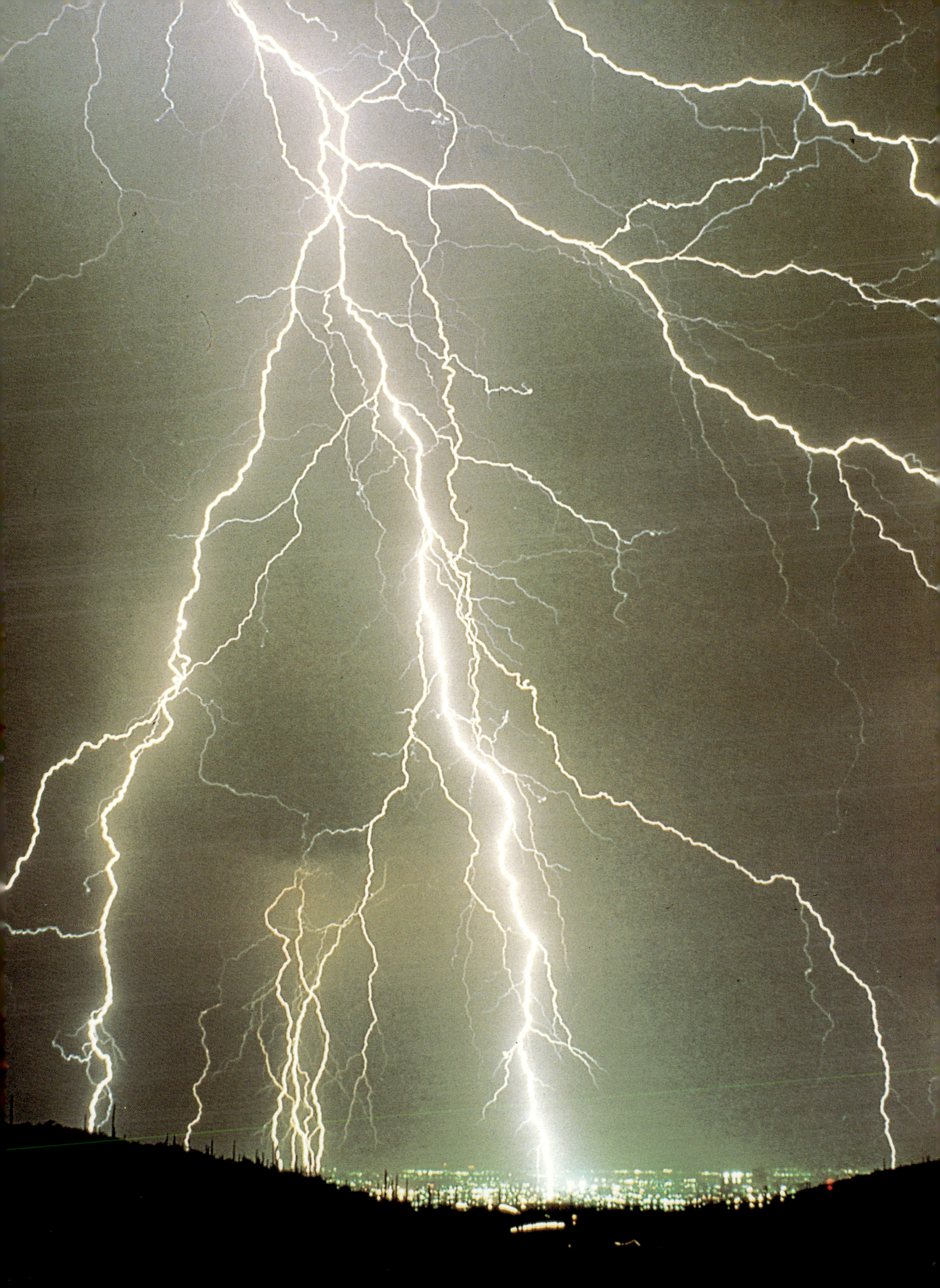

Electrons on the move

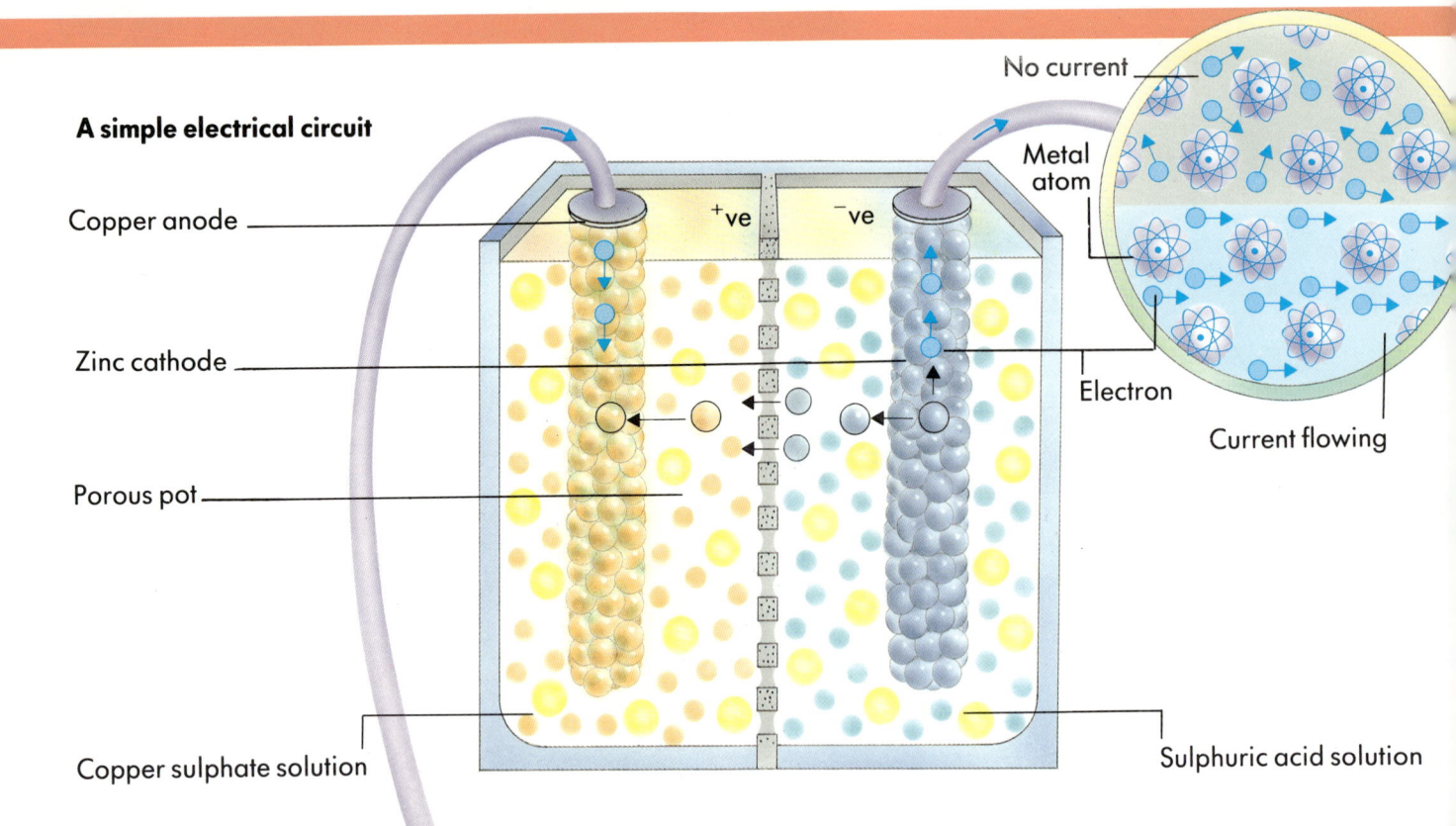

A simple electrical circuit

- Copper anode
- Zinc cathode
- Porous pot
- Copper sulphate solution
- +ve / −ve
- Metal atom
- Electron
- No current
- Current flowing
- Sulphuric acid solution

When a wire is connected to a battery in a continuous path or circuit, the electrons in the wire move along it. An electric current flows through the wire like water flowing through a pipe. But to make water flow through a pipe, a pump is needed to produce a pressure difference between its ends.

In an electrical circuit, the battery acts like an electron pump and produces an electrical pressure difference. The electrical pressure supplied by the battery is called the potential difference. It is measured in units called volts, which are named after the Italian scientist Alessandro Volta, who invented the electric battery in 1800.

The greater the voltage, the more electrons flow in the wire. We could try to measure the amount of current by counting the number of electrons that pass by. But this would be impossible because there are huge numbers of electrons in most electric currents. Instead, scientists measure electric current in amperes. One ampere is equal to a flow of about 6 million million million electrons every second. The ampere is named after a French professor of mathematics, André Marie Ampere, who did important work on the magnetic effects of electric currents in the early 1820s.

▲ A simple type of cell, or battery, called the Daniell cell. It contains a copper anode (positively charged electrode) in a copper sulphate solution and a zinc cathode (negatively charged electrode) in sulphuric acid. Atoms in the zinc cathode give up electrons, which flow through the circuit and form the current. When the electrons reach the anode, they combine with copper ions from the solution to form atoms of copper. The result is a flow of electrons from cathode to anode.

▼ An ordinary but dangerous light bulb (here shown broken but still working) glows as the electric current flows through the filament. The thin filament has a high resistance and is made from tungsten.

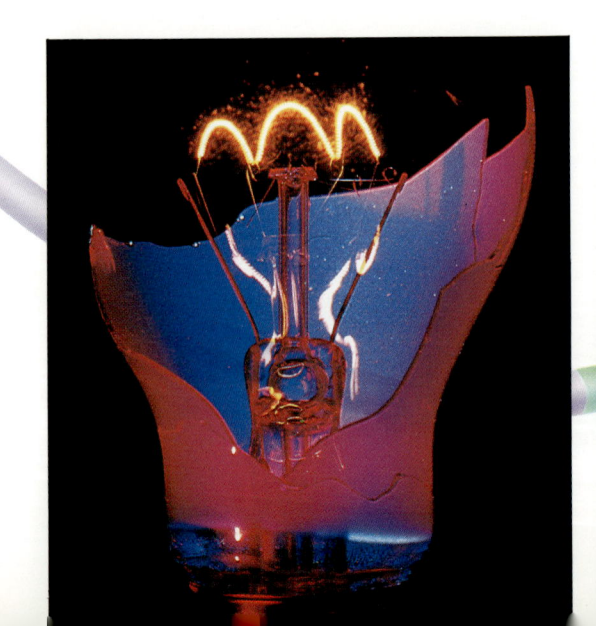

◀ In a metal wire, electrons are free to move about in any direction. As soon as a voltage is applied, as in a circuit containing a battery, the electrons move towards the positive terminal, or anode, of the battery.

▲ A space satellite, such as the *Hipparcos* star-mapping satellite shown here, uses solar cells to generate electricity from sunlight.

▼ Electrical resistance occurs when the flow of electrons is slowed down by collisions with the metal atoms or with impurity atoms in the metal. The electrons lose energy to the atoms. A component with a known resistance is called a resistor.

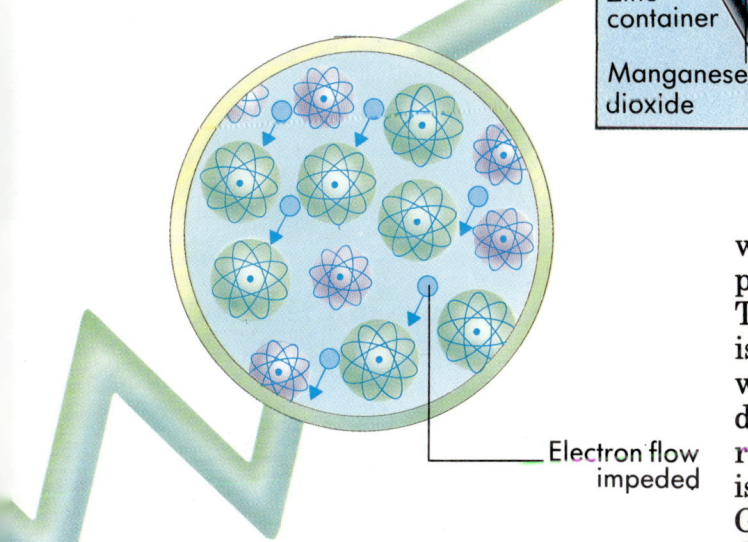

The dry battery

A dry cell, or battery, has a rod of carbon down its centre, acting as the anode, or positive terminal. The cathode, or negative terminal, is the zinc casing. Paper soaked with ammonium chloride solution lines the casing. A chemical reaction causes the zinc atoms to produce electrons, which are able to flow around a wire connected between the terminals. A black powder of manganese dioxide surrounds the carbon anode.

Electrons slow down as they travel through a wire, just as water slows down when it flows in a pipe. This slowing effect is called resistance. The more resistance a circuit has, the harder it is to keep the electric current flowing. A battery with a high pressure, or voltage, is needed to drive a current through a circuit with a large resistance. The amount of resistance in a circuit is measured in units called ohms, after the German scientist Georg Simon Ohm. In 1827 he discovered that the resistance of a wire is equal to the voltage divided by the current. This relationship is called Ohm's law.

Using electricity

When an electric current flows through a wire, the wire heats up. This is because of the resistance of the metal in the wire. The electrons jostle against the atoms of the metal, causing them to move. This raises the temperature of the wire because higher temperatures are linked to faster movements of atoms. The greater the resistance of the metal in the wire, the more energy the electrons lose to the atoms, and the greater is the heating effect of the electric current.

In an ordinary electric light bulb, the filament is made of thin wire, because the resistance of a thin wire is greater than that of a thick wire. The filament is also made in the form of a coil. This allows a greater length of wire to be used. A longer wire has greater resistance, and therefore gets hotter and brighter.

The heating effect of an electric current is used in a fuse. A fuse is a material of fixed low-resistance so that it stops conducting if an excessive current is passed through it. If a fault develops in an electrical circuit, too much electricity may flow along the wires and heat them up. Without a fuse this could start a fire.

▼ Many children's toys use simple electric motors powered by batteries. These convert electricity into mechanical energy for movement.

Bright lights

But if there is a fuse in the circuit, the fuse quickly melts. This breaks the circuit and stops the electricity flowing. The wiring cools down before a fire can start.

Electricity is especially useful and convenient because it can be made to do so many things. It can easily be converted into other forms of energy. A loudspeaker, for example, converts electricity into sound. Electricity can keep us cool when the weather is hot, and warm us

Coloured city lights are made from long, thin tubes that glow when electricity passes through them. They are called discharge tubes. A discharge tube is filled with a vapour or gas, such as neon, at very low pressure. When the tube is switched on, electrons are emitted by electrodes at the ends of the tubes. The electrons travel along the tube, striking the atoms of the gas and causing them to emit light. The colour of the light depends on the gas in the tube. Neon tubes glow bright red; argon tubes glow blue.

Fluorescent tubes are a type of discharge tube. The tube contains mercury vapour, which produces invisible ultraviolet rays when electricity flows through it. The rays fall on a fluorescent coating on the inside of the tube. This material absorbs the ultraviolet rays and re-emits them as white light.

Discharge tube

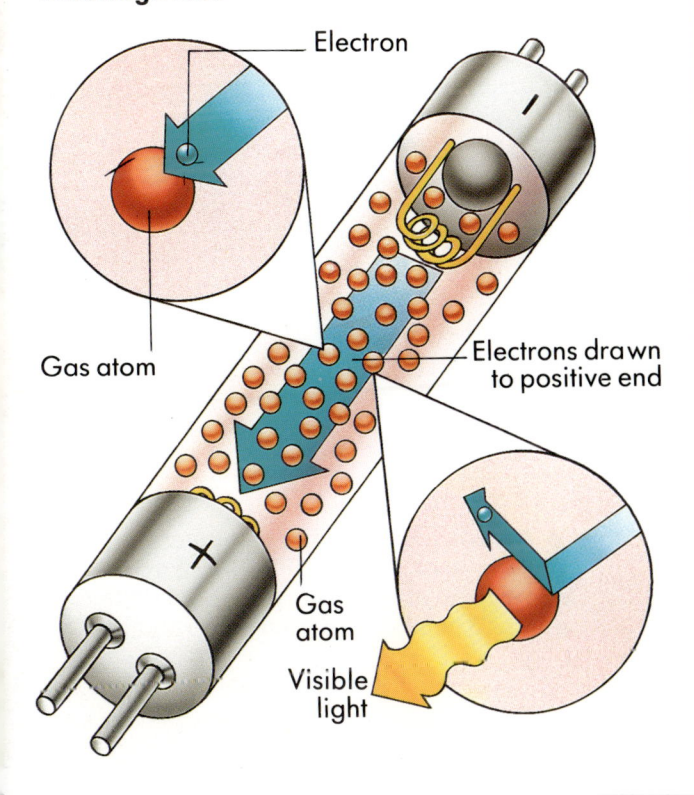

▼ Electroplating is used to coat a metal object with a thin layer of another metal. The object is hung in a tank holding a solution of the salt of the metal with which it is to be coated. A sheet of a metal is also hung in the tank. The object is connected to the cathode (negative terminal) of a battery, and the metal sheet is connected to the anode (positive terminal). The metal at the anode slowly dissolves, and the object becomes coated with the metal. All sorts of objects are electroplated, from silver teapots to chromium car bumpers.

Electroplating

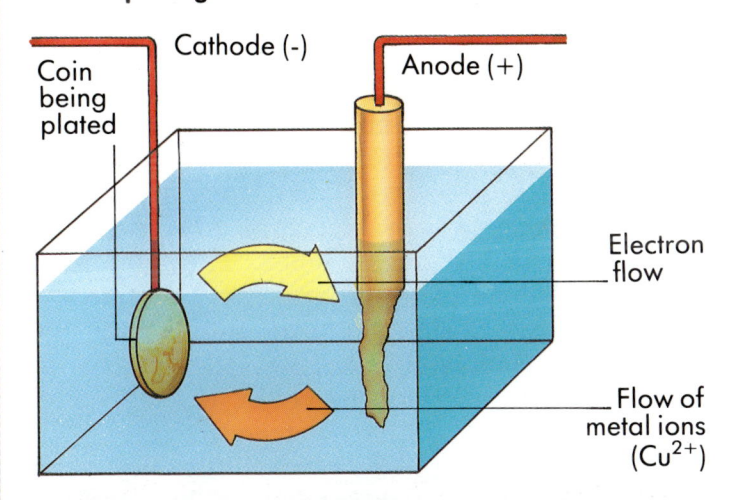

when it is cold. Electricity is vital to everyday life. Hospitals, schools, offices and shops would come to a standstill if it were not for electricity.

Electricity can be used to break a substance down into other substances. This process is called electrolysis. Electrolysis is used in industry to extract metal from mineral ores. For example, aluminium is prepared by passing an electric current through molten aluminium oxide, prepared from the mineral bauxite.

The current is passed via electrodes through a liquid called an electrolyte. If a current is passed through a solution of common salt (sodium chloride), chlorine gas is produced at the positive electrode, or anode. Sodium is released at the negative cathode. But it immediately reacts with the water in the solution to form hydrogen gas and sodium hydroxide. Chlorine, hydrogen and sodium hydroxide are all valuable substances made in this way.

Semiconductors

Semiconductors are materials that conduct electricity only with difficulty, unless they have been treated in some way. The most important semiconductor is silicon. Silicon is a successful conductor of electricity after minute amounts of other materials are added to it. This process is called doping. Silicon doped by adding minute amounts of phosphorus is called n-type; silicon doped by adding boron is called p-type. The effect of adding phosphorus to silicon is to add extra electrons. The extra electrons carry electricity through the silicon.

Adding boron to silicon also allows electricity to flow, but in a different way. Boron atoms have one fewer electron than silicon atoms. So when boron is added to silicon, there are places in the silicon where electrons are missing. These places are called holes. They act like positive electric charges, and carry electricity through the silicon in the same way as electrons. However, because they have a positive charge, they move in the opposite direction to the negative electrons.

Electronics engineers control the electrical properties of a semiconductor by adding precise amounts of impurities. This enables them to produce "integrated circuits". These have all the parts of an electronic circuit on a tiny silicon chip. Without them miniature televisions and personal stereos could not exist.

Inside a semiconductor

(1) Free electrons are the most important carriers of electricity in an n-type semiconductor. There may be a few holes. In a p-type semiconductor, holes are the main carriers of electricity, with a few electrons present. (2) In an n-type semiconductor, electrons are attracted to the positive terminal of the battery. In p-type material, the holes are attracted to the negative terminal. (3) When a slice of n-type material is joined to a slice of p-type material, only a small current flows when the negative terminal of the battery is connected to the p-type material. (4) When the battery is reversed, a large current flows because both holes and electrons move freely. This set-up acts as a semiconductor diode, which allows current to flow in only one direction.

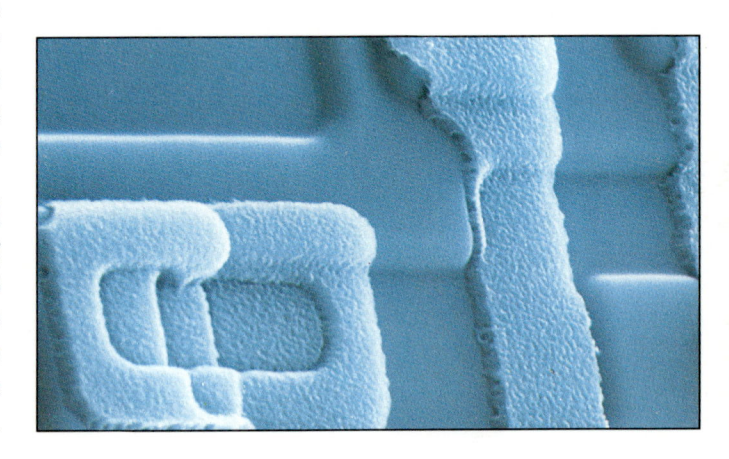

▲ A magnified view of a computer memory chip made of millions of "cells". Each cell can trap an electric charge representing a part of a number and hold it for reading.

◀ Removing silicon wafers from the doping oven. This is one of many stages in the production of silicon chips. During the doping operation a mask is applied over the wafer so that the dopants, such as boron and phosphorus, reach only certain areas.

▶ The chips on a wafer are tested and any faulty ones marked. The wafer is then cut into individual chips, and the faulty ones thrown away. The chips are put into a small plastic box, or case, with "legs", which act as connections to an external circuit. The circuits on the chip are connected to the legs by gold or aluminium wires.

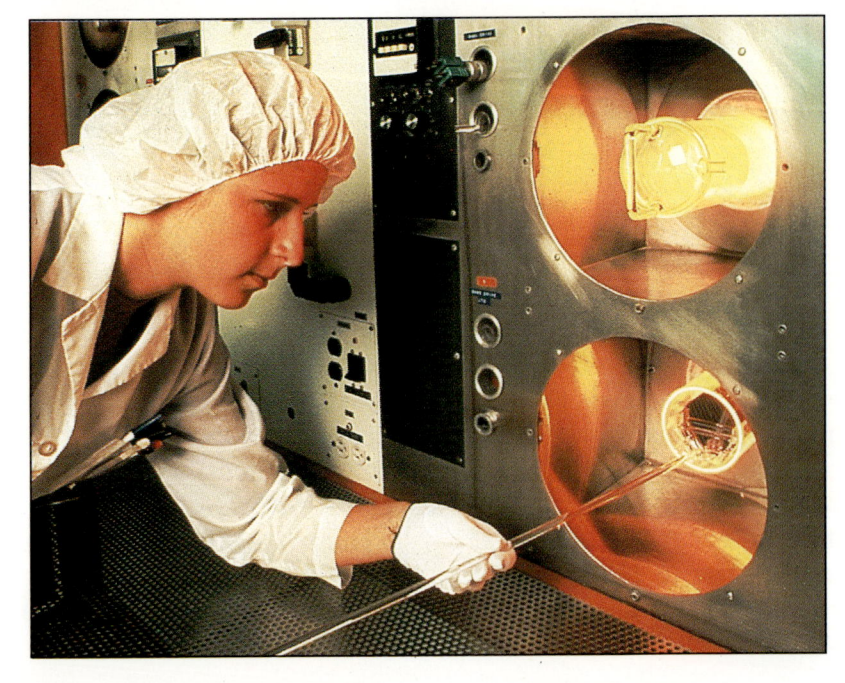

| 1 | 2 | Free electrons | Holes |

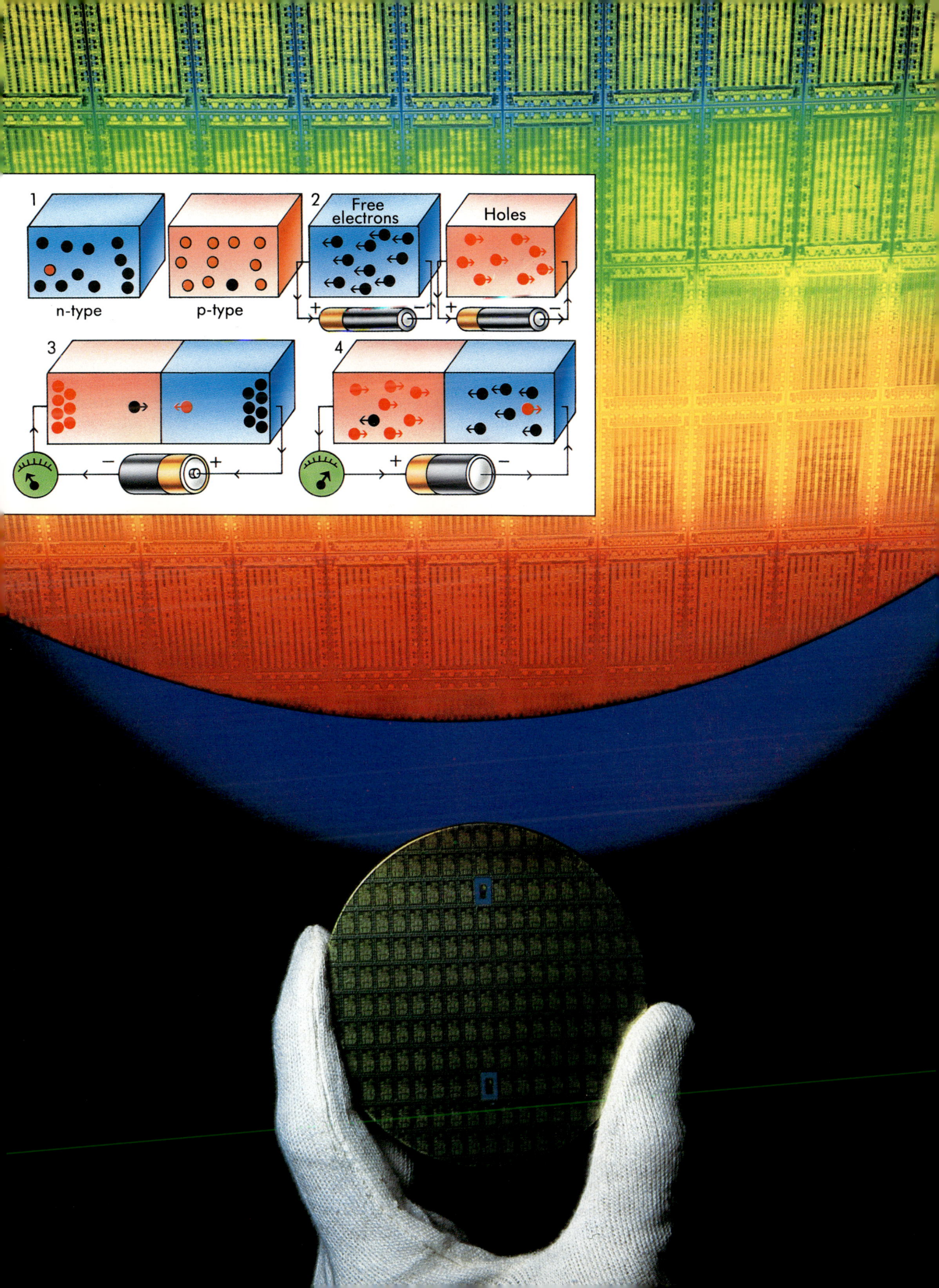

n-type   p-type

3   4

Magnetism

Spot facts

- A large electromagnet at the National Magnet Laboratory, USA, gets so hot that it needs 9,000 litres of cooling water every minute.

- The magnetic force around the giant planet Jupiter is 250,000 times greater than that of the Earth.

- The Earth's magnetic effects can be felt some 80,000 km out into space.

- The Earth's magnetism reverses its direction from time to time. The last reversal took place about 30,000 years ago.

- Migrating animals probably use the Earth's magnetism to help them find the way.

▶ A huge electromagnet shapes a blue haze of hot hydrogen inside a device which attempts to harness the power of nuclear fusion. This is the process which produces the Sun's energy. A strong magnetic field is used to contain the very hot material that is produced by the fusion process.

The ancient Greeks, more than 2,500 years ago, knew about magnets. They had discovered rocks which would attract small iron nails. Greek travellers told stories of mountains that could draw nails out of ships. These tales were untrue, of course, but they do demonstrate that at that time the Greeks knew about magnetism. It was not until the reign of Elizabeth I that the scientific study of magnets began. An English doctor, William Gilbert, pioneered the study of magnets and concluded that the Earth itself was a huge magnet. In 1820 the next advance was made by a Danish scientist, Hans Christian Oersted. He discovered that an electric current could produce magnetic effects. This form of magnetism, called electromagnetism, is used in electric motors and devices such as door bells and telephones.

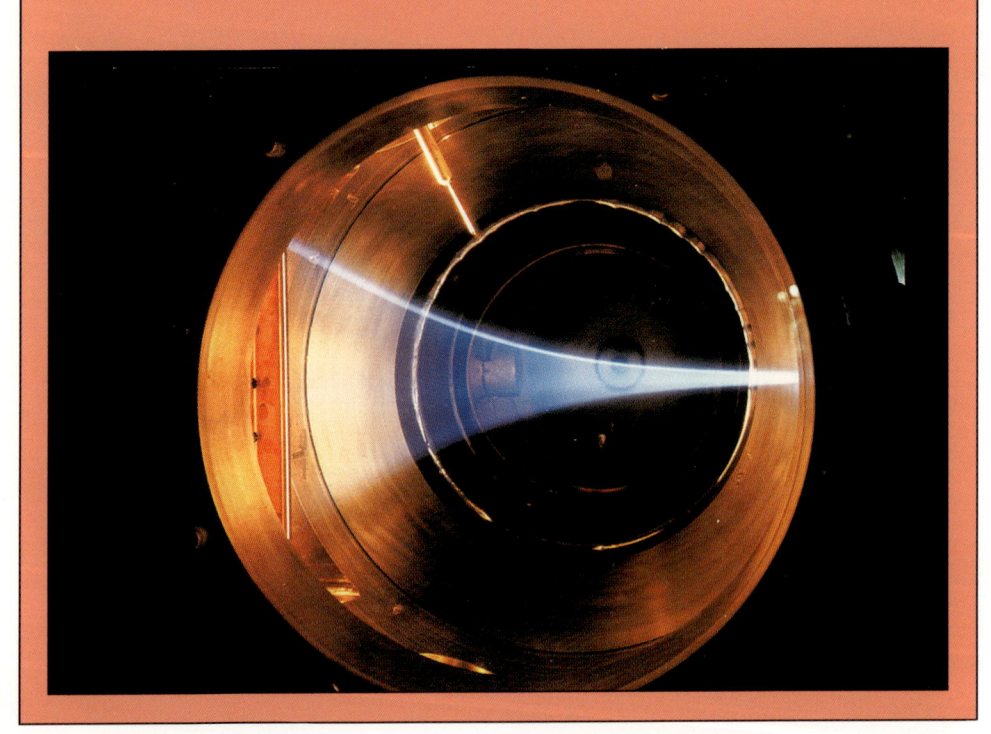

Magnets

Magnets attract things made of iron or steel. Some other metals, such as cobalt and nickel, are also attracted to a magnet. But most metals, such as gold, copper and aluminium, are not attracted in the same way. Neither are plastics, paper, cloth and glass attracted to magnets. These things are said to be non-magnetic.

Iron filings stick to a magnet most strongly at two points, usually near its ends. It is at these points, called the poles, that the magnet's power is strongest. One of them is called the north pole; the other is called the south pole. If a bar magnet is hung at its centre on a thread so that it can swing freely, the north pole points to the north. This effect is used in a compass. All planes and large ships are fitted with a compass as a vital part of the navigation system.

If you put the north pole of one magnet near the north pole of another, the magnets push apart, or repel each other. Two south poles behave in the same way and also repel each other. Magnets attract each other only if different poles are close together. Scientists say: "Like poles repel, unlike poles attract".

A giant magnet

The Earth acts as if it were a giant bar magnet, and has two magnetic poles. The Earth's north magnetic pole is in the Canadian Arctic, about 1,600 km from the true, or geographic, North Pole. The south magnetic pole is in Adélie Land about 2,400 km from the geographic South Pole in central Antarctica. The magnetism of these poles makes the north pole of a compass needle point to the north, and the south pole of a compass point south.

But because the Earth's magnetic and geographical poles are in slightly different locations, a compass does not point exactly due north or south. The slight difference between the magnetic and geographical poles is called the magnetic variation. It changes all the time. Navigators and mapreaders must allow for it when finding their way.

The Earth's magnetism is produced by the molten metal deep within the Earth's core. As the Earth spins, electric currents are created in the molten metal. These currents produce the Earth's magnetic force.

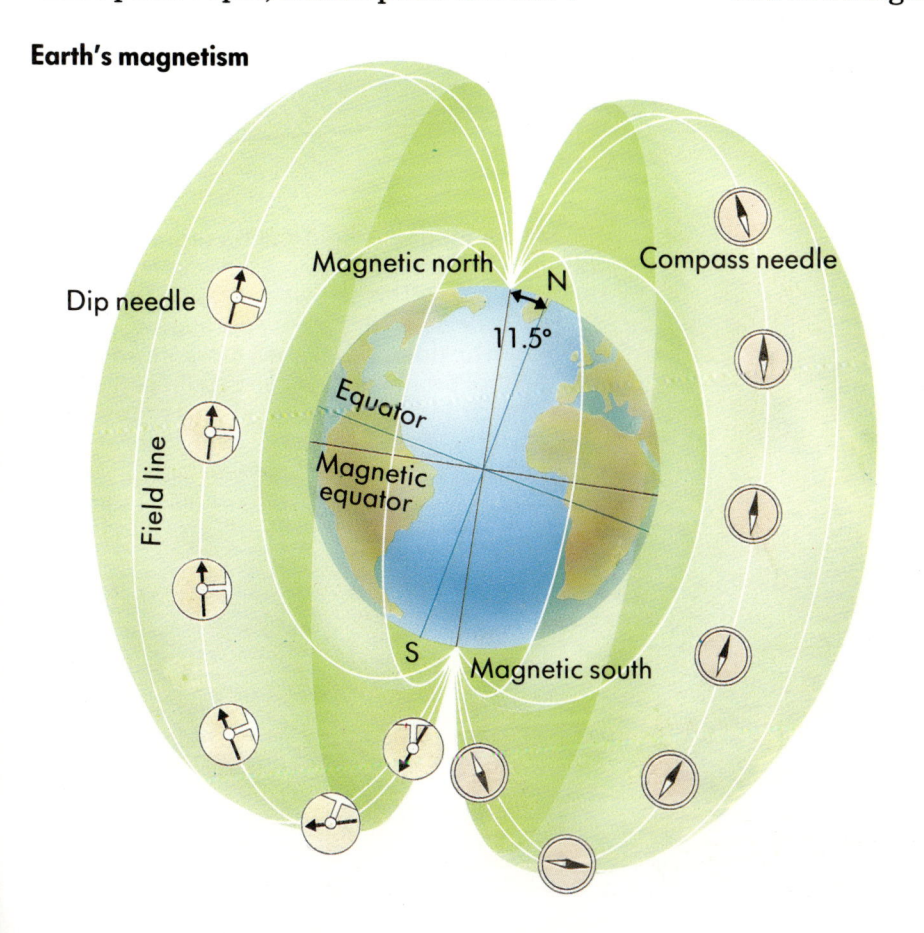

Earth's magnetism

▲ Iron filings cluster around the poles of a magnet, where the magnetic force is strongest.

◀ We live on a huge magnet! Magnetized needles that are free to rotate point more or less towards the Earth's magnetic poles.

Magnetic field

The space around a magnet, where the magnetic force can be detected, is called a magnetic field. A kind of map of a magnetic field can be made with iron filings. Put a magnet under a piece of white paper and sprinkle iron filings on top. Tap the paper gently. You will see the filings form a pattern of lines. These lines follow the lines of force of the magnet.

The lines of force are close together near the poles, where the effect of the magnet is concentrated. Away from the poles, the magnetic effect is weaker and the lines of force are farther apart. The lines show the force a small north pole would experience in the field.

How are magnets made? One way is by stroking a piece of steel, such as a thin needle, with a magnet. To understand why this happens remember that the needle, like all other things, is made up of atoms. In a magnetic material, such as iron or steel, the atoms are like miniature magnets.

The atomic magnets are grouped together into small areas called domains, which are like mini-magnets inside the material. When an iron needle is stroked with a magnet, the domains become lined up so that they all point in the same direction. In this way the needle has become magnetized.

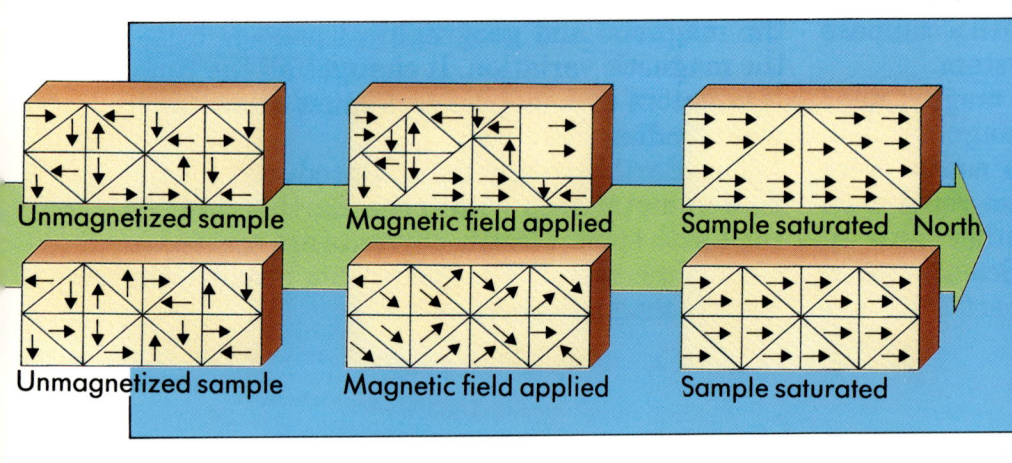

Magnetic domains

In a magnetic material, the atomic magnets form small regions called domains. In an unmagnetized sample, the domains usually point in different directions, producing no magnetic effect. When a magnetic field is applied (top), domains that are aligned with the field grow, and the material becomes magnetized. In strong fields (below) the domains rotate to become aligned.

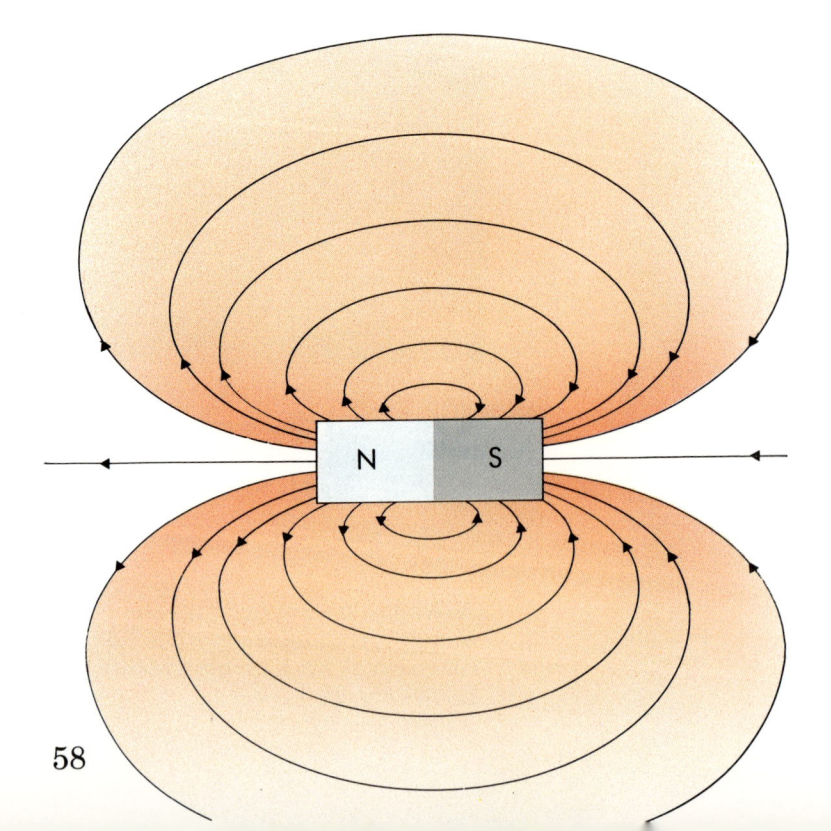

◄▼ A magnetic field can be represented by lines, called lines of force, linking north and south poles. The arrows give the direction of the force on an imaginary north pole. Similar magnetic poles repel each other, giving a "neutral point" between them where there is no force. Unlike poles, on the other hand, are attracted to each other, with lines of force filling the space between them.

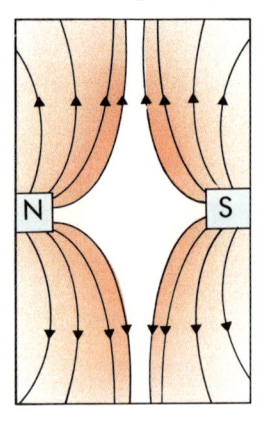

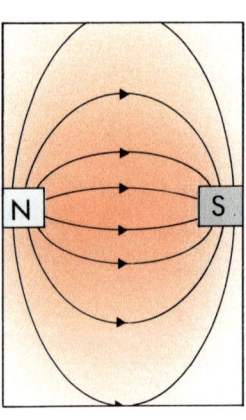

Electricity and magnetism

Moving electric charges – an electric current – causes magnetism. Indeed, magnetic fields are not truly different from electric fields. Rather, both are aspects of the same thing, the "electromagnetic field".

In 1820 the Danish physicist Hans Christian Oersted noticed that a current flowing in a wire deflected a compass needle nearby. It did not take long to discover that the magnetic field around a wire carrying a current is complicated. The strength of the field depends on several things — the strength of the current, the length of the wire and the distance to the wire.

In some instances, the field has a relatively simple form. Around a straight wire, the lines of force are circles. Around a solenoid – a coil with many turns – the field resembles that of a bar magnet. It has a north pole at one end and a south pole at the other.

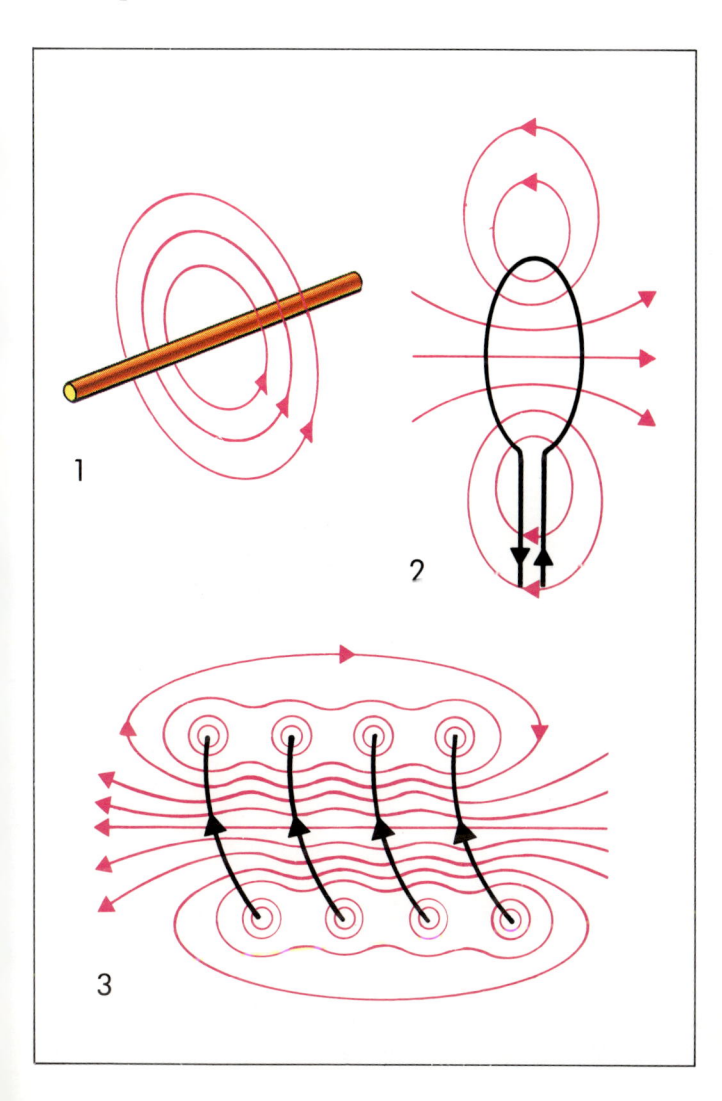

Electromagnetic door bell

▲ A simple application of an electromagnet, the "ding-dong" door bell. When the bell switch is pressed, electricity flows through the coil. The coil becomes magnetic and forces out the iron rod inside. When the rod strikes the chimes, they each produce a different note because they are different lengths.

◀ An electric current flowing along a wire creates a magnetic field encircling the wire (1). In a wire bent into a single turn, the magnetic field consists of many loops (2). In a solenoid (3), the effects of many turns of wire add together to give a field like that of a bar magnet.

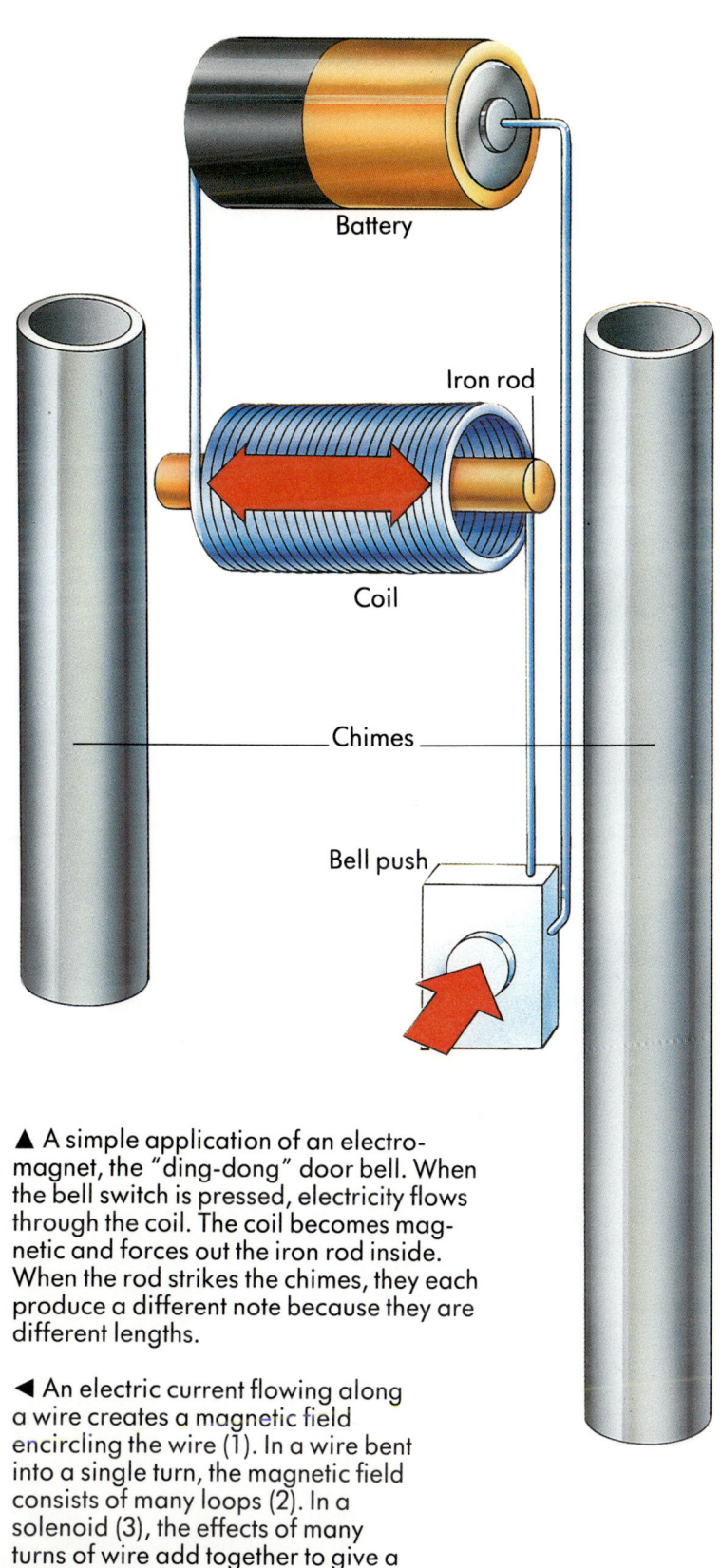

Electromagnetism

The first electromagnet was made in 1823 by an Englishman, William Sturgeon. He wound an insulated copper wire into a coil around an iron bar. When a current flowed through the coil, the bar became a strong magnet. The strength of an electromagnet depends on the number of turns of wire in the coil and on the strength of the current flowing through it. The more turns and the larger the current, the stronger the electromagnet becomes.

Super electromagnets
Today, the strongest electromagnets are made using superconducting coils. The coils are made of materials that lose all electrical resistance when they are cooled to very low temperatures, around −269°C. They can carry the large electric currents needed in powerful electromagnets. Superconducting magnets are used in particle accelerators, which are used for atom-smashing experiments.

Superconducting magnets are also used in medicine. They are an essential part of a type of body scanner, called a NMR scanner. These scanners use a process called nuclear magnetic resonance (NMR) to produce detailed pictures of the inside of a patient's body.

The patient lies in a strong magnetic field produced by an electric current flowing in a superconducting coil. Radio signals are beamed into the area of the body being investigated. The nuclei of the atoms of the body produce tiny magnetic signals, which are picked up by detectors. A computer is used to form a picture of the inside of the body from these magnetic signals.

Electromagnets are found in many household appliances. In a television set, electromagnets are used to control the beams of electrons that form the pictures on the screen. In a telephone, an electromagnet moves the plate in the earpiece that produces the sounds we hear. In a loudspeaker, electric currents produced by a record player or tape recorder are converted into sounds by an electromagnet. Electromagnets in the form of motors are also found in vacuum cleaners, food mixers, hair dryers and washing machines.

▶ A magnetic levitation train is held in the air by superconducting electromagnets. These magnets produce strong magnetic fields because large currents can flow through them without resistance.

▼ Very powerful electromagnets are often used in scrapyards. The magnetism does not exist when the current is switched off, and the magnet releases its load.

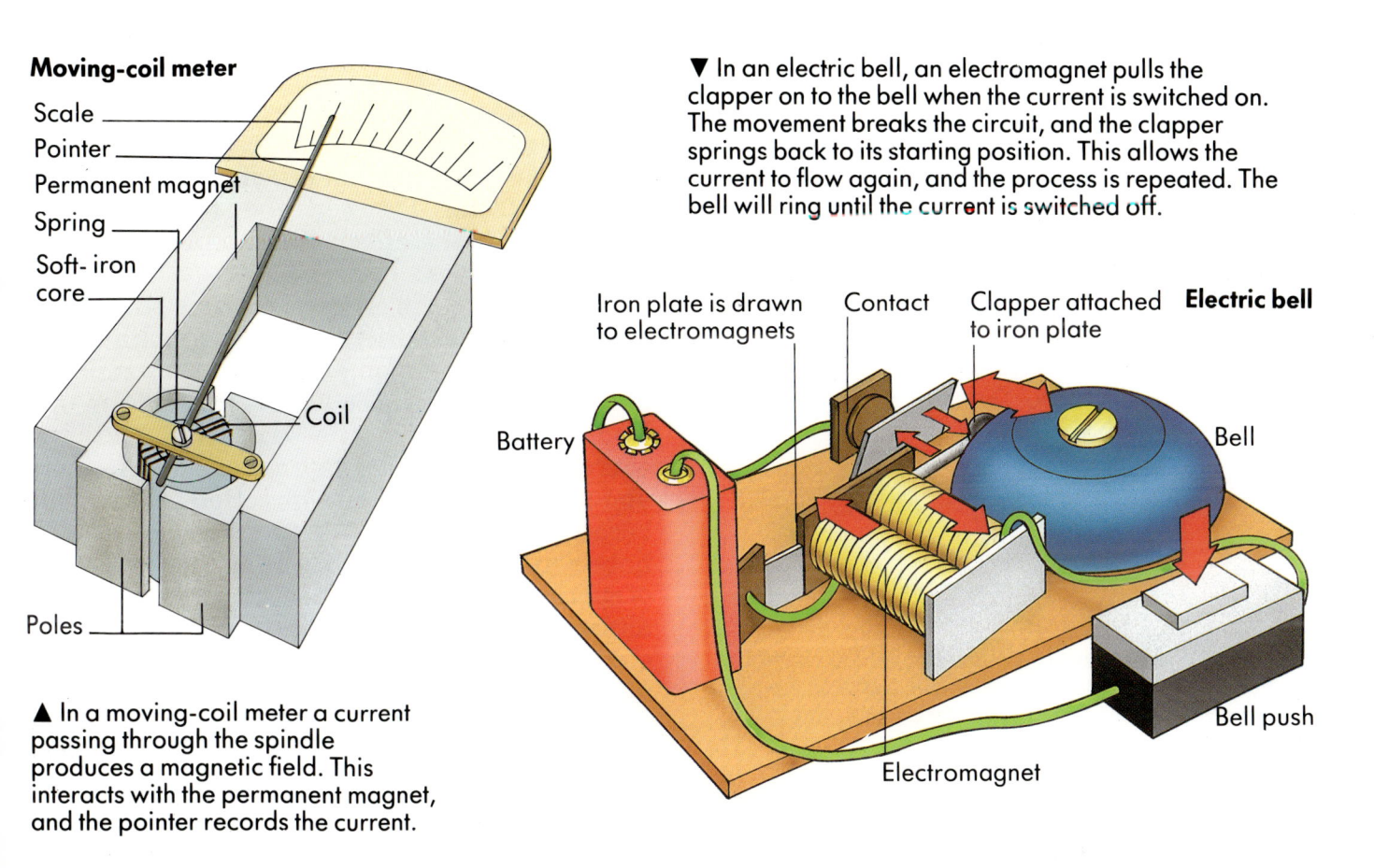

Moving-coil meter
- Scale
- Pointer
- Permanent magnet
- Spring
- Soft-iron core
- Coil
- Poles

▲ In a moving-coil meter a current passing through the spindle produces a magnetic field. This interacts with the permanent magnet, and the pointer records the current.

▼ In an electric bell, an electromagnet pulls the clapper on to the bell when the current is switched on. The movement breaks the circuit, and the clapper springs back to its starting position. This allows the current to flow again, and the process is repeated. The bell will ring until the current is switched off.

Electric bell
- Iron plate is drawn to electromagnets
- Contact
- Clapper attached to iron plate
- Battery
- Bell
- Electromagnet
- Bell push

▼ Circuit-breakers are switches used in power stations. The magnetic effect of an excess current opens the switch, preventing the circuit becoming overloaded.

▶ A transformer consists essentially of two electromagnets. By altering the number of turns in the coils, it can be used to increase or decrease the input voltage.

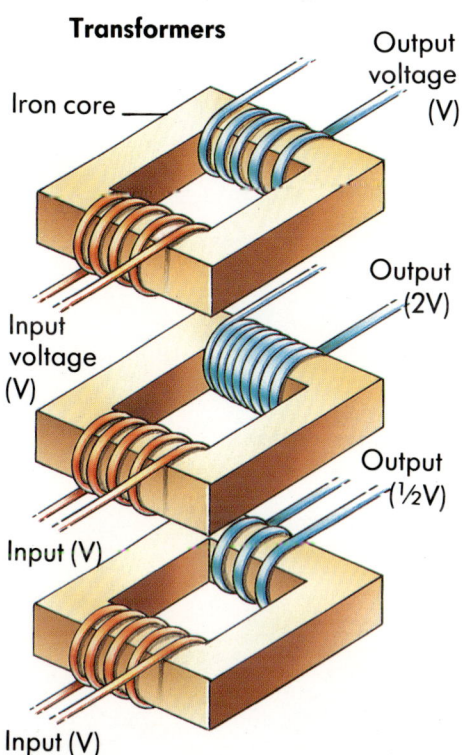

Transformers
- Iron core
- Output voltage (V)
- Input voltage (V)
- Output (2V)
- Input (V)
- Output (½V)
- Input (V)

61

Generators and motors

Principle of the generator and motor

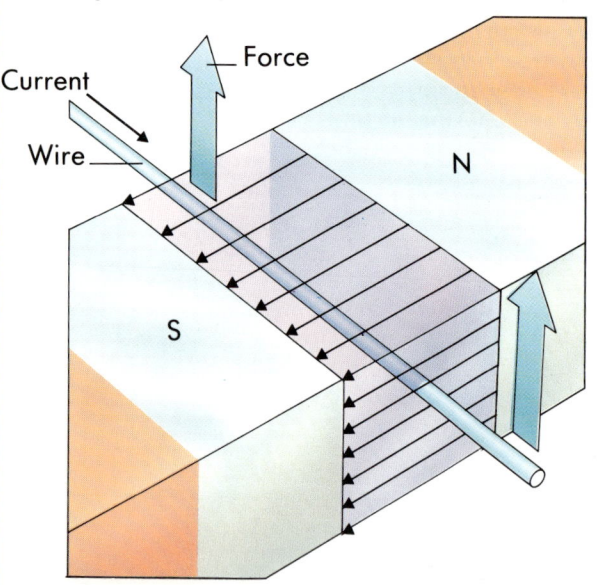

When a wire moves in a magnetic field, a current flows if the wire is part of a circuit, because the electrons in the wire experience a force. This is the principle that underlies the operation of a generator. When a current flows in a wire in a magnetic field, the wire experiences a force. This is how a motor works.

Electrical generators and motors are related in the way they work. A generator converts energy of movement into electrical energy. An electric motor does the opposite; it converts electrical energy into energy of movement. The common factor between the two pieces of equipment is the effect of a magnetic field on moving electrons.

The electrons in a wire moving through a magnetic field experience a force, which sets the electrons moving along the wire. This is how a generator works. If the wire is in the form of a loop being spun in the magnetic field, the current produced moves first in one direction and then in the opposite direction. It happens because the two sides of the coil move alternately up and down through the field. This sort of current, which moves back and forth, is called alternating current (AC). It is the kind of electric current supplied by the electricity mains found in our homes.

It is possible to produce one-way electric current, or direct current (DC) from a rotating coil. A split ring, called a commutator, is attached to the ends of the coil. The commutator connects the coil to the circuit.

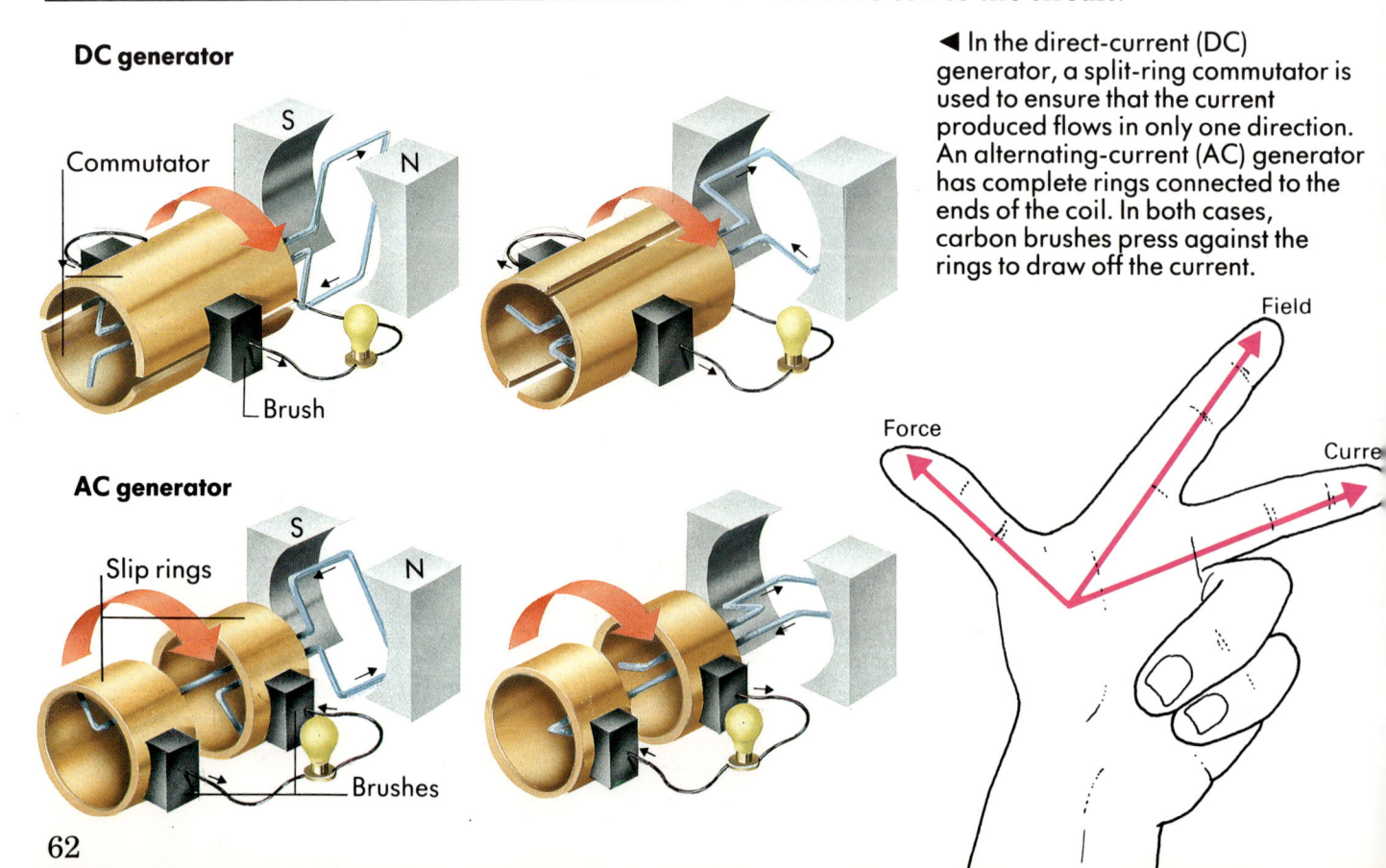

DC generator

AC generator

◄ In the direct-current (DC) generator, a split-ring commutator is used to ensure that the current produced flows in only one direction. An alternating-current (AC) generator has complete rings connected to the ends of the coil. In both cases, carbon brushes press against the rings to draw off the current.

62

In an electric motor, a current is set up through a wire placed in a magnetic field. The moving electrons then experience a force due to the magnetic field, which makes the wire move. If the wire is in the form of a loop, the forces acting on the two sides of the loop make it spin.

A simple motor requires an alternating current to make it work. As the coil turns, the current is reversed at the right moment to continue the rotation. In a direct-current motor, a commutator reverses the current through the coil at each half-rotation, and in this way keeps the rotation going.

There are various kinds of electric motors, used in factories, electric railways and household appliances. Most household appliances use a motor in which the magnetic field is produced by an electromagnet. The electromagnet is connected to the same electrical supply as the rotating coil. To increase the power, there are many coils in these motors. Each coil is in a slightly different position from its neighbours. The commutator is split into many segments, one for each coil. The motors are called universal motors, because they can run on both direct and alternating current.

▲ A powerful electric motor is used to drive a winch at a mine in Zimbabwe. The winch winds in a steel cable to raise a lift carrying rock from deep down the mine.

▼ Fleming's left- and right-hand rules. They indicate the directions of current flow, magnetic field and movement in motors and generators. The left-hand rule applies to motors, the right-hand rule to generators. The forefinger points in the direction of the field, the thumb of the motion, the middle finger of the current.

▼ In a direct-current motor, the commutator, which connects the coil to the current, reverses the direction of the electric current after the coil has turned half a turn. As a result, the coil keeps turning in the same direction. Without a commutator, the coil would come to rest after half a turn, with the coil horizontal.

▼ An alternating-current motor does not need a commutator because the current in the coil is continually reversing in direction. Instead it uses a pair of slip-rings, one connected to each end of the coil. The coil rotates at a speed which keeps it in step with the changes in the direction of the current.

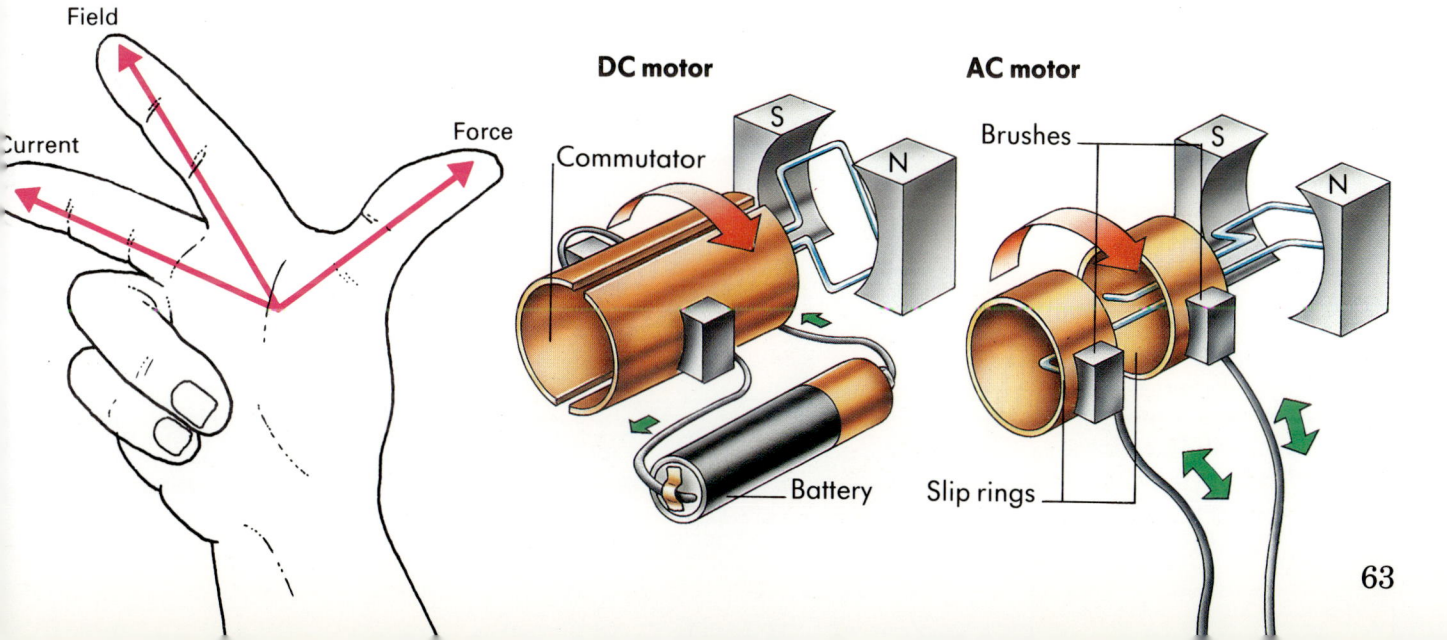

Light and radiation

Spot facts

- Light travels at a speed of 300,000 km/s. It takes one-tenth of a second to travel from New York to London, 8½ minutes to reach the Earth from the Sun, and 4⅓ years to reach Earth from the next nearest star.

- Albert Einstein, that most famous scientist, showed in 1905 that nothing could travel faster than the speed of light.

- Laser light beams shone from the Earth have been reflected off mirrors left by astronauts on the Moon. These experiments have been used to measure the precise distance to the Moon.

Light shining through a hole forms a beam, or ray, that travels in a straight line. Light rays can be seen to bounce off mirrors, and bend when they enter a transparent material. But this is not the whole story. Light is also a form of wave, an electromagnetic wave. Furthermore, modern developments in the use of light, such as the laser, can be explained only if light also consists of packets of energy, or particles of light, called photons.

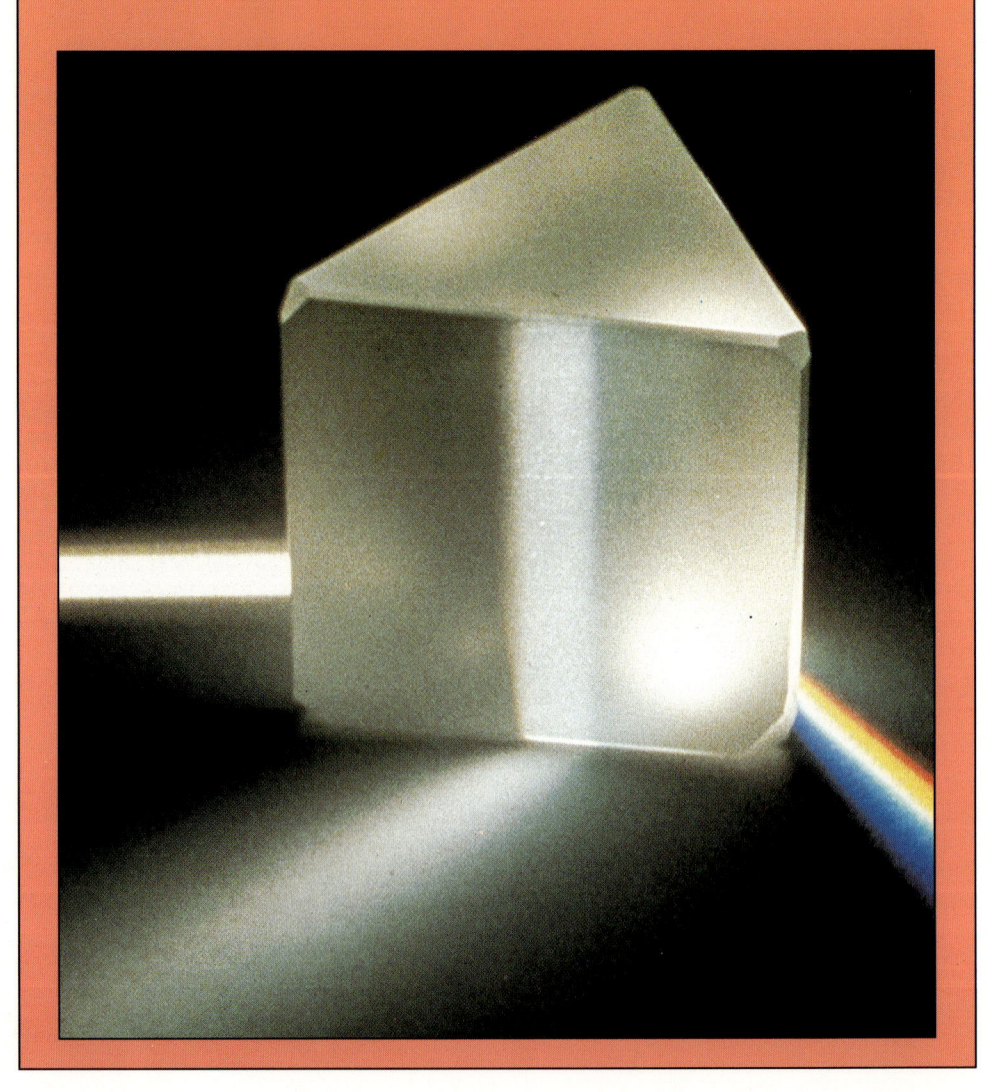

▶ A simple experiment that reveals much! When light passes through a prism, two phenomena occur. The beam bends away from its original direction. This bending is called refraction. The beam also disperses into a band of colours known as a spectrum. This shows that white light is a mixture of these colours.

Reflection and refraction

One of the earliest discoveries about light was that it seems to travel in straight lines. Other discoveries concerned what happens to a light ray when it meets a surface, for example the surface of a mirror, or a glass surface.

When a beam of light falls on a shiny surface, it bounces off, or is reflected. The law of reflection states that the angle at which the ray strikes the surface, called the angle of incidence, is always equal to the angle at which the light beam leaves it. Simple diagrams using this law demonstrate how a mirror forms an image. The reflected image appears to come from behind the mirror.

Light rays bend when they enter a transparent material. This bending is called refraction. Refraction explains why it is so hard to spear a fish from the bank of a river. The fish is not where it seems to be, because the light from the fish is refracted as it leaves the water.

Refraction also explains why a drinking straw appears to be bent where it dips into a glass of water. Refraction occurs because light travels more slowly in water and glass, which are denser than air. This causes a light ray to swerve as it enters these materials, in the same way that a racing car swerves if it drives off the track on to a rougher surface.

◀ The John Hancock Tower in Boston, USA. The glass in many modern buildings acts like a giant mirror, often creating startling effects. The images in a mirror or reflected in a glass window are reversed left-to-right.

▼ The image of a tree reflected in a lake is upside-down and the same size as the real tree appears to be.

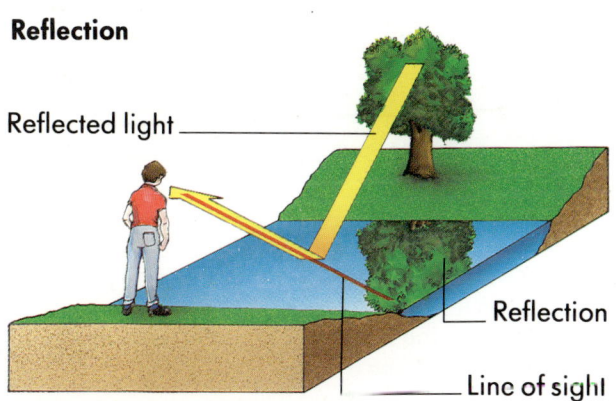

Reflection

▼ A mirage of a tree in a desert is formed because light from a distant tree is refracted when it passes through the hot air near the ground.

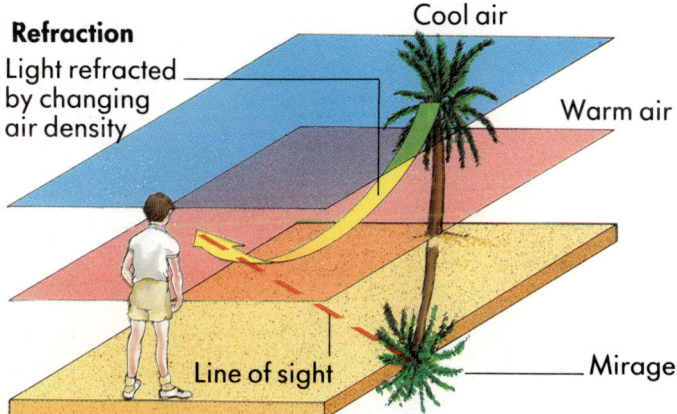

Refraction

Curved mirrors and lenses

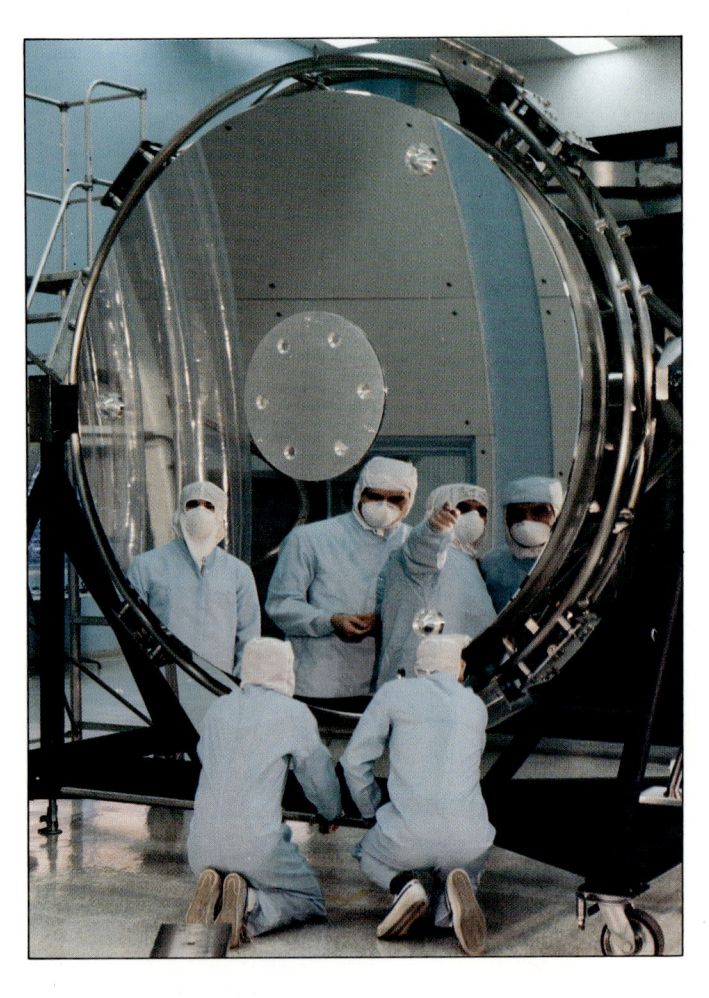

▲ Large astronomical telescopes generally use curved mirrors. This one is for the Hubble Space Telescope, to be launched into orbit on the Space Shuttle. Early telescopes used lenses, but produced poor images. Isaac Newton built the first reflecting telescope in 1671. Mirrors can be made much larger than lenses.

Mirror images

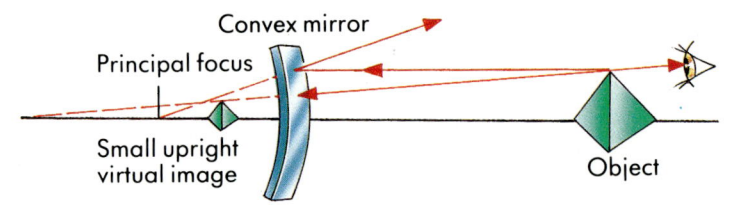

There are two kinds of curved mirrors: concave mirrors that bend inwards, and convex ones that bulge outwards. Concave mirrors are used in astronomical telescopes, shaving mirrors and in car headlamp reflectors. Convex mirrors are used for car rear-view mirrors and in supermarkets and shops to give the staff a good view of the shop floor.

If a beam of parallel light rays falls on a concave mirror, the rays are reflected so that they all pass through a single point in front of the mirror. This point is called the principal focus. Light falling on a convex mirror is spread out as if it were coming from behind the mirror.

Curved mirrors can produce images, or pictures, of objects placed in front of them. The size of the image and its position depends upon where the object is placed. If the object is far away from a concave mirror, no image can be seen in the mirror. However, if a piece of paper is held in front of the mirror, a small, upside-down image can be seen on the paper. This is called a real image. If an object is close to a concave mirror, a large upright image of it can be seen in the mirror. This kind of image, which can be seen in the mirror but not focused on to a piece of paper, is called a virtual image.

Lenses are pieces of glass or other transparent material with curved sides. They are used in cameras, small telescopes and spectacles. Convex lenses have sides that bulge outwards. Concave lenses have sides that bend inwards. Convex lenses can bring parallel light to a focus at one point, the principal focus. A concave lens makes parallel light diverge away from its focal point. Like mirrors, lenses can form images. The type and size of image depends on which type of lens is used and where the object is placed.

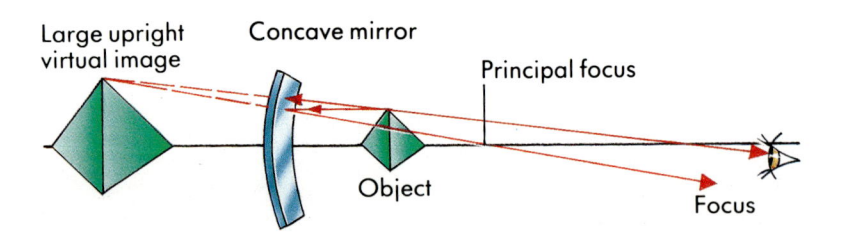

◀ A convex mirror produces a small, upright virtual image behind the mirror. A concave mirror produces a large upright, virtual image if the object is between the focus and the mirror. If the object is farther away, a small inverted (upside-down) real image is formed in front of the mirror.

▲ A magnifying glass is a convex lens. An object between the lens and its focus appears larger, or magnified, but can appear farther away from the lens than it really is.

◀ A lighthouse uses both mirrors and lenses to produce a powerful beam that can sweep the horizon. A solid lens would be too big and too heavy. Instead, Fresnel lenses, named after a French scientist, are used. In such a lens the surface is divided into a series of circles. Relatively thin ridges of glass are set in each circle at angles that would be found in a solid lens at that point.

Convex and concave lenses

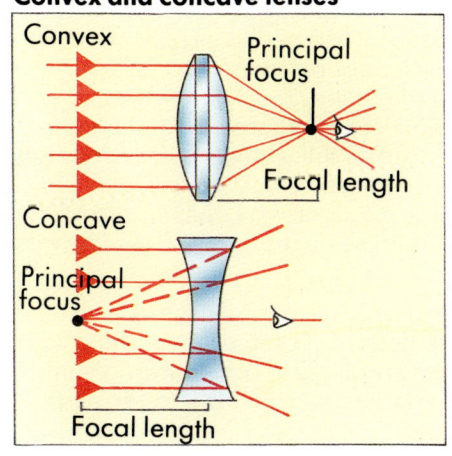

Convex lens image

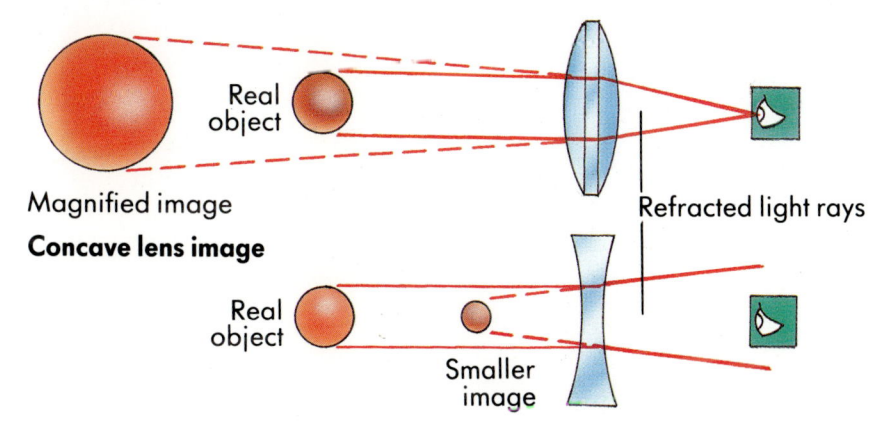

Concave lens image

▲ A parallel beam of light entering a concave lens is bent to a single point known as the principal focus. On leaving a concave lens parallel light rays spread out as if from a point behind the lens.

▲ If an object is placed between a convex lens and its principal focus, an enlarged virtual image is seen through the lens. A concave, or diverging lens, always produces a smaller virtual image between the lens and the focus. A real image is one where the light rays actually come from the object. With a virtual image the rays only appear to come from the object and cannot be focused on to a screen.

67

Waves of light

What is light? Many scientists have studied light and tried to answer this question. In the late 17th century, the great English scientist Isaac Newton suggested that a light beam consisted of a stream of tiny particles, which he called corpuscles.

In about 1690 a Dutch physicist, Christiaan Huygens, put forward another idea. He thought that light travelled in waves. If you drop a pebble into water, you will see waves, or ripples move out from the point where the pebble hits the water. Huygens believed that light travelled in a similar way, but that the light waves were very small. He thought that the distance between the tops of any two neighbouring light waves was only a few ten-thousandths of a millimetre.

Although Huygens's theory explained the properties of light, such as reflection and refraction, it took many years for this theory to win support. In 1801 the Englishman Thomas Young performed an experiment which showed that light consisted of waves.

In Young's experiment, light from a small point source was shone on to a screen through two fine slits side by side. The light source was a sodium lamp, which gave light of a pure colour. Young saw a pattern of alternating light and dark bands on the screen. He realized that this result is similar to that seen when two stones are thrown into water at the same time. A series of waves is sent out by each stone, and where the waves meet, they combine. Where two wave peaks overlap, there is a peak of double height. Where two troughs overlap, there is a deep trough. This is called interference of waves. Interference could happen in Young's experiment only if light were a form of

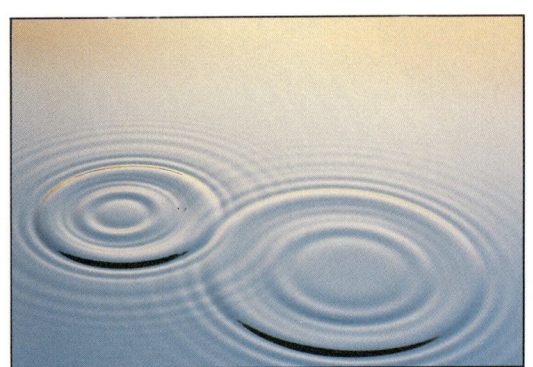

▲ Two stones dropped into a still pond produce circular patterns of spreading ripples. Where they meet, the two wave patterns interfere. Where the ripples cancel out, the result is still water. Where two crests combine, a large ripple results. Light behaves in the same way, but instead of ripples, light produces light and dark bands.

◀ Clear plastic rulers and protractors show coloured patterns in some lights. The colours are caused by the plastic splitting light into different beams, which then interfere with each other.

▶ Coloured patterns form in soap bubbles when light reflected from the top of the soap film interferes with light reflected from the bottom. The delicate, swirling rainbow colours seen in thin oil films on water are produced in the same way.

wave. With white light, Young's experiment produced a pattern of coloured bands. This is because the different colours that make up white light have different sized waves, which cancel out in different places.

The modern idea is that light waves consist of packets of light energy called photons. In some situations the photons are important, and light behaves like a stream of particles. In others the wave properties are more important, and light behaves like waves. So the two explanations by Newton and Huygens each provide a part of the answer to the question "What is light?"

▶ Interference patterns are produced when two beams of light, with their waves in step, overlap. Bright bands are seen where the waves reinforce each other. Dark bands are seen where the waves cancel out.

Colour

The colours of light correspond to waves of differing lengths. Red light, for example, has a wavelength – the distance between successive crests of a wave – of about 700 nanometres. One nanometre is equal to one thousand millionths of a metre. Violet light has the shortest wavelength in visible light. It is about 400 nanometres. Yellow, green and blue light have wavelengths between these values. White light is a mixture of light of all wavelengths.

When coloured lights are shone together onto a white surface, the colours are added together. Any colour can be produced by mixing combinations of red, green and blue light. These are the three primary colours of light and mixed in equal proportions produce white light. Two primaries, for example red and green, make a secondary colour, yellow.

Coloured objects or paints absorb, or subtract, certain colours from light and reflect the rest. Our eyes see the reflected light only, and so the object appears to be the colour of the reflected light. For example, red paint absorbs the green and blue colours in white light, and reflects only the red light. The secondary colours of light, yellow, magenta and cyan, are the primary colours of paint. Many colours can be made by mixing them; for example a mixture of magenta and yellow make red. All three mixed equally together make black.

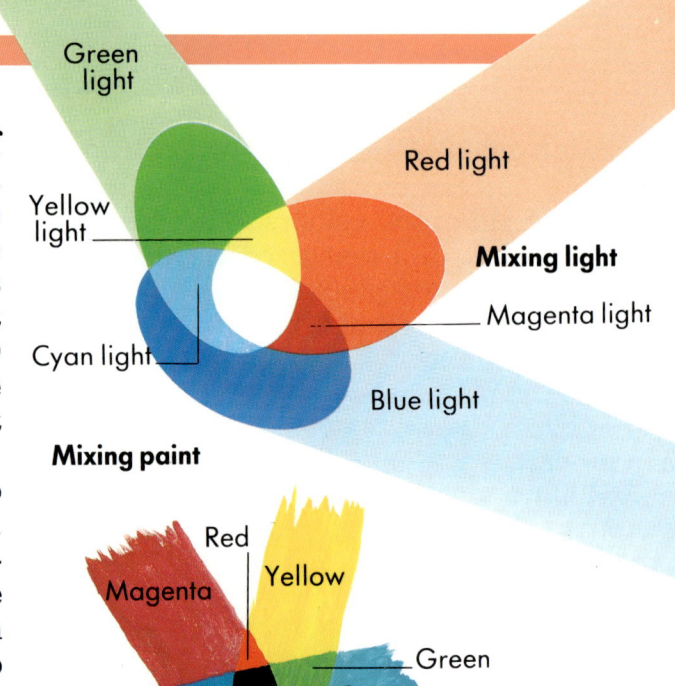

▲ Mixing paints is different from mixing coloured lights, like those found in a theatre or disco. Many colours of paint can be produced by mixing magenta, yellow and cyan paints. With coloured lights, any colour can be produced by mixing red, green and blue lights.

▼ The lights of a disco help to enhance the music and create an exciting atmosphere. The lights are usually wired up to the loudspeakers so that they flash in time with the music. This is known as sound-to-light.

Primary rainbow
Light ray
Total internal reflection
Refraction
Raindrop
Water vapour
Refraction

Secondary rainbow

▲ The rainbow is a spectacular demonstration that shows white light to be a mixture of colours. Rainbows occur when sunlight from behind the observer is refracted and reflected by water droplets in the air. Often a fainter "secondary" rainbow is seen outside the bright "primary" one. The colours in the secondary bow are in the reverse order to those in the primary.

◀ In forming the primary rainbow, light from the Sun is first refracted as it enters a raindrop, then instantly reflected from the back of the drop. Finally it emerges, spread into a band of colours. In forming the secondary bow, the light is reflected twice within the raindrop before it emerges. In the primary rainbow the light is reflected only once.

Lasers

An American, Theodore H. Maiman, invented the laser in 1960. A laser is a device that produces a very powerful beam of light of a single colour. The word laser comes from a set of initials that stand for "light amplification by stimulated emission of radiation". This name was chosen because a laser persuades, or stimulates, its atoms to amplify, or make stronger, a flash of light.

Lasers are used in the home, factory, and hospital. A compact-disc player contains a low-power laser. The laser beam "reads" the music on the disc, just as the stylus reads the music on an ordinary disc. Low-power lasers are also used at supermarket checkouts to read the barcodes on packets of food. Surgeons use high-power lasers to carry out delicate eye operations. In industry, lasers are used to cut and weld metal sheets. When the Apollo astronauts were on the Moon, they set up a mirror pointing at the Earth. Later, scientists shone a laser at the mirror. By measuring how long it took the laser beam to travel to the Moon and back, the scientists were able to measure the distance to the Moon very accurately.

Lasers are used to create three-dimensional photographs called holograms. To produce a hologram, laser light is shone on the object being photographed. The light is reflected off the object on to a photographic plate. At the same time, some light from the laser is shone directly on to the photographic plate. The two beams of light produce a complex pattern on the plate. Later, if laser light is shone through the plate, a three-dimensional picture is formed.

▼ Lasers are key research tools. They provide a source of light of a single wavelength, or pure colour. Furthermore, the waves are all in step with each other. This greatly enhances the intensity of the light.

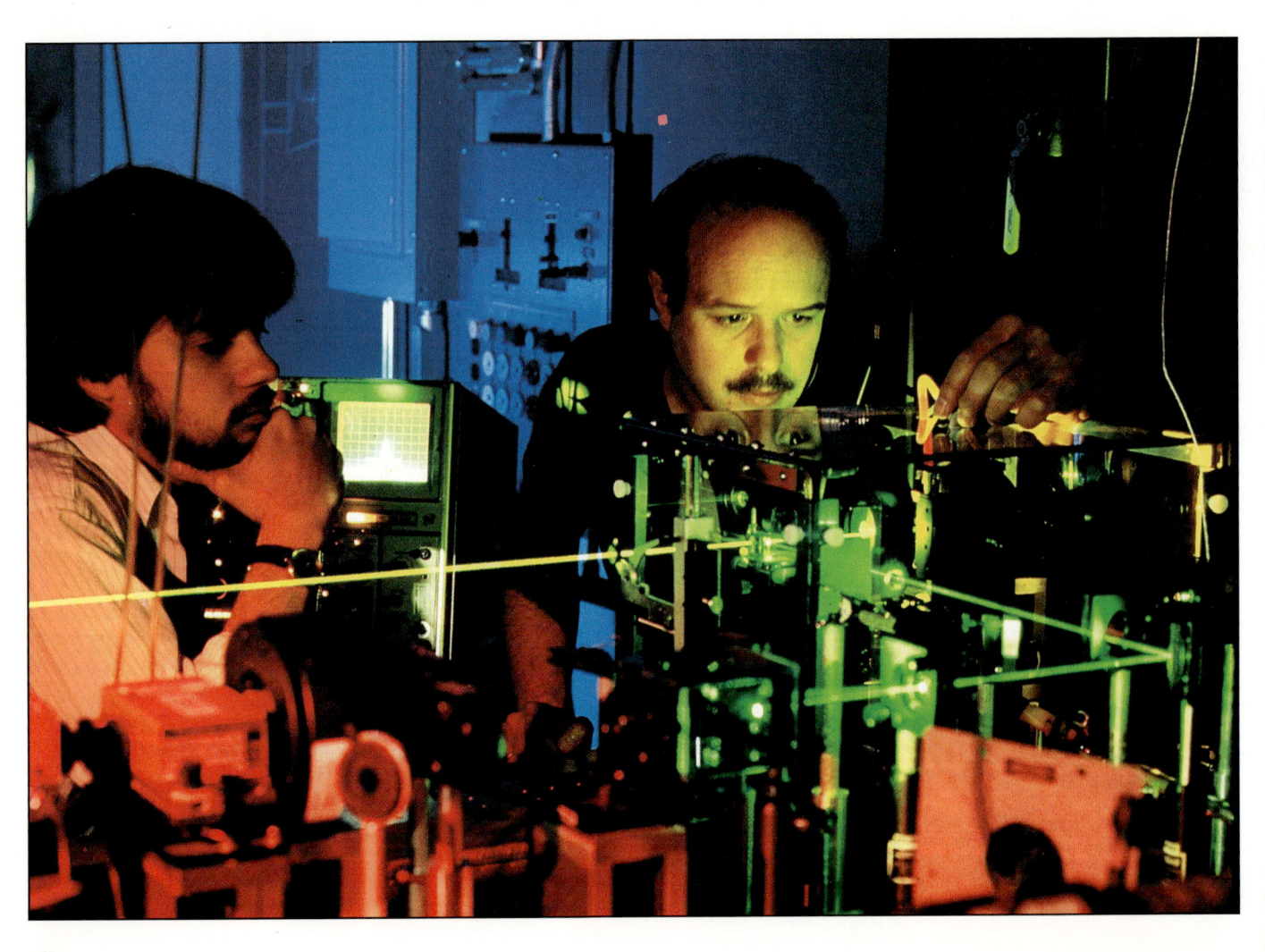

How a laser works

Electrons emit light when they lose energy. This happens when an electron jumps from one orbital to another one of less energy. As this happens, a photon is emitted.

When an electron in an atom is given extra energy, it is raised to a higher-energy orbital. The energized electron can then emit a photon and return to its usual orbital spontaneously. However, it can also be stimulated to fall back by a photon passing nearby. This is stimulated emission. The emitted photon and the passing photon move away in step, with their troughs and peaks matching exactly. The light produced in this way is said to be coherent.

The first lasers used a cylinder of ruby to produce their light. In the ruby laser a powerful flashlight tube is wrapped around the ruby cylinder. When the flash tube is turned on, the ruby becomes bathed in light. The ruby absorbs the light and its electrons move to high-energy orbitals.

One light photon is released, which then stimulates other energized electrons to release more photons. There is a rapid build-up of photons, until the light becomes bright enough to pass through a partly silvered mirror at one end. Ruby lasers are still used. Other types use gases or liquids.

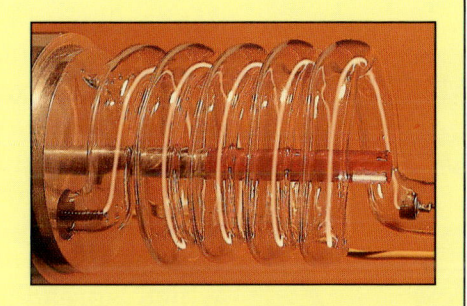

▲ The crystal and flashlight tube in a ruby laser. Crystals other than rubies have been developed for use in lasers. Most common are yttrium-aluminium-garnet (YAG) crystals, which allow the laser to operate continuously.

▼ When an atom absorbs light, electrons are raised to higher-energy orbitals. They can be stimulated to fall back to their usual orbital by a photon. When this happens, another photon is emitted, which moves away in step with the stimulating photon.

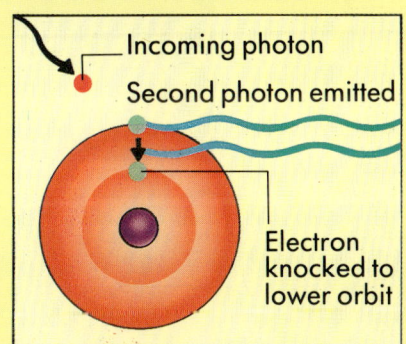

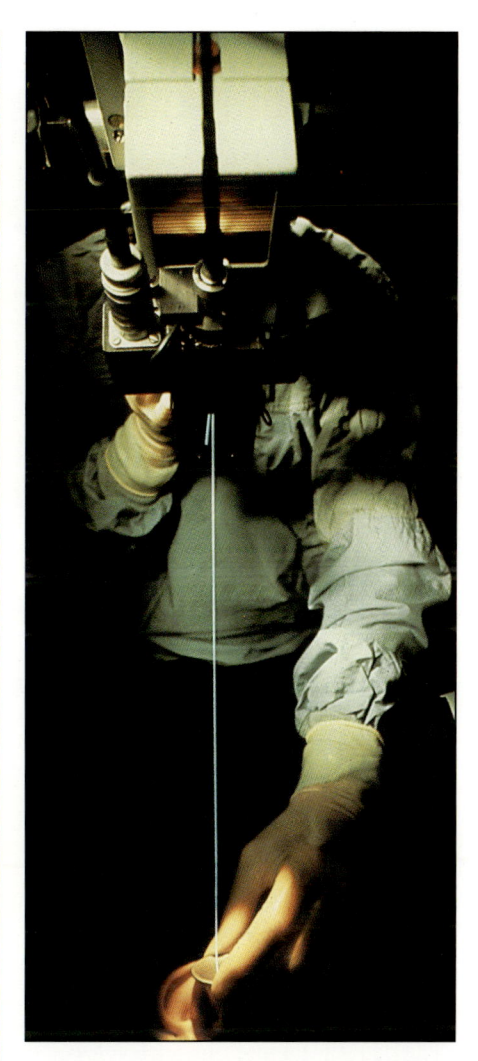

▲ Using a medical laser. Here a surgeon directs the beam from an argon laser through a small funnel into a patient's ear in order to remove a tumour between the ear and the brain.

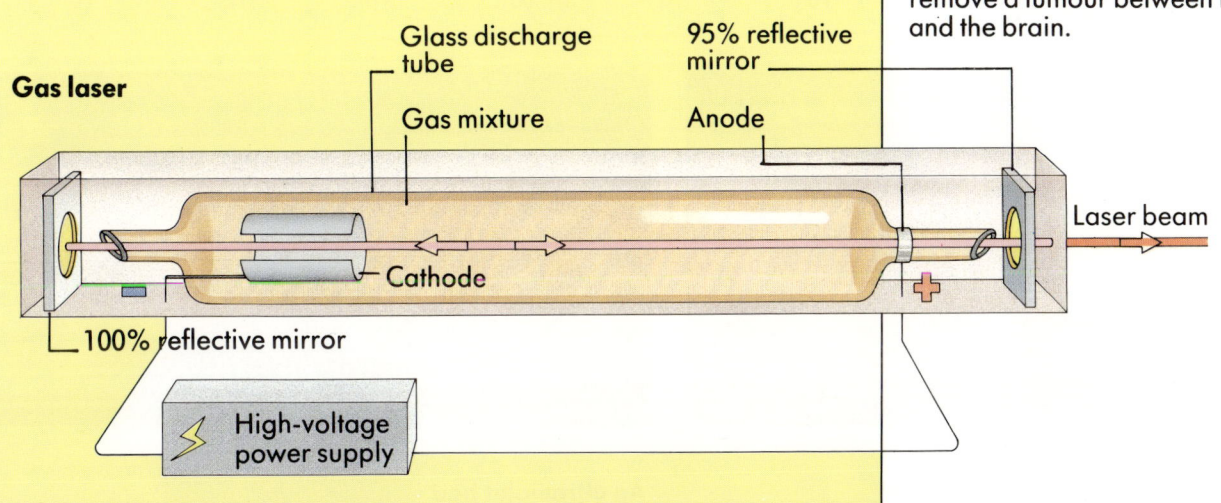

Gas laser

The electromagnetic spectrum

What bees see

The eyes of animals and insects are often sensitive to wavelengths we cannot see. The bee responds to ultraviolet light. A flower that looks a uniform colour to us, seems to have a dark centre to a bee, enabling it to find the pollen.

In 1865 the Scottish physicist James Clerk Maxwell used mathematics to show that waves which were a combination of electricity and magnetism, could spread through space. He called these waves electromagnetic waves.

You can visualize these waves by thinking about what happens when an electric charge is moved rapidly up and down. When still, the charge is surrounded by lines of electric force, which spread straight out from the charge. When the charge is moved up and down, the lines of force wiggle, in the same way that a stretched rope wiggles when one end is waved back and forth.

The wiggles in the lines of force move out from the charge in the same way that the wiggles move along the rope. However, the charge also generates a magnetic field as it moves, because it is a small electric current.

X-ray of a fractured leg

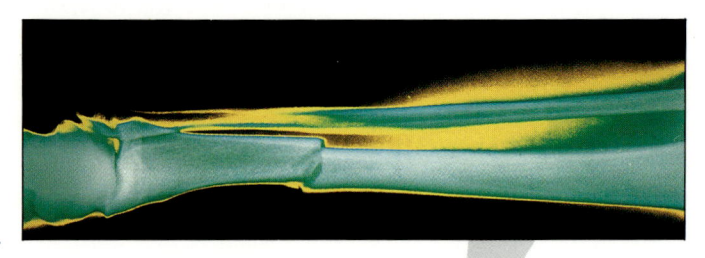

▼ The electromagnetic spectrum includes all forms of electromagnetic waves from gamma rays at the short-wavelength end to radio waves at the long-wavelength end. Visible light falls in about the middle of the spectrum.

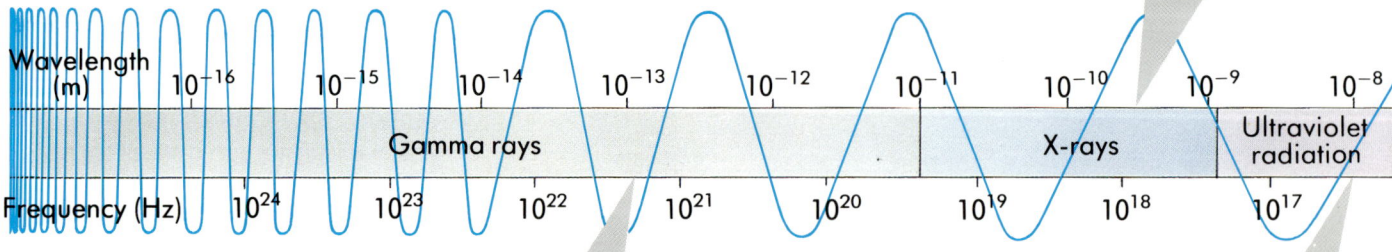

Wavelength (m): 10^{-16}, 10^{-15}, 10^{-14}, 10^{-13}, 10^{-12}, 10^{-11}, 10^{-10}, 10^{-9}, 10^{-8}

Gamma rays | X-rays | Ultraviolet radiation

Frequency (Hz): 10^{24}, 10^{23}, 10^{22}, 10^{21}, 10^{20}, 10^{19}, 10^{18}, 10^{17}

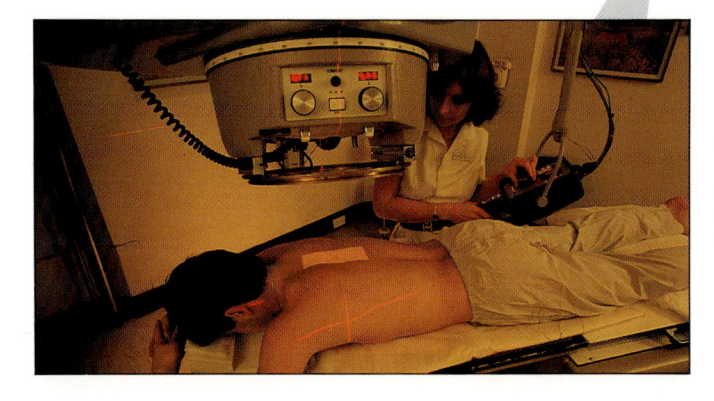

Gamma-ray therapy

An ultraviolet bed

The lines of magnetic force form circles around the moving charge. They move out from the charge like ripples moving across a pond. They move outwards at the same time and at the same speed as the electric wiggles. And so they form a wave that is a combination of changing electric and magnetic fields.

Maxwell calculated the speed of these electromagnetic waves to be the same as the speed of light, so he suggested that light consisted of electromagnetic waves. In 1889 the German physicist Heinrich Hertz produced radio waves and showed that they were electromagnetic waves too. They differed from light waves only in having longer wavelengths, or smaller frequencies. We now know that gamma rays, X-rays, microwaves, and infrared rays are all electromagnetic waves. They make up the electromagnetic spectrum.

Electromagnetic waves

James Clerk Maxwell described how electromagnetic waves consisted of electric and magnetic fields lying across the direction in which the wave is travelling. The electric field is at right-angles to the magnetic field.

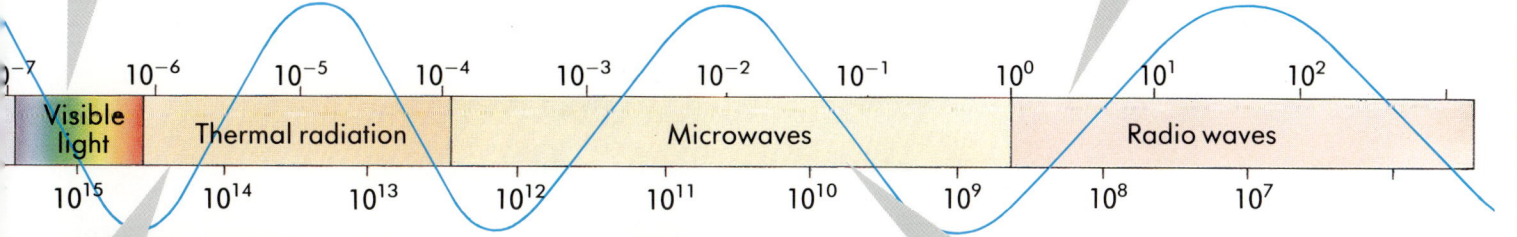

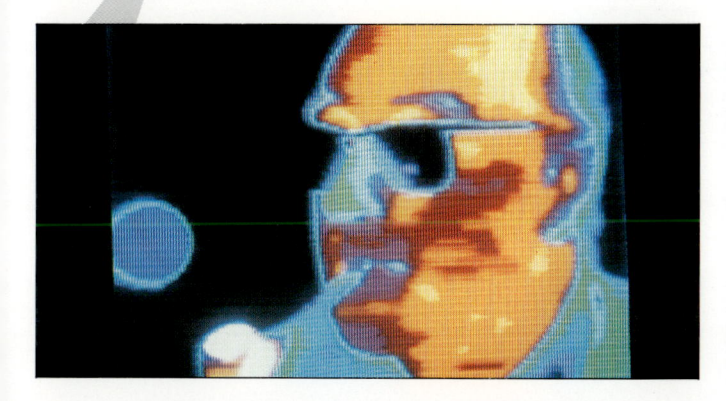

Visible light fibre optics

A short-wave CB radio

An infrared thermograph of a man smoking a pipe

A microwave radar aerial

75

Forces, energy and motion

Spot facts

- On a 16th-century battleships, the guns were never fired at once. This would have created such a big reaction force that the ship would have overturned.

- A flash of lightning releases up to 3,000 million joules of energy. It needs 100,000 million joules to send a rocket into space. A tropical hurricane releases about 100,000 million million joules. Earthquakes can release up to 10 million million million joules.

- If you can jump 1 m high on Earth, you would be able to jump to a height of 6 m on the Moon.

▶ The Panavia Tornado fighter-bomber. As a jet aircraft moves through the air, it is acted on by a number of forces. The force of gravity pulls it down. However, its wings provide an upwards force, called lift, which keeps it in the air. The engines provide a forward force, which overcomes the resistance, or drag, of the air.

Imagine a ball being hit by a golf club. It is obvious that the force of the club on the ball starts the ball moving. But why does the ball continue to move after it has lost contact with the club?

The answer to this question can be found in the three laws of motion proposed by Isaac Newton in 1665. The three laws of motion explained how forces make objects move. These laws are still used today for tasks such as calculating the paths taken by spacecraft. Such calculations also make use of another great discovery by Newton: gravity. This is one of the great forces of the Universe, which attracts objects and makes them move.

Force and movement

Nothing starts moving by itself. A push or a pull is needed to start any object moving. These pushes or pulls are called forces. When you kick a ball, the force which starts the ball moving is provided by your foot. As well as starting things moving, forces can stop moving objects.

Friction
When a ball rolls across the ground, a force called friction acts on it, and eventually stops it moving. Without friction, the ball would go on moving at the same speed and in the same direction for ever. If a football bounces off a wall, the ball changes the direction in which it is moving because of the force exerted by the wall. Forces can speed up, slow down or change the direction of a moving object.

A seat on a fairground roundabout is continually changing its direction, so there must be a force acting on it. This force acts through the chain that holds the seat to the roundabout. If the chain were to break, this force would cease to act. The seat would fly off and continue in a straight line. Any force that produces circular motion is called a centripetal force. Centripetal forces act towards the centre of the circle.

Scientists believe that there are only four basic types of force. One is the electrical and magnetic force and another is gravity. The two other types of force, called the weak and the strong forces, are found only inside the atomic nucleus. All other forces are derived from these basic four.

▼ The people on a roundabout feel a centripetal force, acting towards the centre of their circular path. This force is a combination of their weight and the tension in the chain holding the chair. Confusingly, it is popularly called centrifugal force, which means one acting away from the centre.

Laws of motion

In 1687 the English scientist Isaac Newton set down three laws of motion, which show how forces affect moving bodies. The first law says that an object at rest will stay at rest unless a force acts on it, and an object moving at a constant speed in a straight line will continue at the same speed and in the same direction unless a force acts.

The second law says that when a force acts on an object, the object changes its speed or direction of motion in the same direction as the force that has been applied. The change of speed or direction is called the acceleration of the object. The greater the force acting on a body, the greater is the acceleration produced. The greater the mass of the body, the greater is the force required to move it or change its direction.

The third law says that for every force, there is an equal force acting in the opposite direction. Newton illustrated this law with the example of a horse pulling a stone tied by a rope. While a forwards force acts on the stone, the horse feels an equal force backwards.

◀ The launching of the Space Shuttle illustrates all of Newton's laws of motion. As predicted by the first law, before the motor fires, the Shuttle is still. In line with the second law, with the motors firing, the Shuttle lifts off the launch pad. The third law indicates that the upwards force on the Shuttle is equal and opposite to the force acting on the hot gases streaming from the engines.

▼ Some dragster engines deliver an enormous force, accelerating the car to speeds of over 400 km/h in as little as 6 seconds.

Using these laws, scientists are able to explain how objects move and what happens when they collide. The first law explains a simple party trick. If a tablecloth is pulled quickly and firmly enough, it can be taken off the table without disturbing the dishes on it. The point is that the dishes do not experience the force that pulls away the cloth, and so they stay undisturbed. Unfortunately, as many tricksters have discovered to their cost, there is another force which should be taken into account, friction. Friction is a force which acts when two surfaces rub together. It slows down the movement and may cause the dishes to move and crash to the floor.

One result of the second and third laws is that when objects collide, their total momentum does not change. The momentum of a moving object depends upon both the speed and the mass of the object. When a car is moving at a high speed, it has more momentum than when it is moving at a slower speed. Also, a heavy lorry has more momentum than a small car moving at the same speed.

Sometimes if a moving ball in a pool or snooker game collides with a similar but unmoving ball, the first one stops. The second ball moves off with the same speed that the first ball had before the collision. The momentum of the first ball has been completely transferred to the second ball. But the total momentum after the collision is the same as the momentum before the collision. When you jump up and down, you are like a small ball banging into a very large ball, the Earth. After the collision, you stop moving, and the Earth absorbs your momentum. You make the Earth move. But because the Earth is 100,000 million million million times heavier than you are, the movement is very tiny and you will not notice it.

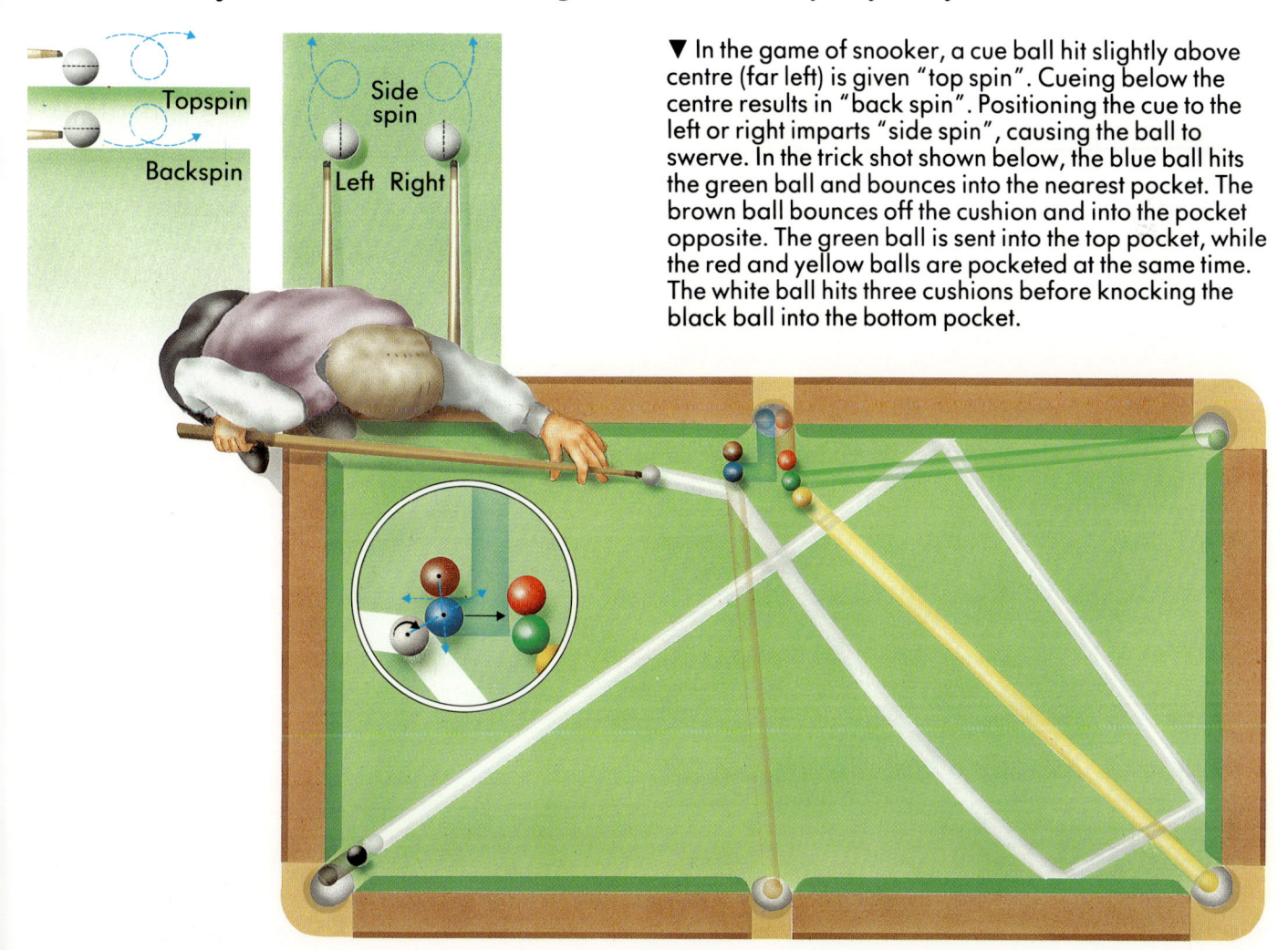

▼ In the game of snooker, a cue ball hit slightly above centre (far left) is given "top spin". Cueing below the centre results in "back spin". Positioning the cue to the left or right imparts "side spin", causing the ball to swerve. In the trick shot shown below, the blue ball hits the green ball and bounces into the nearest pocket. The brown ball bounces off the cushion and into the pocket opposite. The green ball is sent into the top pocket, while the red and yellow balls are pocketed at the same time. The white ball hits three cushions before knocking the black ball into the bottom pocket.

Energy and work

What is energy? We cannot touch it, see it, or weigh it. However, its effects are sometimes very obvious. When a lightning flash strikes, a bomb explodes, or a speeding train rushes by, it is clear that much energy is being used. The movement of the train shows one of the effects of energy. Nothing can move without energy. And because doing work involves movement, no work can be done without energy.

To a scientist, work is done whenever a force moves something. The greater the distance moved, and the greater the force involved, the more work is done. The more work done, the more energy is used. Work and energy are measured in units called joules, named after the British scientist James Prescott Joule, who lived in the 19th century. He did experiments to measure the heating effect of friction.

One joule is the work done when a force of one newton moves through a distance of one metre. A newton is a force that gives a mass of one kilogram the acceleration of one metre per second per second. This is equivalent to lifting a bag of sugar from one shelf to another in a cupboard.

It is clear that energy comes in different forms. Some forms of energy are obvious, others are more difficult to spot. A wound clock spring has energy because it can make the hands of the clock move. This form of energy is called stored or potential energy. Moving objects also have energy, called kinetic energy, because of their movement. Food contains energy in chemical form, which allows children to run about and play.

The different forms of energy can interchange. Electrical energy changes into heat energy in an electric fire. The chemical energy in food changes into energy of movement or heat in our bodies. The kinetic energy of the wind can be changed into electrical energy using a wind-driven generator. But when one form of energy changes into another, the total amount of energy remains the same. This is a statement of the law of conservation of energy.

▼ In 1977 a solar car, drove more than 3,000 km across Australia in six days. The car had 7,200 solar cells arranged around its body, which converted energy in sunlight into electrical energy to power electric motors to drive the wheels. The top speed was 72 km/h.

Potential energy

Linear kinetic energy

Rotational kinetic energy

Electrical energy

Changing energy

(1) A hydroelectric power station taps the store of potential energy held in a water reservoir. As the water is released, its potential energy is converted into kinetic energy as it runs downhill. (2) Below the reservoir, the water drives round a turbine. (3) In the turbine, some energy is lost in doing work against friction as the turbine shafts rotate. This "lost" energy is converted to heat and sound. (4) The turbines drive generators, which convert the kinetic energy of the rotating shafts into electrical energy. (5) The electrical energy created by the generator is in the form of low-voltage alternating current. This is converted by transformers to high voltage for transmission.
(6) Overhead transmission lines carry current at high voltage to reduce losses caused when electrical resistance creates heat. (7) Once the electricity reaches the consumer, it is converted to other forms of energy, such as light and heat. In the home, mechanical energy is produced in devices such as washing machines and lawnmowers.

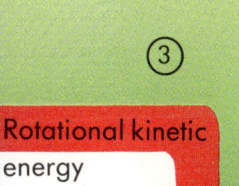

James Prescott Joule carried out experiments to prove that heat and mechanical work are equivalent. They are both different forms of energy

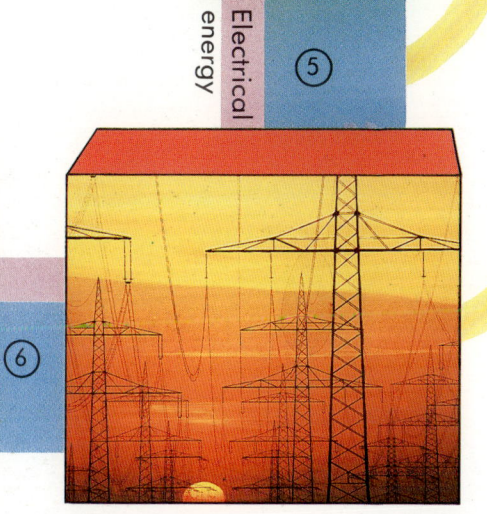

Gravity

The first person to realize why things fall to the ground was Isaac Newton. It is said that he was sitting in the orchard of his house at Woolsthorpe, in Lincolnshire, England, in 1665. He saw an apple fall from a tree. He realized that the Earth must be pulling, or attracting, the apple. He went on to discover that all objects attract each other. The attractive force between objects is called gravity. You can feel this force if you try to lift anything. The weight of an object is due to the force of gravity between the object and the Earth.

Newton realized that gravity was a long-range force. The force of gravity of the Earth reaches beyond the Moon. It stops the Moon from flying off into space. In turn, the Moon's gravity pulls the Earth's seas towards it, causing tides. The force of the Sun's gravity reaches far out into space and controls the movements of the planets.

Newton's studies of gravity revealed that the force of gravity gradually got weaker away from the Earth. Because of this, a person is attracted less in a high-flying aircraft than on the ground. However, the change in weight is very small and we do not notice it. At 25,000 km above the Earth, you would weigh only about one-tenth what you do on the ground.

The force of gravity is smaller on the Moon than on Earth. This is because the force of gravity depends upon the amount of matter in the objects being attracted together. The more matter, or mass, the objects have, the stronger the force of gravity between them. Because the Moon has only one-sixth the mass of the Earth, its gravity is only one-sixth that of the Earth. A person who weighs 60 kg on Earth would only weigh 10 kg on the Moon. Astronauts can throw things much farther and jump much higher on the Moon because of the weak gravity.

◀ An astronaut in orbit above the Earth appears to be floating motionless in space. But gravity has not ceased acting. He is falling freely and, at the same time, moving forward at great speed. The combination of the two movements produces a circular path that keeps him at the same distance above the Earth.

▼ According to modern ideas, the gravitational force of a large collection of stars, such as a galaxy, bends the space around it. This causes light rays to bend rather than follow a straight path. If a bright object, such as a quasar, is behind the galaxy, two slightly separated images of the quasar can be seen from Earth. One image is the direct view of the quasar; the other is due to the light bent around the galaxy. This effect is called a gravitational lens.

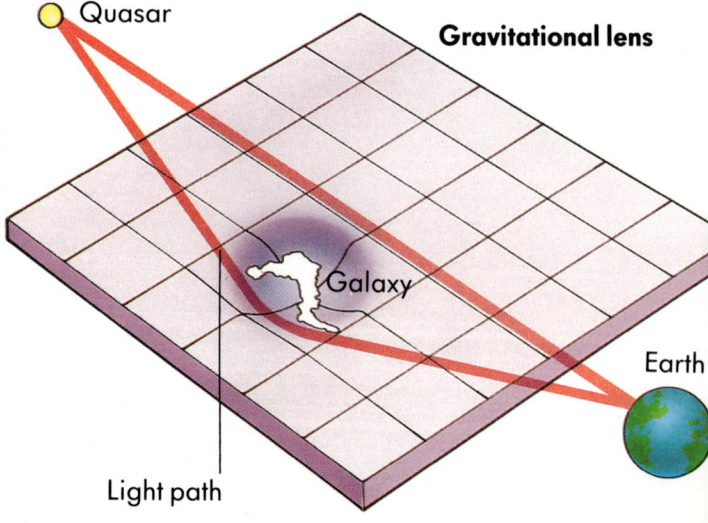

▲ Skydivers experience a force due to air resistance as they fall. This increases with speed, and at a certain speed becomes equal to the force of gravity, which is accelerating them downwards. When this happens, the divers fall at a constant speed, called the terminal velocity. Skydivers can reach a speed of 298 km/h in a head-first position in the lower atmosphere.

Galileo's experiment

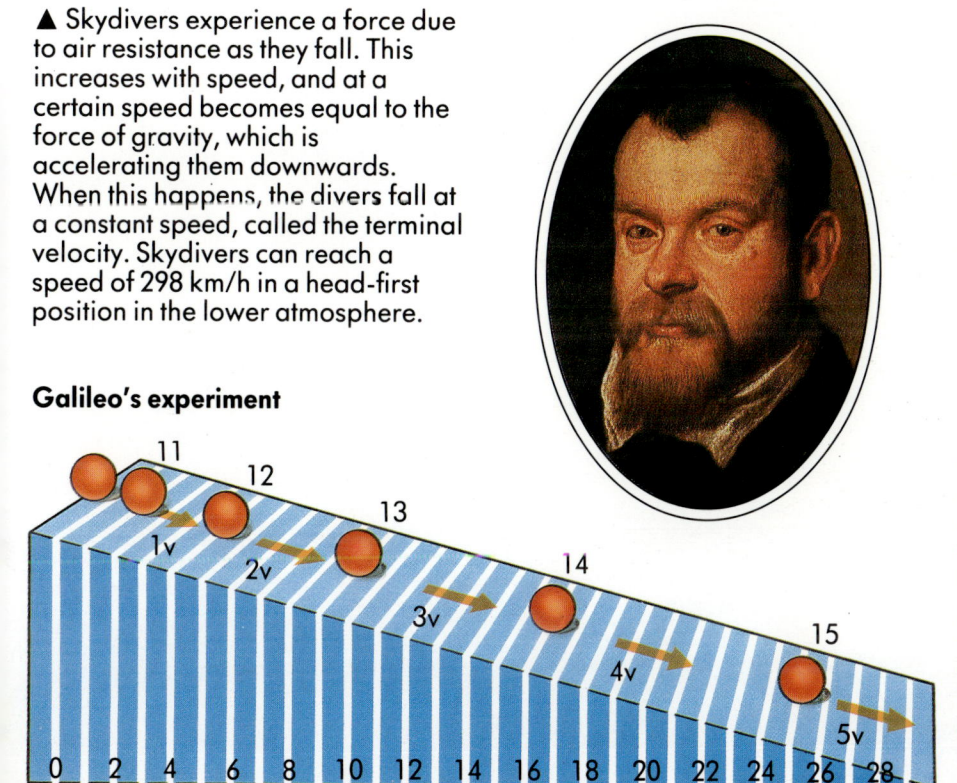

◀ The Italian scientist Galileo began the scientific study of moving objects in about 1590. He is said to have dropped objects of different weights from the Tower of Pisa, to show that all objects fall at the same rate. He also made many important astronomical discoveries using a telescope, which he constructed in 1609. He was the first to see the moons circling around the planet Jupiter.

◀ An experiment performed by Galileo involved rolling balls down a gently sloping plank and measuring the distance moved in equal intervals of time. Unfortunately, Galileo did not possess an accurate clock; he used a water clock. Nevertheless, he was able to show that the speed increased steadily as the ball moved down the slope. In other words, the force of gravity produced a steady acceleration on the ball.

83

Measuring

Spot facts

- The basic SI unit of length, the metre, is defined as the distance travelled by light in 1/299,792,458th of a second.

- The second is defined as 9,192,631,770 cycles of the radiation emitted by atoms of the caesium-133 isotope.

- The basic SI unit of mass, the kilogram, is the mass of a small cylinder made of platinum-iridium alloy. Called the international prototype kilogram, it was made in 1889 and is preserved at the International Bureau of Weights and Measures at Sèvres, near Paris.

- The length of the sides of the square base of the Great Pyramid of Khufu (completed in about 2580 BC) vary by less than 15 cm from the mean length of 230.36 m.

▶ Weighing mussels (moules) in a shop in L'Aiguillon-sur-Mer, on the west coast of France. The woman has scooped up mussels in a measuring container and is pouring them into a bag to be weighed before sale.

Measuring is as old as civilization. It became necessary as societies became more organized. Times and dates had to be fixed for religious and civic functions, goods had to be weighed in trading, and dimensions had to be marked out in building construction. Without accurate measurements, our modern technological civilization of ingenious structures, machines, instruments and devices could not exist. Scientists could not carry out their work in such depth or give accurate values to their results. Science and technology would founder.

Standards

Some 5,000 years ago, the Ancient Egyptians had various weights and measures. For example, they used different-sized stones for weights, and measured lengths in a unit called the cubit. This was the distance from a person's elbow to the tip of their middle finger. But everybody's cubit was different. So a "royal cubit" was established as a standard and used to mark measuring sticks for use in construction.

The Romans adopted a different cubit, dividing it into 2 feet, and each foot into 12 unciae (inches). For distance measurement they used *mille passus* – one thousand paces. On this, the English mile was based. The Romans also established a standard pound (libra) for weight. The abbreviation for pound (lb) is based on the Latin word.

The relationships between smaller divisions of the standard weight, length, distance, and so on, were quite arbitrary. In the old English, or Imperial, system of length measurement, for example, there are 12 inches in a foot, 3 feet in a yard, 22 yards in a chain, and so on.

It was to simplify matters and establish new standards that a commission of French scientists in the 1790s developed the metric system. It was based on a length unit of the metre: one ten-millionth of the distance between the poles and the Equator, measured along a meridian. Larger and smaller units were derived from it as multiples or subdivisions of 10. Other basic units were established: for example, gram for weight and litre for capacity, with derived units again as multiples or subdivisions of 10. In 1960 scientists adopted the present standard unit system based on the metre, kilogram and second: the Système International (SI).

▼ Early means of measurement. Ancient devices like the water clock (clepsydra) were accurate enough for their age. But precision instruments such as the micrometer and sextant became necessary as technology progressed.

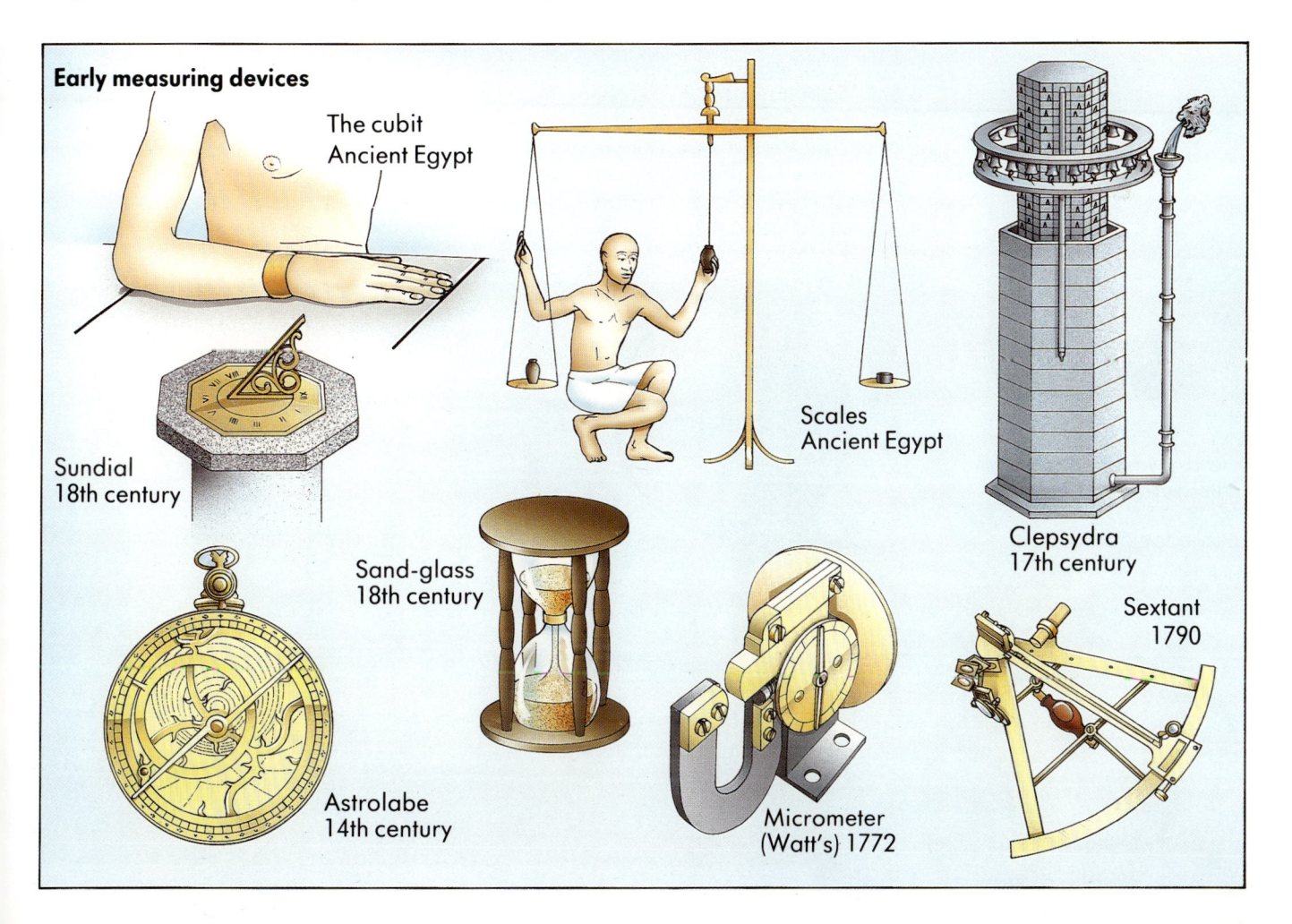

Early measuring devices

The cubit — Ancient Egypt
Sundial — 18th century
Scales — Ancient Egypt
Clepsydra — 17th century
Astrolabe — 14th century
Sand-glass — 18th century
Micrometer (Watt's) 1772
Sextant — 1790

85

Weight

Scales to measure weight were first introduced by the Ancient Egyptians in about 3500 BC to weigh gold. Illustrations in the famous papyrus known as the *Book of the Dead* show what these scales were like. They consisted of two pans hung from the ends of an arm suspended in the middle. The object to be weighed was placed in one pan, and weights were added to the other pan until the object and weights balanced each other and the arm was again level.

Scales like this are still in use today in markets and shops in many parts of the world. They are what we call equal-arm balances. The same principle is also used in the traditional precision balances once used in every chemical laboratory. In these, the balance arm pivots on a sharp knife edge made of agate or similar material. The pans also hang from knife edges. A pointer attached at right-angles to the arm indicates on a scale when the arm is in balance.

▼ This chemical balance was made by Jesse Ramsden in England in 1787. It can measure to the nearest one-thousandth of a gram. The cone-shaped beam pivots on a steel knife-edge when the beam support is raised. Knife-edge pivots were used in most balances before electronic ones came into use.

▲ Strain gauges are not only used in balances to measure light weights in the laboratory. They are also used to measure heavy weights as well, from elephants to 350-tonne jumbo jets. Aircraft are weighed every few years to check for changes in weight from paint, dust and modifications to on-board equipment.

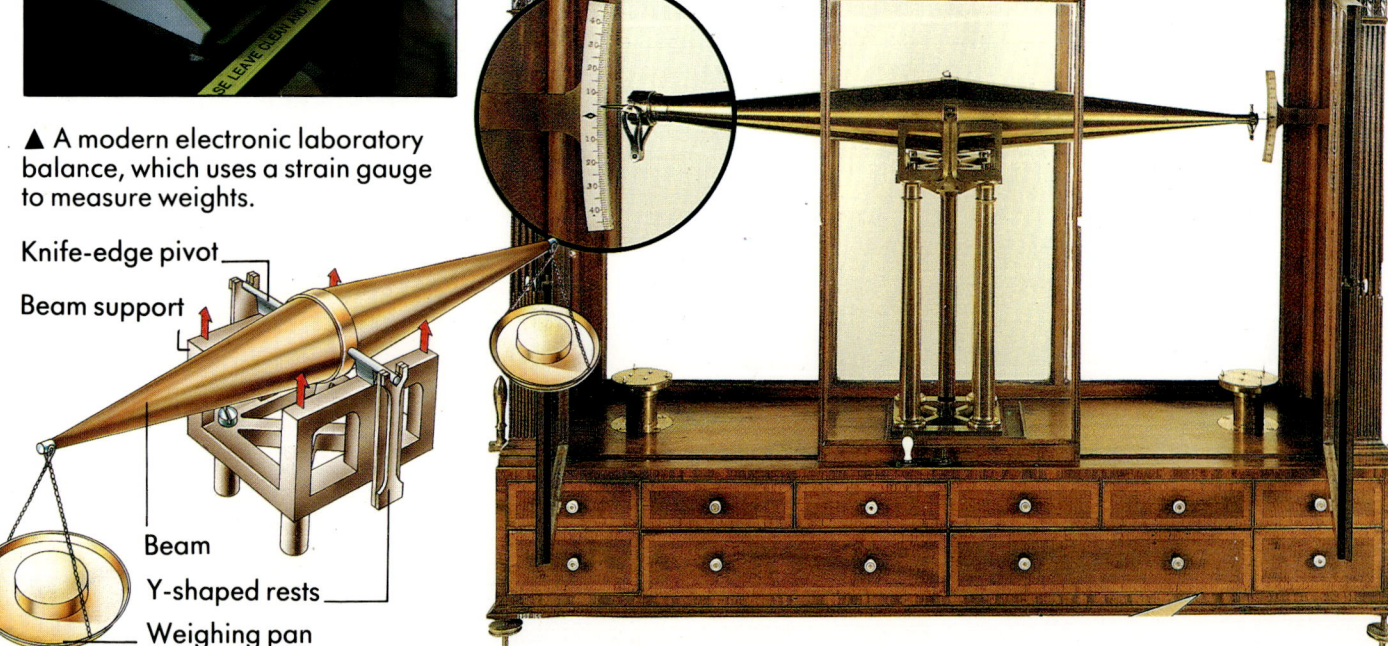

▲ A modern electronic laboratory balance, which uses a strain gauge to measure weights.

Knife-edge pivot
Beam support
Beam
Y-shaped rests
Weighing pan

86

Also common are single-pan balances, in which the pan and a cradle of weights are balanced by a counterweight. The object to be weighed is placed in the pan and weights are removed from the cradle until the arm is in balance again.

Such balances, enclosed in a glass case to prevent disturbance by air currents, can weigh to an accuracy of up to one-thousandth of a gram. Much more accurate are the so-called microbalances, which can measure with an accuracy of up to one-millionth of a gram. Some, called torsion balances, employ the twisting action of quartz fibres to restore balance. Others use the electromagnetism set up by passing electric current through a coil as the restoring force.

The latest laboratory balances are electronic, and are based on a strain gauge. This is a ribbon, usually of metal, which is stretched by the applied load. Stretching slightly changes the ribbon's electrical resistance. The changes are detected by microcircuits and can be interpreted as weights.

Other everyday balances work on different principles. Spring balances measure weight by the extension of a spring under load. A rack-and-pinion system moves the pointer over a scale. The load on a letter balance is balanced by a weighted pointer.

Spring balance

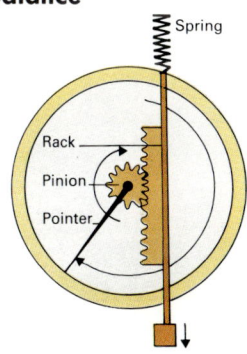

▲ ▶ A spring balance being used at a salmon fishery in Scotland to weigh the day's catch. The balance measures weight as the increase in length of a spring. A toothed rack attached to the spring turns a toothed pinion as it is pulled downwards. And a pointer attached to the pinion moves over a scale.

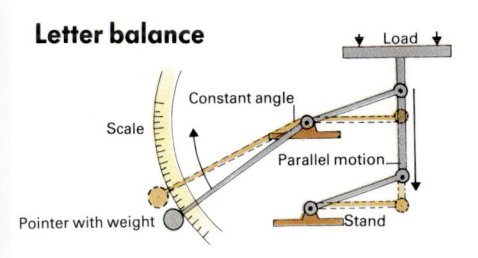

Letter balance

▲ In the letter balance no spring is involved. When a letter is placed on the load pan, it causes a weighted arm to move upwards to restore balance. This causes an attached pointer to move over a scale.

Length and distance

Measuring sticks, or rulers, for measuring length date back 5,000 years or more. And, along with flexible measuring tapes, they are still in widespread use in everyday life.

In industry the need for more accurate measurement than that afforded by rulers became pressing as the Industrial Revolution got underway in the late 1700s. Machines just had to fit together accurately. To make accurate measurements, engineers like James Watt, the Scottish steam-engine manufacturer, built micrometer gauges. These relied on the movement of a spindle driven by the rotation of a screw along an accurate screw thread. An accuracy of fractions of a millimetre became achievable.

Later, using precision screw-threaded machines, Joseph Whitworth in England produced sets of gauges of standard dimensions for

Triangulation

The classical method of surveying land to make maps, called triangulation, makes use of the geometry of the triangle. The surveyor first measures out a baseline. Then from each end he or she measures with a theodolite (picture) the angle between the baseline and a distant point. From these measurements the distances to the point from each end of the baseline can be simply calculated.

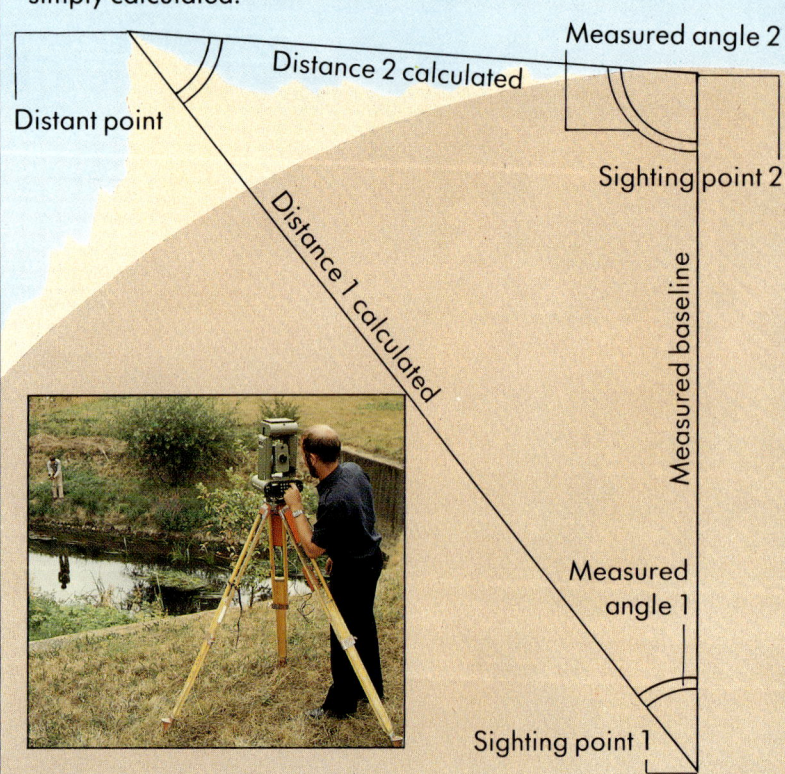

▲ A micrometer screw gauge being used to measure accurately the external diameter of steel tubing. Rotating the thimble (handle) of the gauge drives forward a spindle until it touches the tubing. The diameter is then read from a fixed and moving scale.

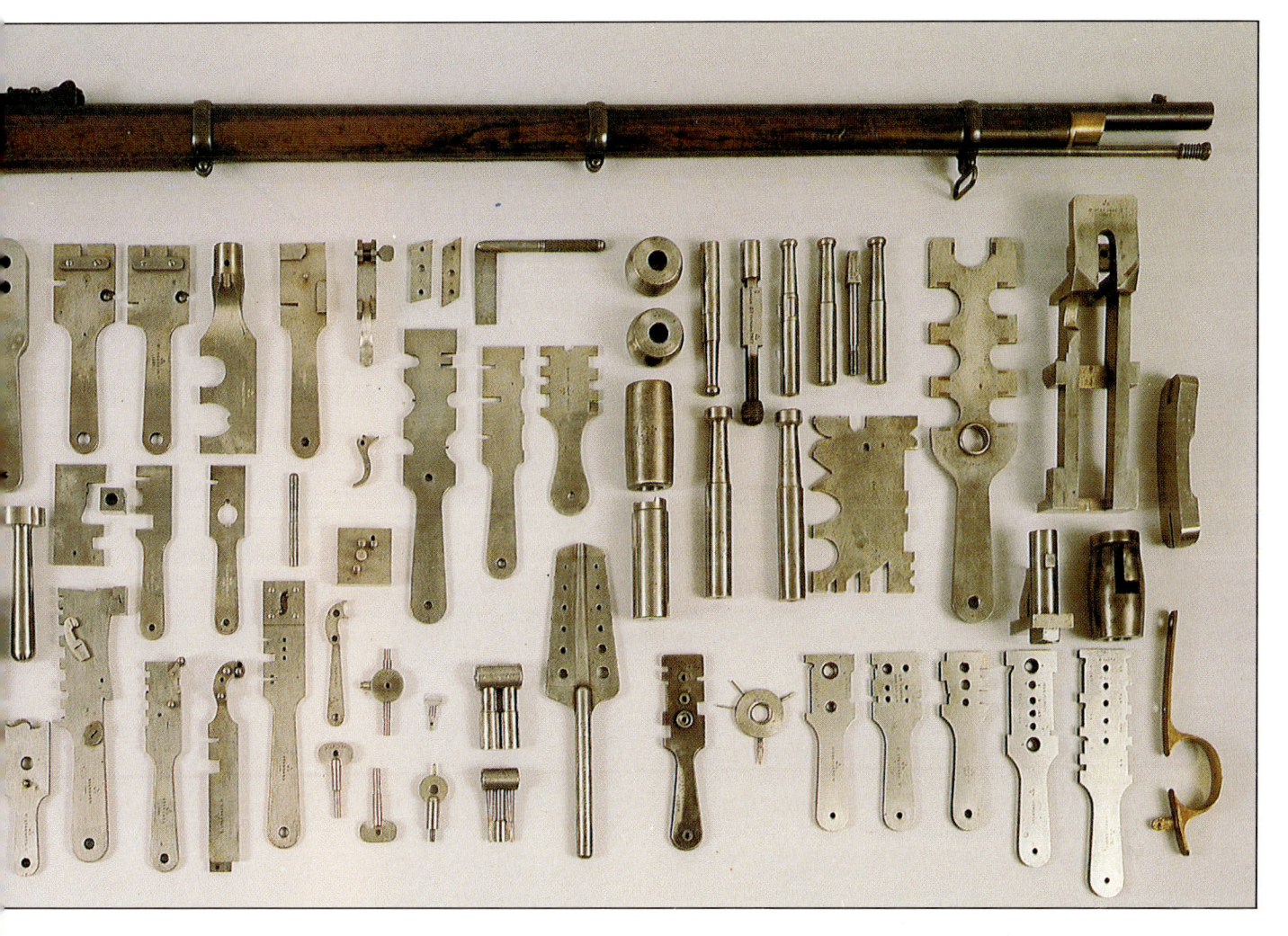

▲ The mass-produced Royal Enfield rifle used by British infantry from the mid-1850s. It was assembled from standardized parts made using this set of accurately measured "slip" and "go-no-go" gauges.

engineering use. Armed with a suitable set of gauges, workers could turn out nearly identical, or interchangeable, parts. The assembly of such parts made mass-production possible.

For measuring distance, surveyors traditionally use a long steel tape or a measuring chain. The chain is made up of standard-sized links: the 25-m long metre chain is widely used. In recent years surveyors have begun using electronic rangefinders. Beams of infrared light, for example, are reflected off a target object. The time interval between transmission and the reception of the reflection provides a measure of the distance. Rangefinders working on the same principle but using laser beams are now widely used by the military, for fitting to rifles and artillery.

The vast distances to the stars presents a greater problem. Distances to some of the nearest ones are measured by exploiting parallax, the apparent shift in position of an object against a more distant background when viewed from two different positions. The distance of far-off galaxies can be gauged from the spectrum of their light. Astronomers know that the more distant a galaxy is, the faster it is travelling away from us. This causes a shift in dark lines in their spectrum towards the red end. From the amount of red-shift, the distance to the galaxy can be estimated.

To express astronomical distances, ordinary terrestrial units of length are inadequate. And astronomers often express distances in terms of the distance light travels in a year (10 million million km), calling this a light-year.

Time

Unlike the metre and the kilogram, the SI unit of time, the second, does not fit into a decimal (10-based) system. Sixty seconds make one minute; 60 minutes make one hour; and 24 hours make one day. It was Babylonian astronomers more than 5,000 years ago who divided the day in this way. They selected these divisions because their number system was based on 60.

The early peoples told the time by day from the position of the Sun in the sky, and by night from the position of the stars. They devised shadow clocks and sundials to indicate the passage of time by day, and water clocks the time by night.

Not until the 1300s did more accurate mechanical clocks appear in Europe. The first ones used a swinging arm called a foliot as a regulator, a device that repeats its action in a standard time. On each swing, the foliot let escape one tooth of a gearwheel (escape wheel) to move the clock hand. Later clocks used a pendulum as a regulator.

The early clocks were driven by falling weights. They became portable, as watches, when spiral springs (mainsprings) were used instead from 1500. The watch was later improved by the introduction of a hairspring as regulator. In the modern mechanical watch, this is now linked with a balance wheel.

In the modern world mechanical clocks and watches have been largely overtaken by electronic ones, which use a quartz crystal as regulator. Precision quartz-controlled clocks are used for accurate time measurements in scientific work, but they are calibrated against atomic clocks. These clocks are regulated by the frequency of radiation from atoms, often caesium atoms. The standard atomic clock varies by only one second in 1,000 years.

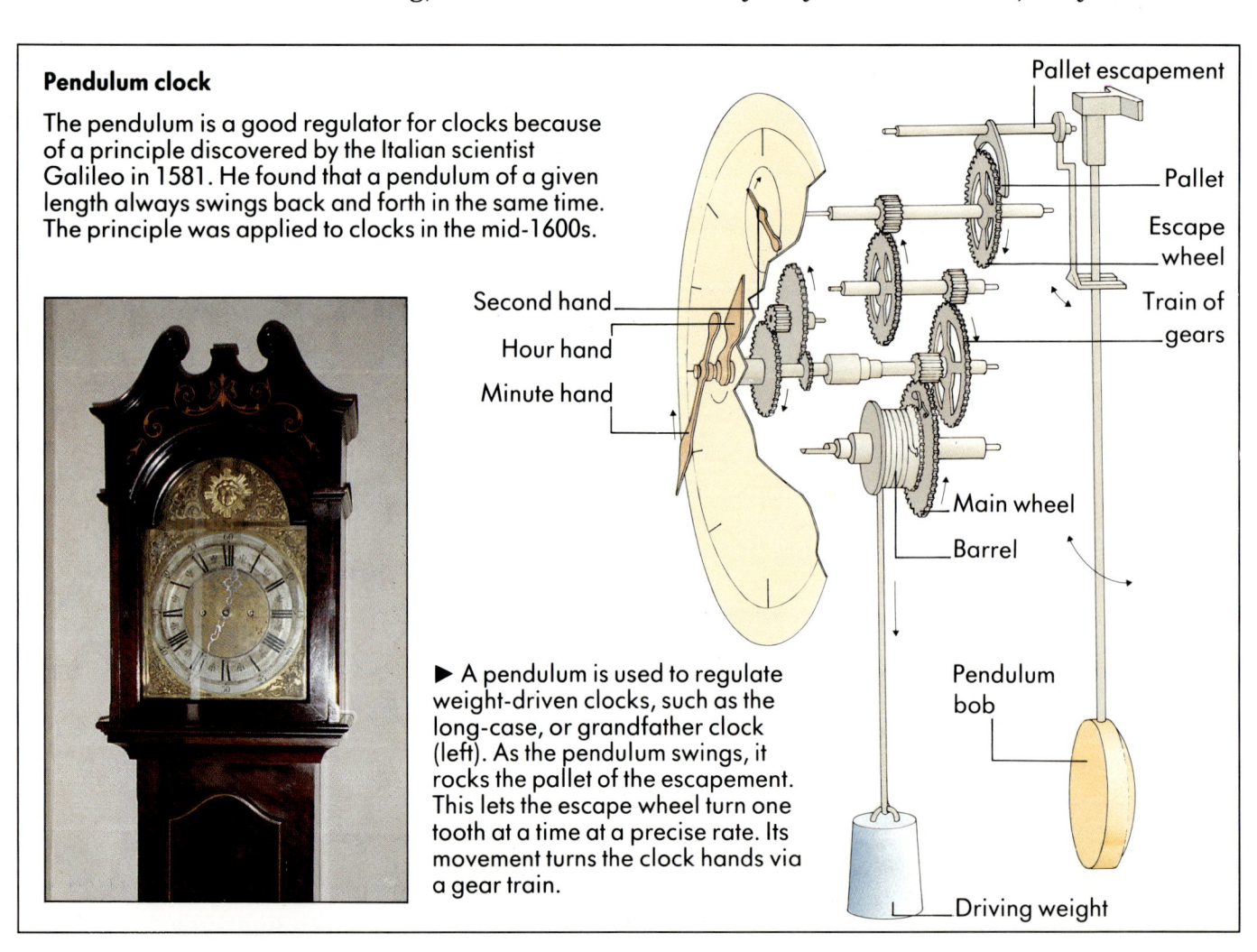

Pendulum clock

The pendulum is a good regulator for clocks because of a principle discovered by the Italian scientist Galileo in 1581. He found that a pendulum of a given length always swings back and forth in the same time. The principle was applied to clocks in the mid-1600s.

▶ A pendulum is used to regulate weight-driven clocks, such as the long-case, or grandfather clock (left). As the pendulum swings, it rocks the pallet of the escapement. This lets the escape wheel turn one tooth at a time at a precise rate. Its movement turns the clock hands via a gear train.

Digital watch

Quartz crystal is a piezoelectric material. This means that if it is deformed, it produces a tiny electric current. The reverse is also true: if an electric current is applied to it, it will deform. A quartz crystal is used as the regulator in digital watches. When a tiny electric current from a battery is applied to it, it is made to vibrate at a precise frequency of 32,768 hertz (cycles per second). Electronic circuits on a chip reduce the frequency in stages to one vibration per second, and this is then used to drive a digital display that shows the time. Most watches these days have a liquid-crystal display (LCD). Early ones used light-emitting diodes (LEDs).

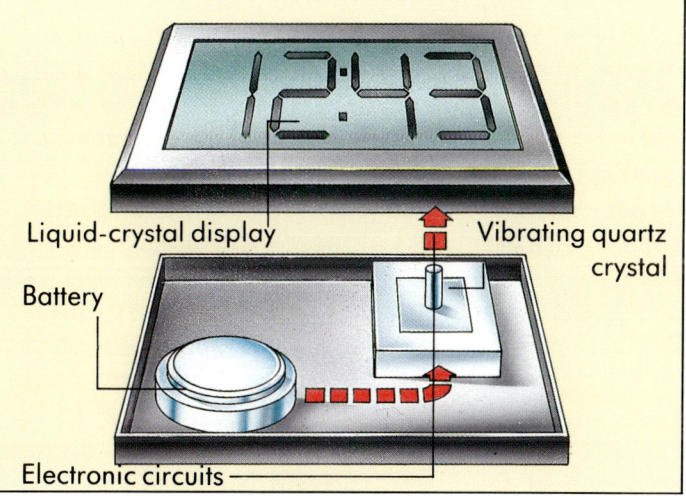

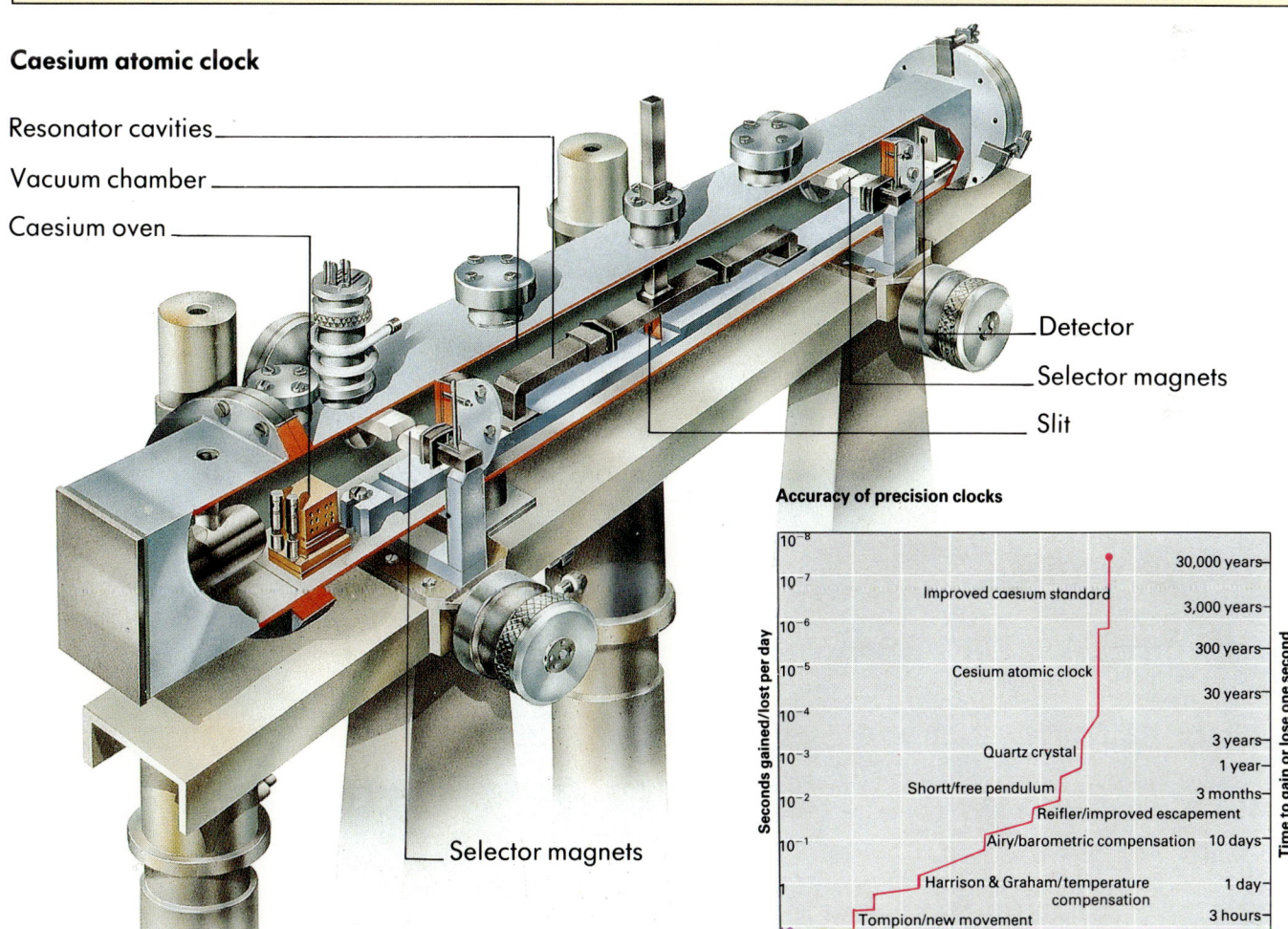

Caesium atomic clock

▲ In the caesium atomic clock a sample of caesium metal is heated and emits atoms, which travel down the tube. Half-way along they pass through a cavity fed with microwave radiation vibrating at a frequency set by a quartz clock. When this frequency matches the frequency of the caesium atoms, which is 9,192,631,770 hertz, the detector registers maximum.

▲ The accuracy of clocks has improved with each new development. Atomic clocks have brought about the most dramatic improvement. US scientists are now developing one that should be accurate to one second every 10 billion years.

Electricity

Scientists and engineers use a wide variety of electrical instruments for measuring, and for displaying and recording experimental data. In essence most electrical instruments measure either current (rate of flow of electricity) or voltage (electrical "pressure difference").

An ammeter ("amp-meter") measures current in units named amperes (amps), after the French physicist André-Marie Ampère. The standard instrument is a moving-coil meter. The current to be measured is fed through a wire coil located within the poles of a permanent magnet. The passage of the current sets up in the coil a magnetic field. This interacts with the permanent field, and the coil is forced to turn. Attached to the coil is a pointer, restrained by a spring, which indicates the level of current on a marked scale.

A voltmeter is a similar instrument used to

▼ The diagram shows the essential features of a moving-coil meter. The photographs show two views of a type widely used in industry, which can be set to measure electric current, voltage and resistance.

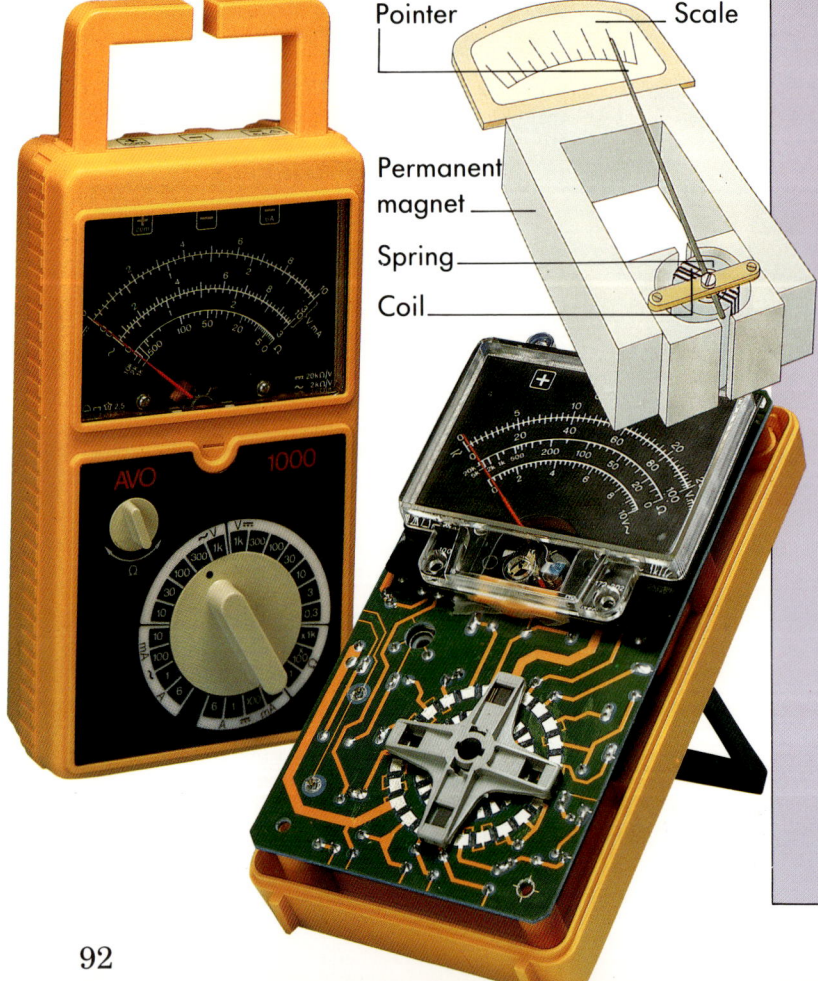

Liquid-crystal display (LCD)

Liquid crystals twist light passing through them. In an LCD that shows blank (left), light passes through a polarizing sheet, which lets through light vibrating in one plane only. It passes through the liquid crystal and then through another polarizing sheet before being reflected back by a mirror. But when current is applied (right), the crystal no longer twists the light. And no light is reflected because it cannot reach the mirror. As a result, the display shows black.

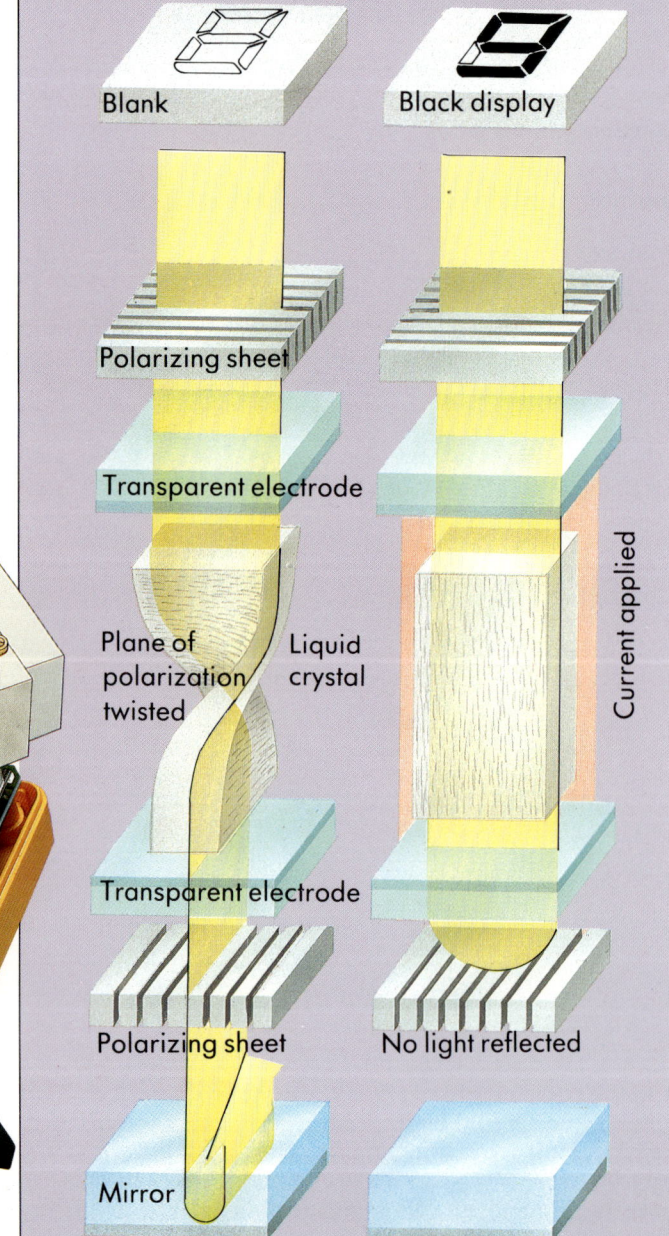

92

measure voltage in volts, units named after the Italian inventor of the electric battery, Allesandro Volta. It measures the current flowing through a resistor of known electrical resistance, and this is directly related to the applied voltage.

In this type of voltmeter, however, some current is lost in the meter itself. To avoid this inaccuracy, a potentiometer is used. This is a device which balances the unknown voltage against a standard voltage source by means of a sliding resistance. When the voltages are equal, no current flows through the circuit meter, and so no inaccuracy is introduced.

Ammeters and voltmeters can also be used indirectly to measure non-electrical quantities, such as pressure, rates of flow and temperature. They can do so if these quantities can be suitably converted, or transduced, into electric current and voltage. For example, temperature can be measured electrically with a resistance thermometer. This uses a coil of platinum wire as a transducer. The resistance of platinum alters as the temperature alters, so current passed through the wire will vary according to the temperature and be a measure of that temperature.

Electrical measurements can also be displayed, notably on an oscilloscope. This is a cathode-ray tube, like the tube in a television set. In the tube a cathode emits a stream of electrons which are focused in a narrow beam on to a fluorescent screen. When electrical voltage is applied to magnetic coils around the tube, the beam is deflected up and down. Usually it is also swept from side to side across the screen, and this gives a visual image of how the applied voltage is varying.

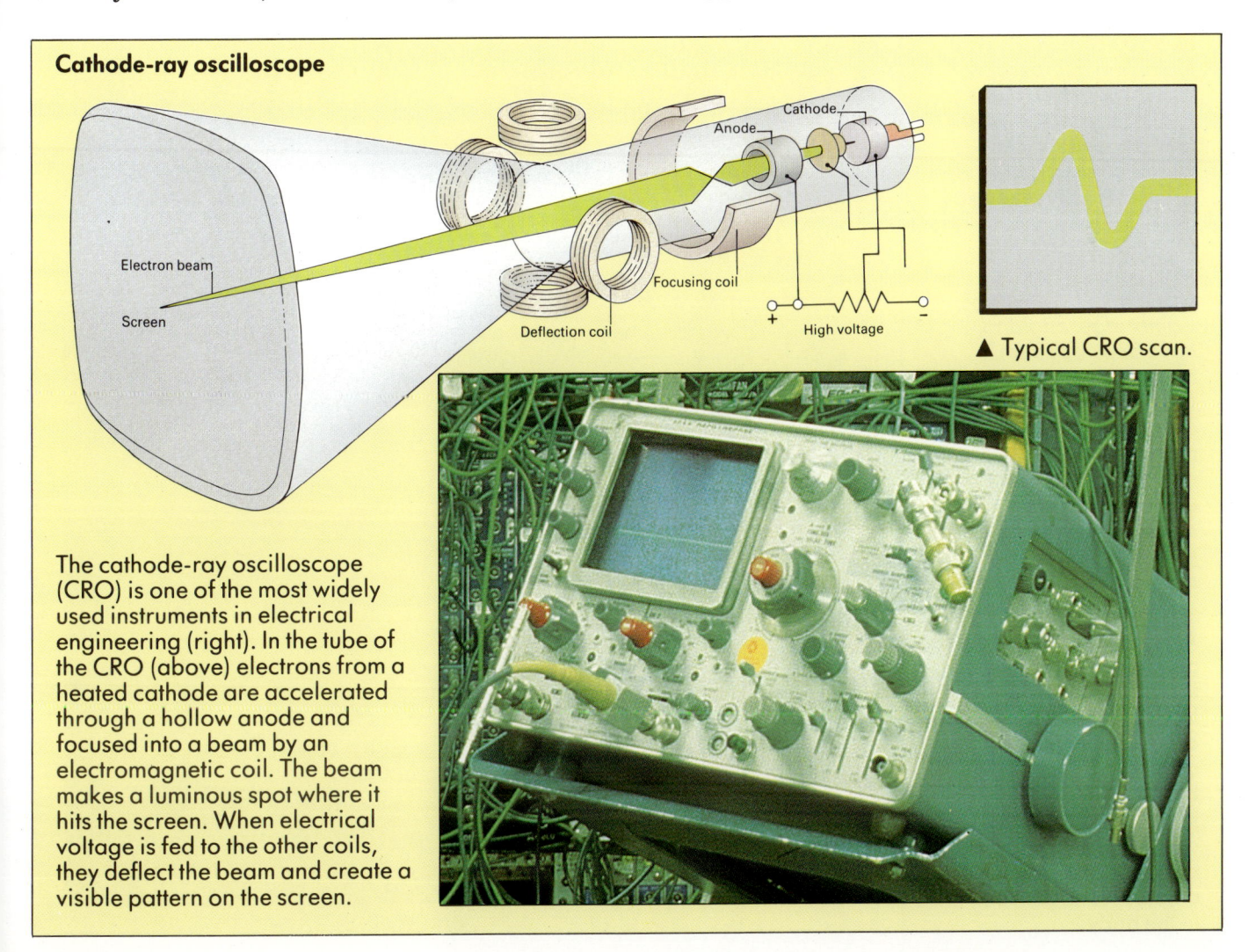

Cathode-ray oscilloscope

▲ Typical CRO scan.

The cathode-ray oscilloscope (CRO) is one of the most widely used instruments in electrical engineering (right). In the tube of the CRO (above) electrons from a heated cathode are accelerated through a hollow anode and focused into a beam by an electromagnetic coil. The beam makes a luminous spot where it hits the screen. When electrical voltage is fed to the other coils, they deflect the beam and create a visible pattern on the screen.

Seeing near

Spot facts

• Microscopes have been developed that use sound waves to examine materials. These acoustic microscopes are very suitable for examining biological specimens because sound waves do not damage living tissue.

• In 1990 IBM engineers, using a scanning tunnelling microscope, manipulated 35 atoms of xenon to form the letters IBM. The letters were "written" a million times smaller than the type you are reading.

• Scientists at the Cavendish Laboratory, Cambridge, England, have developed an electron-beam technique for the microstorage of information. They reproduce the information as a pattern of dots on aluminium fluoride. They are able to cram up to 10 million words in each square millimetre.

▶ A microscope photograph showing the tubes radiating from a spiracle, or breathing hole, of a caterpillar. All insects breathe through spiracles, and the tubes carry oxygen to, and bring back carbon dioxide from, the cells in the body.

One of the smallest things our eyes can see is a pinprick. If we prick our skin with a pin, a drop of blood oozes out. We can see the drop, but not what it contains — as many as five million tiny saucer-like bodies called corpuscles. To see such bodies we need to use a microscope. We can see blood corpuscles in an optical, or light microscope, which uses glass lenses to magnify objects. To see very much smaller things, such as bacteria and viruses, we must use an electron microscope, which works with a beam of electrons instead of light. Other ingenious microscopes are even able to picture atoms.

Early microscopes

Magnifying lenses made of glass came into use in the 1200s. Like modern cheap magnifying glasses, they magnified about three to five times. In the 1670s the Dutch biologist Anton van Leeuwenhoek began grinding more accurate lenses with much greater magnification (up to 300 times). With these lenses he began observing the tiny animals in pond water and even some bacteria. In so doing he helped pioneer the science of microscopy.

A few years earlier in England, the English physicist Robert Hooke had begun making observations with a compound microscope. This used two lenses to achieve a two-stage magnification. The compound principle had been discovered at the turn of the century by a Dutch spectacle-maker, Zacharias Jannsen.

The early lenses suffered from many defects, which caused progressive blurring of the image as magnification increased. The main defects were spherical and chromatic aberrations. The first was caused by the lens surface not being quite spherical. It could be reduced by accurate grinding. The second was caused by the glass lens acting like a prism and splitting up light into a spectrum. And the various colours were brought to a focus at different points, causing colour blurring of the image. Chromatic aberration was cured in the 1830s by the use of achromatic lenses, following the practice in telescope design.

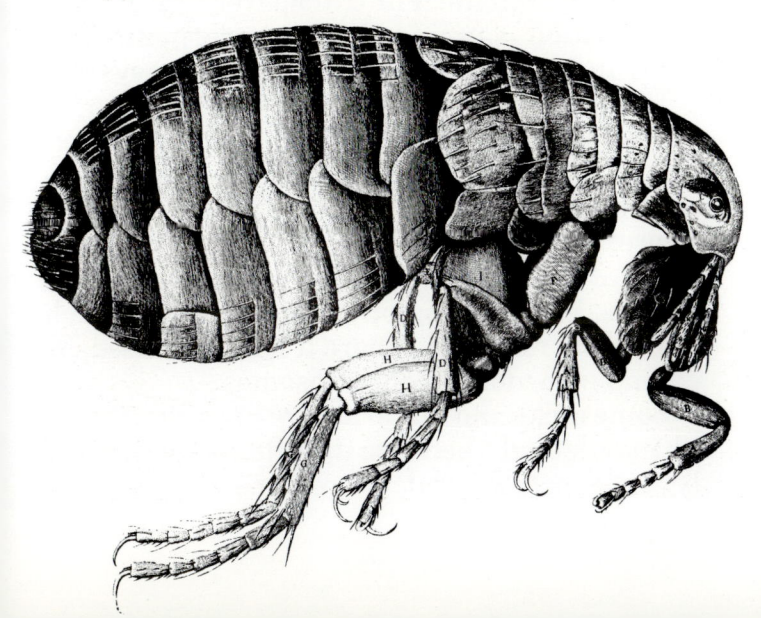

▼ This compound microscope was made by the London instrument-maker Christopher Cock in the late 1600s. It has ornate decoration typical of the period.

▲ The first microscope, made by Anton van Leeuwenhoek in the Netherlands more than three centuries ago. It was a simple microscope. It had a single glass bead lens set between brass plates. The specimen was placed on a pointer close to the lens.

▼ Robert Hooke pioneered microscopy with a compound, or two-lens instrument. This sketch of a flea appeared in his book *Micrographia* of 1665.

Optical microscopes

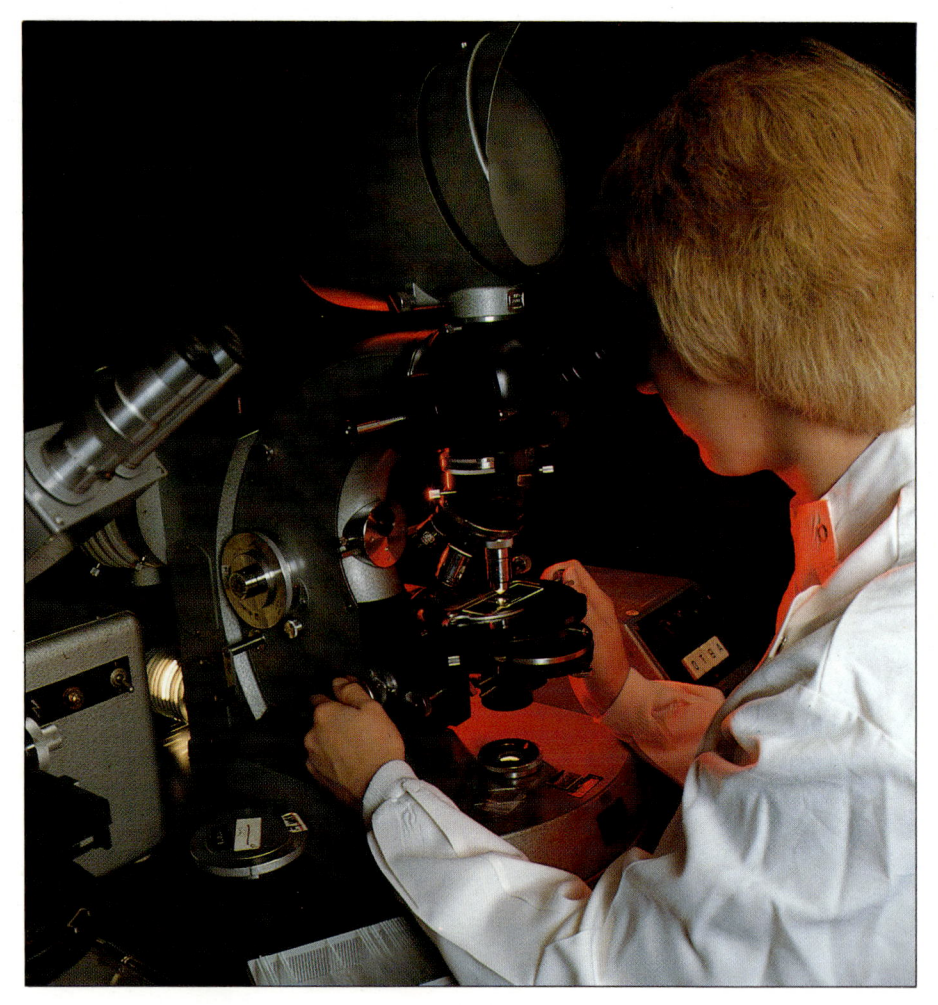

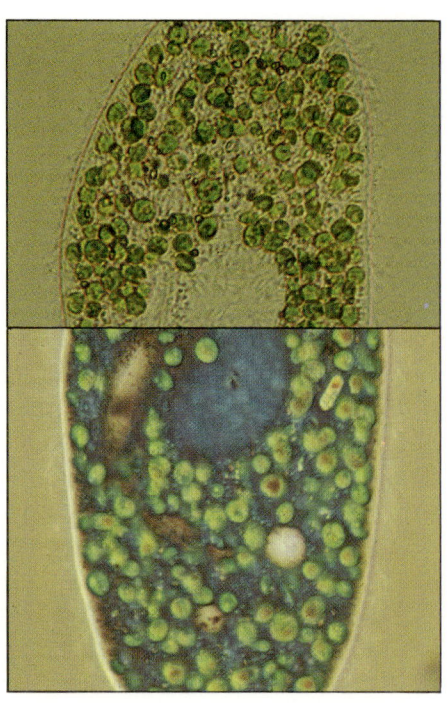

▲ A researcher looks through the binocular eyepiece of one of the latest microscopes. It has facilities for phase-contrast and polarized-light microscopy. It is equipped with a camera for photomicrography, and it has a screen for general viewing.

▲ Pictures of a simple invertebrate (*Paramecium bursaria*), taken at a magnification of 150 times using two different microscopes. In an ordinary microscope (top), the internal structure of the creature is indistinct. In an interference-contrast microscope (bottom), features show up much more clearly.

The ordinary, or optical microscope uses two lenses, or rather combinations of lenses, to magnify an object. The magnification is achieved in two stages. The first magnification is brought about by a lens positioned close to the object under study. This objective lens produces a magnified, real and upside-down image of the object. The second lens, the eyepiece, is positioned close to the eye. It views the image and magnifies it further, producing a virtual image – one that can be seen by the eye but cannot be displayed on a screen.

The diagram shows the main features of a practical microscope. This one has a number of objective lenses mounted on a rotating turret, each with a different power of magnification. Specimens to be examined are placed on a slide, and mounted on the stage. They are often thin transparent slices of material, cut by a device called a microtome. For transparent specimens, illumination is provided from underneath by a lamp and a mirror and lens (condenser) system.

Various techniques are used in microscopy to bring out details in a specimen. Many biological materials are stained by dyes to make their structure stand out. Some are dyed with fluorescent substances, which show up when they are illuminated with ultraviolet light. Fluorescence microscopy has proved valuable in medical research, where it has been used, for example, to identify chromosomes.

Some rocks and minerals are examined under ultraviolet light because they fluoresce naturally. In a petrological microscope they are examined in polarized light, which is light that vibrates in only one plane. Using this technique, crystals in a rock that look much the same in ordinary light, appear quite different and often brilliantly coloured in polarized light.

Another technique used in research work to bring out details in an image is phase-contrast microscopy. It does not involve staining or other techniques that could harm living things, so it can be used to record such events as cell division. The phase-contrast microscope exploits the fact that different parts of a specimen slow down light passing through them by differing amounts. This results in a difference in phase between the light waves. In other words, they get out of step and interfere with one another. The result is a pattern of light and dark, which emphasizes details of structure. The interference-contrast microscope works on the same principle.

Compound microscope

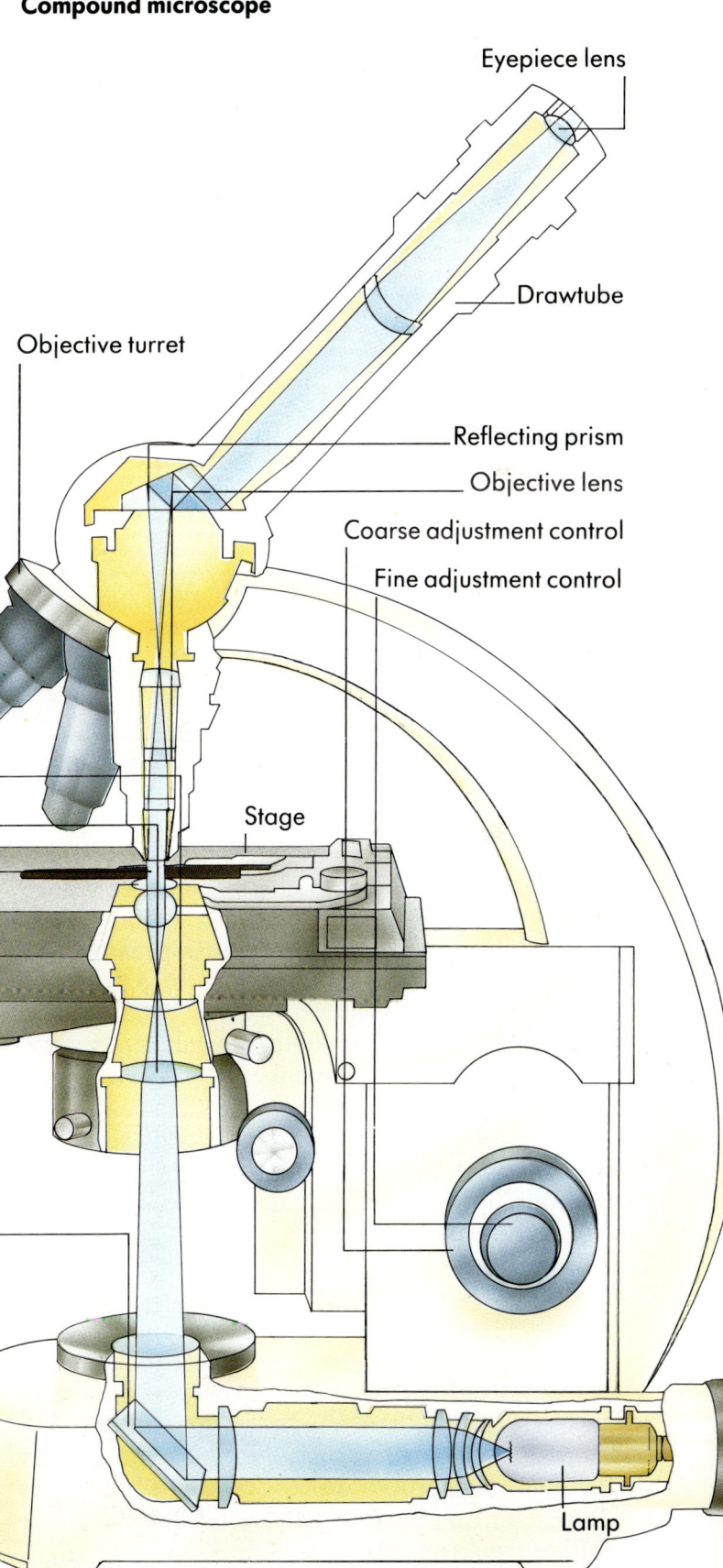

▲ Rock crystals can be studied under a microscope in ordinary light (top), but show up better in polarized light (bottom).

▶ The main features of a compound microscope. The optical system below the stage provides even illumination of the specimen. The objective forms a magnified image, which the eyepiece then views, via a reflecting prism.

97

Electron microscopes

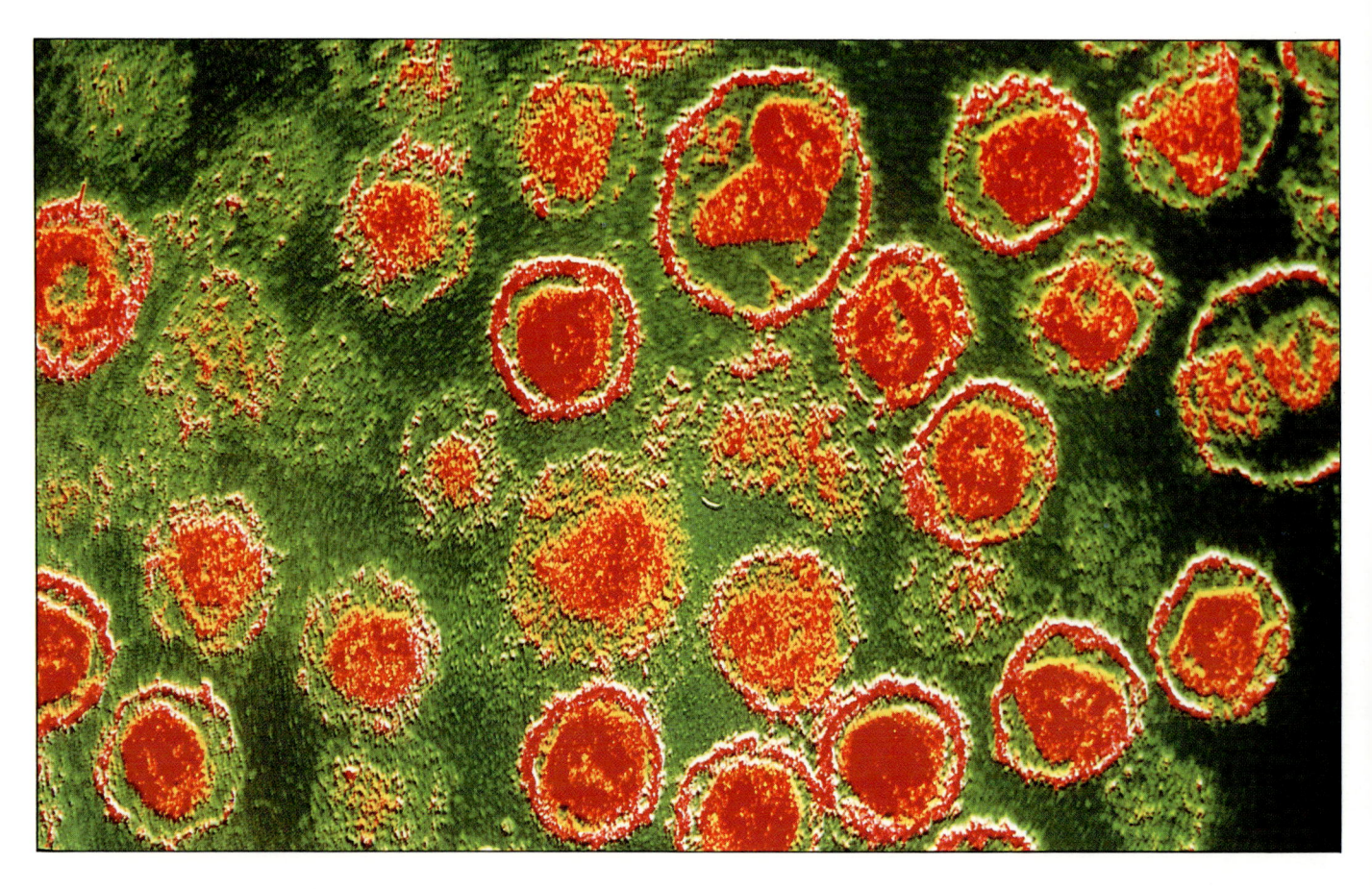

The most powerful ordinary microscope can resolve, or show as separate, objects that are up to about 200 millionths of a millimetre apart. They can magnify up to about 2,500 times. Their resolving power, or resolution, is limited by the length of the light wave – it is too big a yardstick.

The shorter the wavelength used, the greater is the resolution. So using shorter-wave ultraviolet light, for example, twice the resolution (100 millionths of a millimetre) is possible. But for very much higher resolution, we must abandon light and turn to the electron.

Electrons, when they are produced in a beam, behave like a wave motion. And it is a wave motion with very short wavelengths indeed – down to less than one-millionth of a millimetre.

Experiments in using an electron beam to create a high-resolution microscope began in the early 1930s. And the first practical electron microscope appeared in 1936. Since then many developments have taken place, which now allow scientists to achieve magnifications of one million times or more.

The original type of electron microscope is the transmission electron microscope (TEM). The essential features of the instrument are shown in the diagram. It uses an electron "gun" to produce a beam of electrons. Then it uses a series of magnetic "lenses" to focus the beam and form a magnified image when it passes through a specimen. They act like the objective and eyepiece lenses of an optical microscope. Finally, the beam holding the magnified image is projected on to a fluorescent screen, which makes the image visible. This can be viewed through a binocular eyepiece, or photographed.

The scanning electron microscope (SEM) produces images by scanning the surface of a specimen in a series of lines with a fine electron beam. Where the beam strikes it, the surface gives off secondary electrons. The pattern of electrons given off holds an image of the surface. These electrons are collected, line by line, and form the brightness signals for a beam scanning in a cathode-ray tube. An image is built up on the screen, line by line, in much the same way as a television picture is made.

◀ This is a picture of viruses that cause cancer in birds, taken in a transmission electron microscope. False colours have been added to make the virus particles (red) stand out. Viruses are so tiny that they can be seen only in electron microscopes, which can provide magnifications of a million times or more.

▶ Features of a transmission electron microscope (TEM). The main body consists of a column from which the air has been removed. An electron beam is produced at the top of the column by an electron gun, similar to that in a cathode-ray tube. At intervals lower down are sets of magnetic coils, which act as lenses to focus the beam and direct it through a thin slice of specimen. The beam that emerges carries an image of the specimen, which is magnified by another set of lenses (objective). A final set of lenses projects the magnified electron image on to a fluorescent screen, where it becomes visible, and can be photographed.

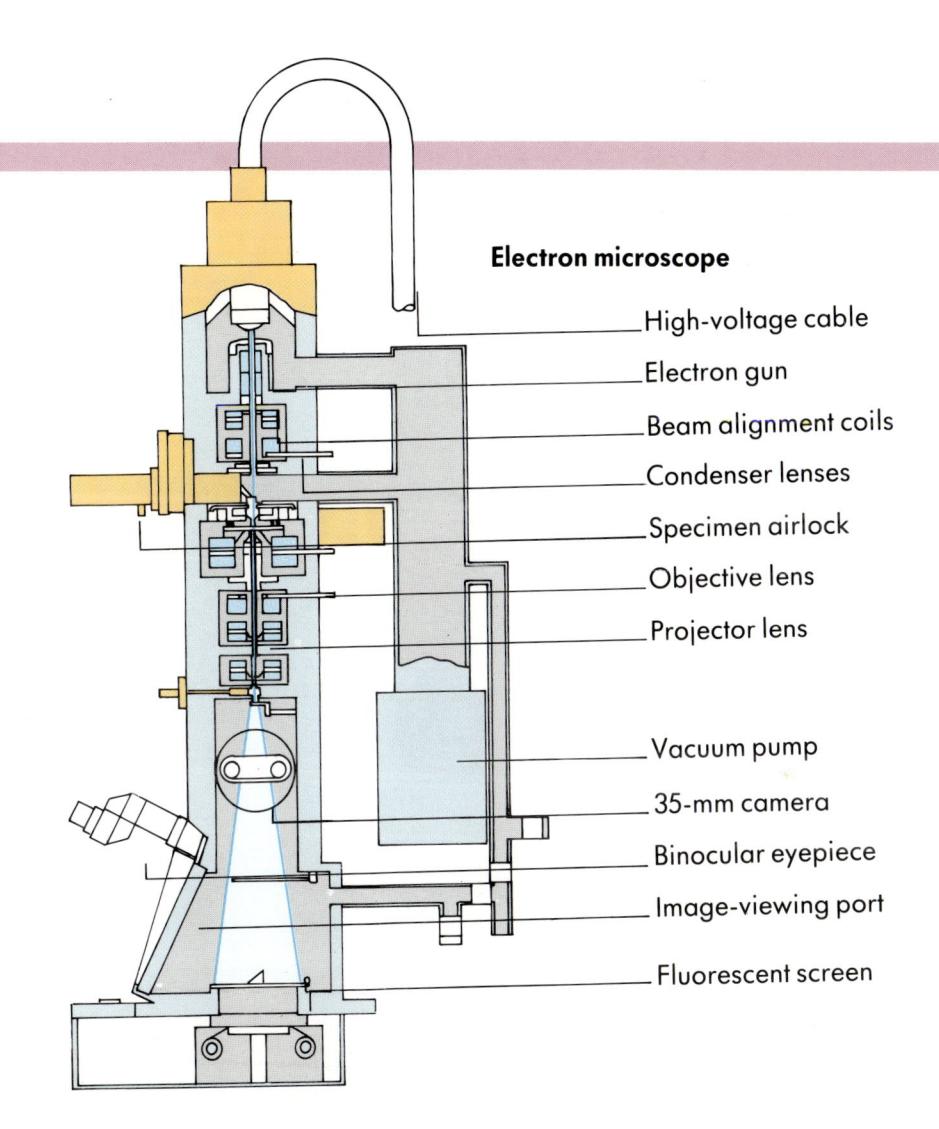

Electron microscope

- High-voltage cable
- Electron gun
- Beam alignment coils
- Condenser lenses
- Specimen airlock
- Objective lens
- Projector lens
- Vacuum pump
- 35-mm camera
- Binocular eyepiece
- Image-viewing port
- Fluorescent screen

◀ A Bird cherry aphid next to the exoskeleton it has just cast off, pictured by a scanning electron microscope (SEM). Vivid three-dimensional images like this make the SEM one of the most fascinating tools in scientific and medical research.

▼ A researcher at the operating console of an SEM. In the instrument the specimen is scanned by a pencil-thin electron beam.

Seeing far

Spot facts

- The world's largest optical telescope is the 6-m reflector at Zelenchukskaya in the Caucasus Mountains, Russia, which was completed in 1974.

- Radio telescopes can detect faint traces of the radiation produced when the Universe came into being some 15,000 million years ago.

- Astronomers fleetingly thought that the pulsing radio waves from pulsars, first detected in 1968 by Jocelyn Bell at Cambridge Radio Astronomy Observatory, England, might be signals from an alien civilization. They called them LGM, for Little Green Men.

- The infrared astronomy satellite IRAS spied stars being born during its survey of over 200,000 infrared sources in the sky in 1983.

▶ Telescope domes at Kitt Peak Observatory in the Arizona Desert, USA. The largest dome houses the Mayall reflector, which has a light-gathering mirror 4 m in diameter.

Just as they cannot easily see very tiny objects, our eyes cannot easily see very distant or very faint objects, such as the heavenly bodies. To see such objects more clearly, we must use telescopes, whose lenses or mirrors can gather more light than our eyes.

Using powerful telescopes and a variety of other instruments, astronomers can tell us what stars are made of and how they live and die. They also build telescopes to collect radio waves from the heavens, and send telescopes into space on satellites to observe our mysterious Universe at other wavelengths as well.

Early telescopes

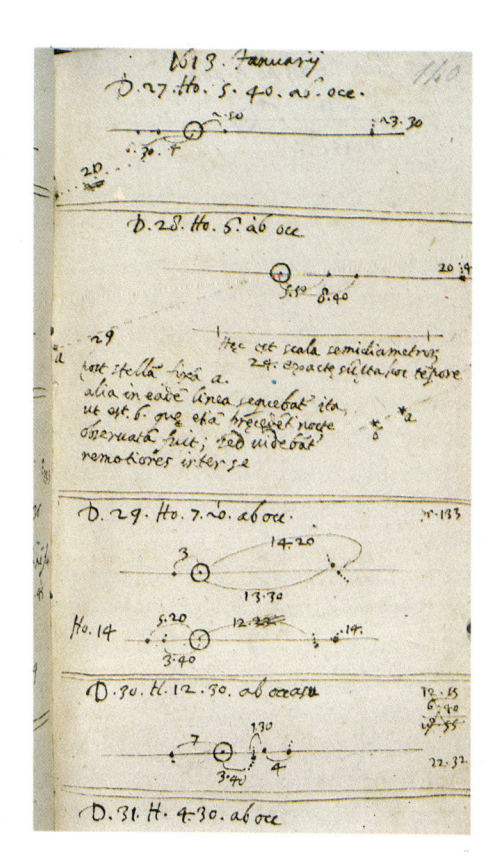

The birth of modern astronomy, the scientific study of the heavens, can be dated at 1609. In the winter of that year in Padua, Italy, a professor of mathematics, Galileo Galilei, made a telescope and trained it on the heavens. He became the first person to see the mountains on the Moon, the four large moons of Jupiter, and the phases of Venus. His observations helped convince him and other astronomers that the Sun and not the Earth was the centre of our planetary system.

Galileo's telescope used a pair of lenses to gather light and produce a magnified view of the heavenly bodies. It was a type called a refractor, because the path of light was refracted, or bent, as it passed through the lenses.

The early refractors, however, suffered from several defects, which blurred the image. To improve image quality, the English scientist Isaac Newton in 1668 built a telescope that used a concave (dish-shaped) mirror to gather and focus light. This type of telescope, called a reflector, is the one used mostly by astronomers today. But whereas the early reflectors were made of polished metal, the modern ones are made of glass coated with aluminium.

▲▶ Galileo pioneered astronomy through the telescope, leaving sketches and detailed notes. His telescopes (shown right) have magnifications of 14 and 20.

▼ Isaac Newton's reflecting telescope of 1668 was a great advance because it did not suffer from lens defects.

101

Refractors and reflectors

Refractors

These days lens-type, or refracting, telescopes are still widely used by amateur astronomers. In small sizes, they can give excellent results. A refractor uses two lenses to form an image – an objective and an eyepiece. The objective is a converging lens, or rather lens combination, which has a long focal length. It gathers the light from a distant object and forms an image. The eyepiece, which is also a converging lens, views and magnifies the image. The objective is mounted at the front of the main telescope tube, while the eyepiece is mounted in a tube that can slide in and out of the main one to bring the image into sharp focus.

Refractors suffer from two main optical defects, or aberrations, which cause blurred images. One is spherical aberration, which results from the lens having slightly the wrong curvature. This defect is reduced by grinding the lens into shape with the utmost accuracy. The other main lens defect is chromatic (colour) aberration. This is a colour blurring of the image caused by the lens refracting light of different wavelengths (colours) by slightly different amounts. This defect is corrected by using achromatic lenses.

Refractors are difficult to make in large sizes because of the way they are constructed. They can be supported only around the edge. Large lenses are heavy and difficult to mount in this way without being distorted. This is why the largest refractor, at Yerkes Observatory in Wisconsin, USA, has a lens only 100 cm across.

Reflectors

Mirror-type telescopes, or reflectors, do not suffer from the chromatic aberration. Also, they can be made in much larger sizes because their mirrors can be supported from behind, which prevents distortion. As a result, reflectors can be built with mirrors several metres across. The world's largest single-mirror telescope has a mirror 6 m across. It is sited near Zelenchukskaya, in the Caucasus Mountains, Russia. However, this 480-tonne reflector is optically not as flexible in use as smaller modern instruments, such as the 4.2-m William Herschel

▲ This refractor has an objective lens of 10 cm in diameter, a popular size with amateur astronomers. It carries a small "finder scope" to help location of the target object. It is mounted on a sturdy tripod.

▶ The Mayall 4-m reflector at Kitt Peak Observatory in Arizona, USA. The telescope is moved bodily to follow the motion of the stars by rotation of the aptly named horseshoe bearing, clearly visible here.

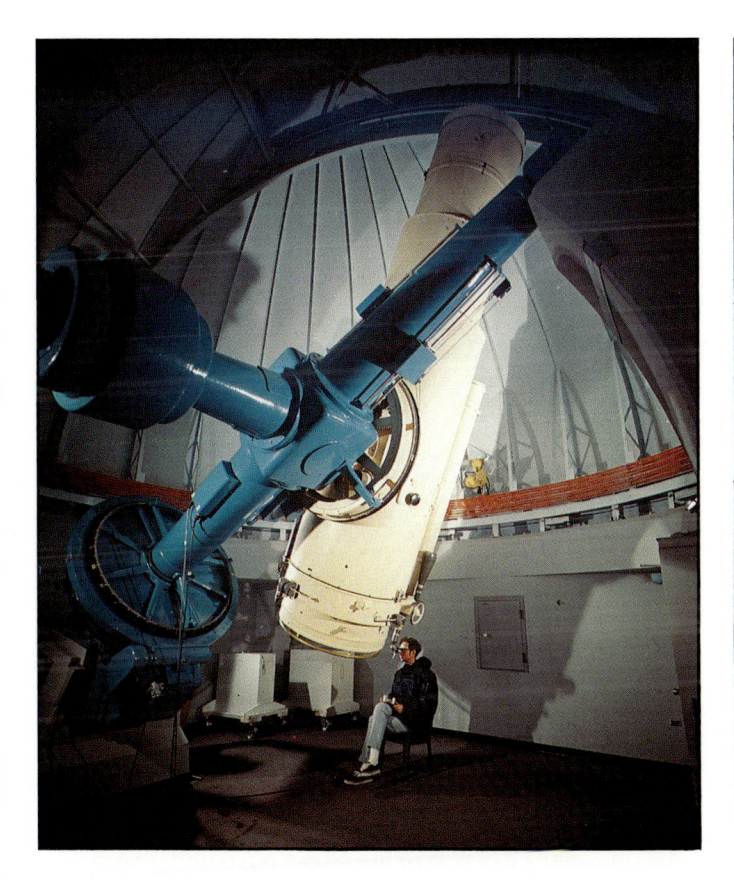

▲ This is another Kitt Peak instrument, a Curtis-Schmidt reflector with a 61-cm diameter mirror. The Schmidt telescope is designed to have a wide field of view. It has a mirror with spherical curvature, which focuses light on a photographic plate. A specially shaped lens is placed in the mouth of the telescope tube to ensure that light reflected from every point on the mirror is brought sharply into focus.

reflector. This telescope, at the Roque de los Muchachos Observatory at La Palma in the Canary Islands, uses superior optics and advanced computer control and electronic instrumentation. It is so sensitive that it could detect the light of a candle 150,000 km away.

There are several kinds of reflectors, which gather and focus incoming starlight in different ways. They all have a concave primary mirror to gather the light. This mirror may focus an image directly on to a photographic plate at the so-called prime-focus position. Or it may reflect the light on to another mirror. This secondary mirror may in turn reflect the light into a viewing eyepiece at the side (Newtonian focus) or back down the telescope tube and through a hole in the primary mirror (Cassegrain focus).

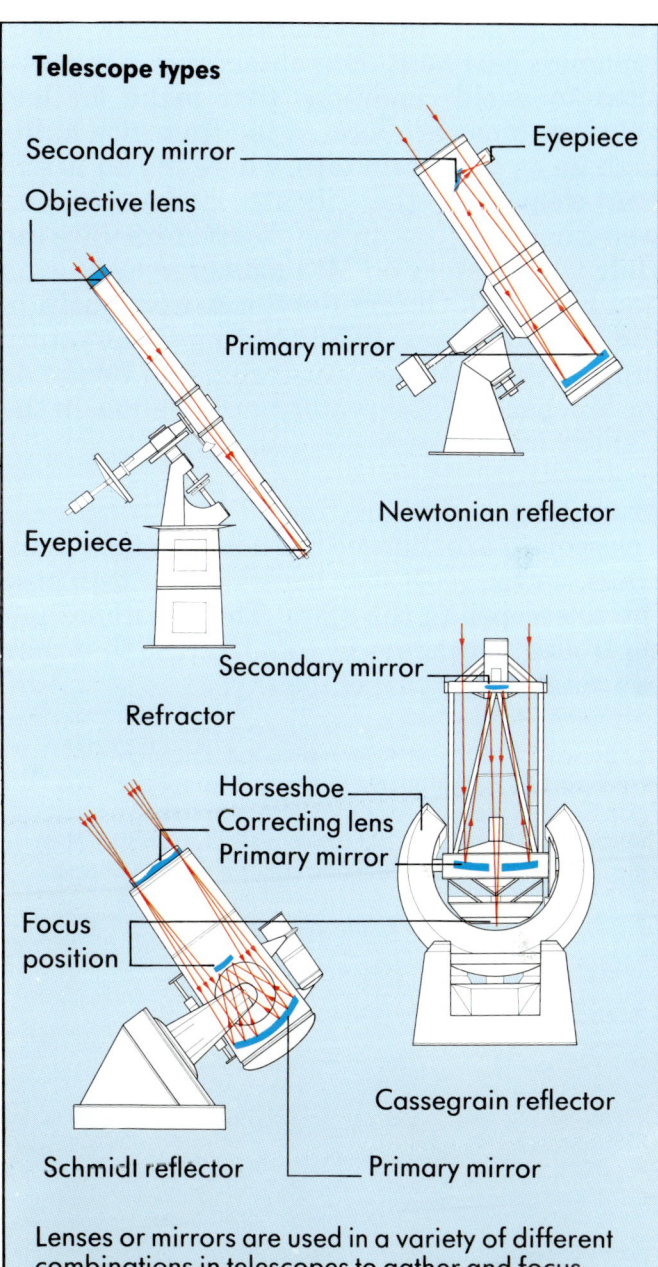

Telescope types

Lenses or mirrors are used in a variety of different combinations in telescopes to gather and focus the faint light from the stars. A refractor uses two sets of lenses (objective and eyepiece) to gather the light and view the image. In the Newtonian type of reflector, light gathered by the primary mirror is further reflected by a plane secondary mirror into an eyepiece in the side of the telescope tube. In a Cassegrain reflector, a curved secondary mirror reflects light back down the telescope tube through a hole in the primary mirror. A Schmidt reflector uses a correcting lens in the telescope tube to ensure sharp focusing by its spherical mirror. It provides a wide-angle view of the heavens.

Astronomical observatories

Astronomical observatories, where astronomers carry out their observations, must be sited to avoid conditions that make for bad viewing. Most are located far from city lights high up on mountain tops, where the air is still and clear and the climate is dry. Among well-known mountain-top observatories are the Hale Observatory on Mt Palomer in California and Kitt Peak Observatory in Arizona, both in the United States; Siding Spring Observatory in New South Wales, Australia; and Roque de los Muchachos Observatory at La Palma, in the Canary Islands.

Mounting and recording

Telescopes are housed in dome-shaped buildings, whose roofs slide back at night to expose the telescopes to the stars. The telescopes are controlled, nowadays by computer, so that they can point to any part of the sky.

▼ The world's largest solar telescope, the McMath, forms part of the extensive telescope complex at Kitt Peak Observatory in Arizona, USA. A mirror (heliostat) reflects sunlight down a sloping shaft 121 m long. Other mirrors reflect and focus it into an image.

▲ One of the newest and finest observatories, the Roque de los Muchachos Observatory on La Palma, in the Canary Islands. It is located at an altitude of 2,400 m, where it is above the cloud base and where the skies are nearly always crystal clear.

▶ Stars trail in circular arcs around the north celestial pole in this long-exposure photograph taken at the Roque de los Muchachos Observatory in the Canary Islands. The two telescopes here are the 1-m Jacobus Kapteyn and the 2.5-m Isaac Newton telescopes. The Isaac Newton telescope was originally located at the Royal Greenwich Observatory in Sussex, England.

▼ An astronomer works at the operating console of the 3.9 m Anglo-Australian telescope at the Siding Spring Observatory, near Coonabarabran in New South Wales. One of the most powerful instruments in the Southern Hemisphere, the telescope is under computer control.

Because the Earth spins on its axis, the stars appear to move through the heavens. So, to follow a particular star for any length of time, the telescope must be driven. This is most simply done if the telescope has an equatorial mounting. This has one axis of movement parallel with the Earth's axis. The telescope is then rotated around this axis at the same speed as the Earth (and the star) is moving.

Astronomers use their large telescopes as giant cameras and record images on photographic film. This is done because film is more sensitive than the eye and can store light. And the longer it is exposed, the more light it stores, which enables it to record very faint stars and galaxies. Light coming through the telescope is also analysed, notably in spectroscopes, which split it into a spectrum. By studying the spectrum, astronomers can discover a remarkable amount of information about the star the light came from, including its composition, temperature and speed.

In recent years electronic methods of light detection have come into use, which are more sensitive than photographic film. They include such devices as charge-coupled devices (CCDs). They are kinds of silicon chips that are light-sensitive. The pattern of incoming light produces on the chip a pattern of charges, which a computer can convert into a visible image.

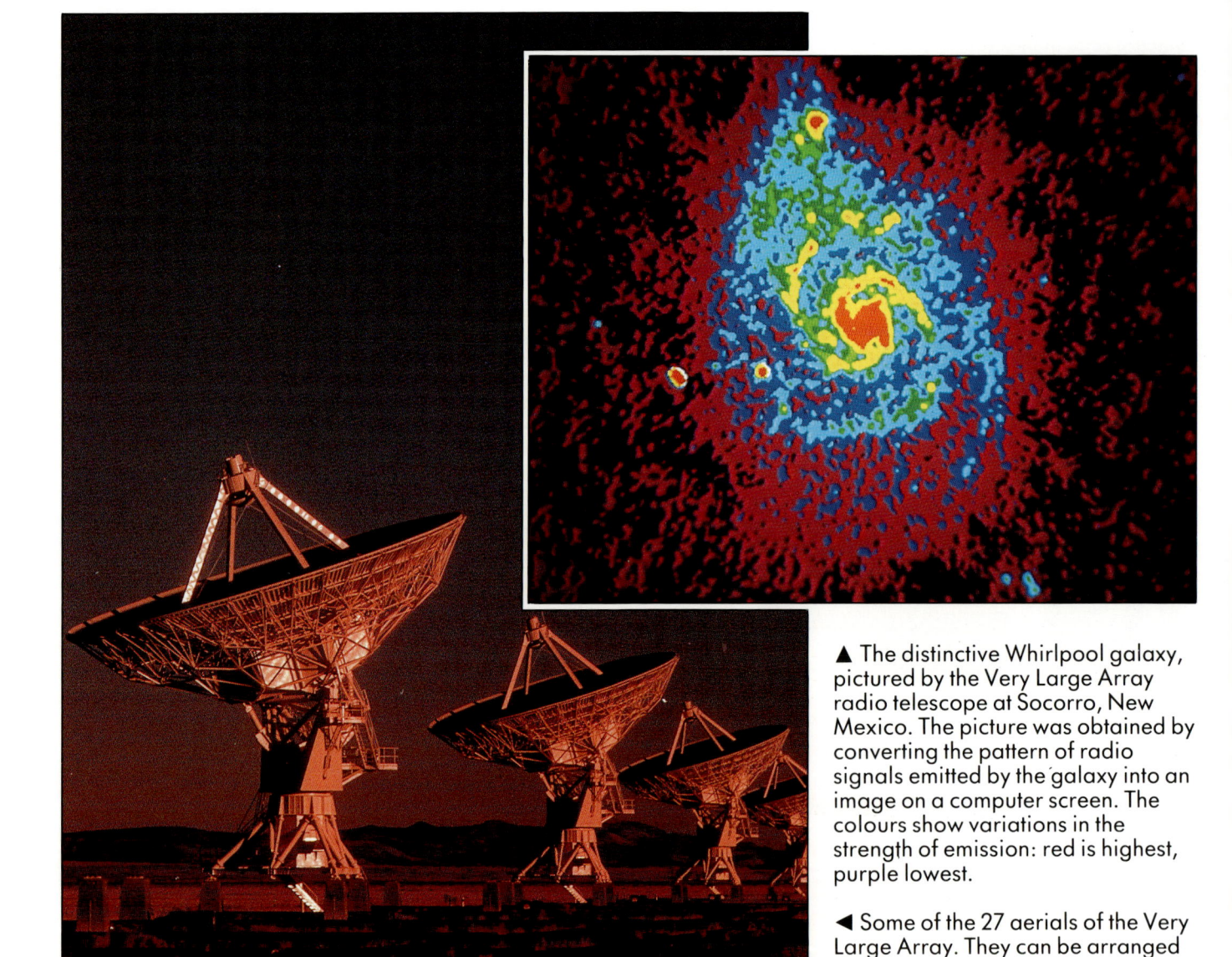

Radio telescopes

▲ The distinctive Whirlpool galaxy, pictured by the Very Large Array radio telescope at Socorro, New Mexico. The picture was obtained by converting the pattern of radio signals emitted by the galaxy into an image on a computer screen. The colours show variations in the strength of emission: red is highest, purple lowest.

◄ Some of the 27 aerials of the Very Large Array. They can be arranged in different positions along a Y-shaped track. Each aerial has a dish 26 m in diameter.

Visible light is not the only means by which stars give off energy. Stars give off many other kinds of radiation as well, such as gamma rays, X-rays, ultraviolet rays, infrared rays and radio waves. Like visible light, these are all electromagnetic waves. They differ only in their wavelength, some having a shorter wavelength, and others a longer wavelength than visible light. Most of these other radiations, however, are blocked by the atmosphere. Only radio waves can get through.

Radio waves were first detected coming from the heavens in 1931. Since that time radio astronomy has grown into one of the most exciting branches of astronomy. It has led to the discovery of hyperactive galaxies that pour out millions of times more energy than normal; pulsating stars made out of solid neutrons; and mysterious quasars, bodies that are hundreds of times brighter than galaxies but millions of times smaller.

Radio telescopes are quite unlike optical telescopes. Some consist of long arrays of wire aerials, and are fixed to the ground. Others are giant dishes, which can be steered to point to different parts of the sky. The dishes act as

▲ American radio engineer Karl Jansky, and the aerial with which he discovered radio waves coming from the heavens. At the time, 1931, Jansky was investigating the source of interference in radio transmissions for Bell Telephone Laboratories. In pinpointing the source as outer space, he launched the science of radio astronomy. In 1937 another American, Grote Reber, built a 10-m dish reflector to gather radio waves, pioneering the most common type of radio telescope today.

reflectors to gather the very faint waves and focus them on a central aerial, or antenna. Electronic circuits then amplify (strengthen) the incoming signals, which are converted into images by computer.

Two of the largest steerable radio telescopes are at Jodrell Bank (76-m diameter) in Cheshire, England, and at Effelsberg (100-m diameter), near Bonn, Germany. The biggest dish, however, is fixed. It is made of aluminium mesh and suspended in a natural bowl in hills near Arecibo, in Puerto Rico. It measures 305 m across. It is used not only in passive mode as a receiving telescope, but also in active mode for radar observations of the nearer planets. It beams signals to a planet and records the "echoes", or reflected signals. In this way it can build up a picture of the planet's surface, which in the case of Venus is permanently hidden by layers of cloud.

Some radio telescopes use several dishes working in concert, a technique called aperture synthesis. Moving the dishes into different positions gives the same results as one very large dish would. Telescopes in different parts of the Earth can be linked in this way.

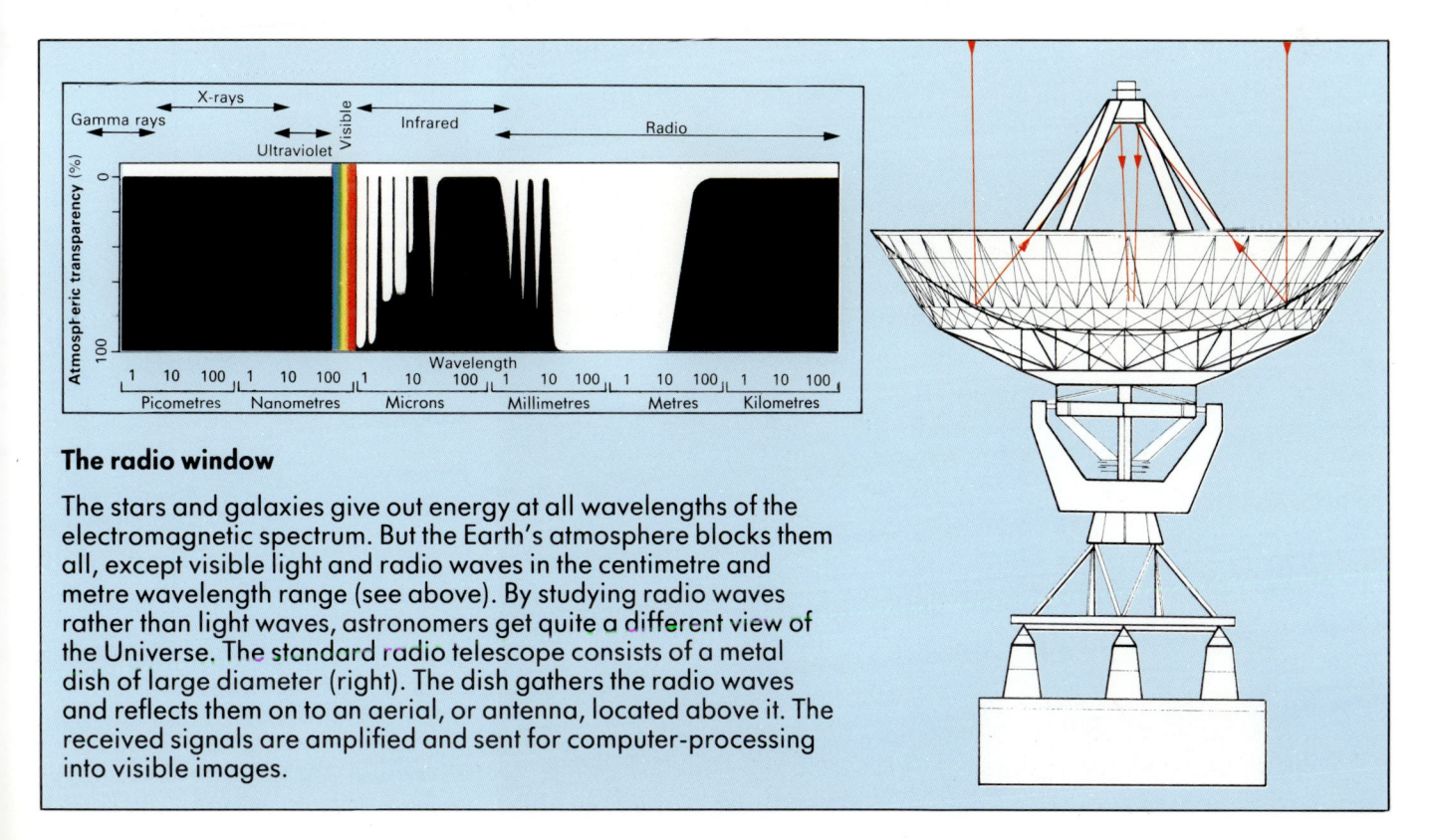

The radio window

The stars and galaxies give out energy at all wavelengths of the electromagnetic spectrum. But the Earth's atmosphere blocks them all, except visible light and radio waves in the centimetre and metre wavelength range (see above). By studying radio waves rather than light waves, astronomers get quite a different view of the Universe. The standard radio telescope consists of a metal dish of large diameter (right). The dish gathers the radio waves and reflects them on to an aerial, or antenna, located above it. The received signals are amplified and sent for computer-processing into visible images.

107

Space telescopes

When the Space Age got into its stride in the 1960s, space scientists began sending instruments into orbit on satellites. Astronomers were not slow to take advantage of this new technology. On Earth their vision of the Universe is distorted by the dust, moisture and air currents in the atmosphere. Sending their telescopes, detectors and other instruments into space on satellites enabled them to view the Universe more clearly. They were also able to view it at wavelengths the atmosphere absorbs: gamma-ray, X-ray, ultraviolet and infrared. Viewed at these other wavelengths, the Universe looked quite different from what it did in visible light. The study of X-rays from the heavens began in the 1960s with instruments carried into the high atmosphere by rockets. X-ray detectors were included on Orbiting Astronomical Observatories launched in 1968 and 1972. *OAO 3*, named *Copernicus*, was particularly successful. Later satellites, such as *Exosat* (1983), mapped the X-ray sky. They

▼ The *Einstein* observatory was an astronomy satellite designed to record X-rays from the heavens. It had a "grazing incidence" telescope, which focused the X-rays by reflecting them through shallow angles. At bottom is an image obtained from *Einstein* of powerful X-ray sources in the Eta Carinae nebula.

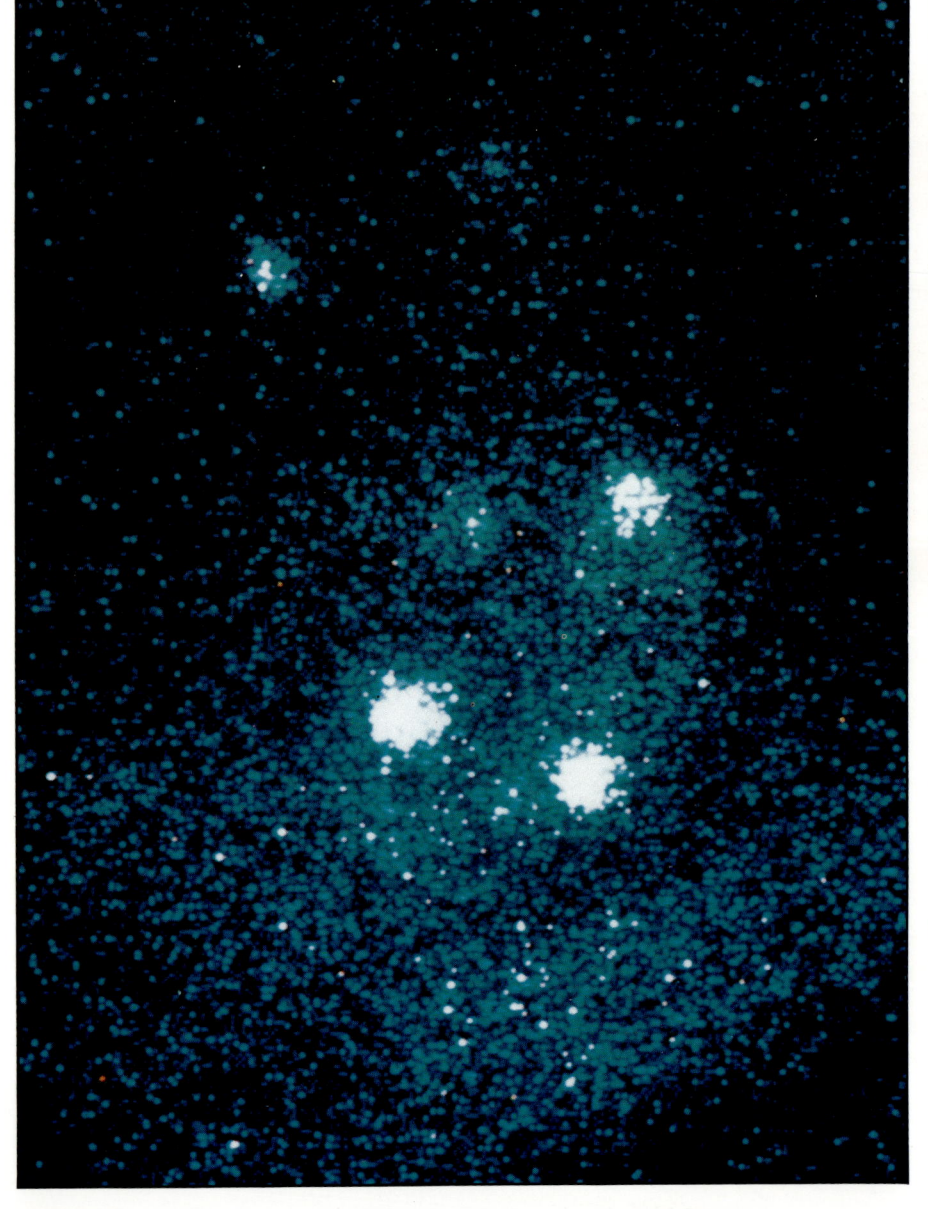

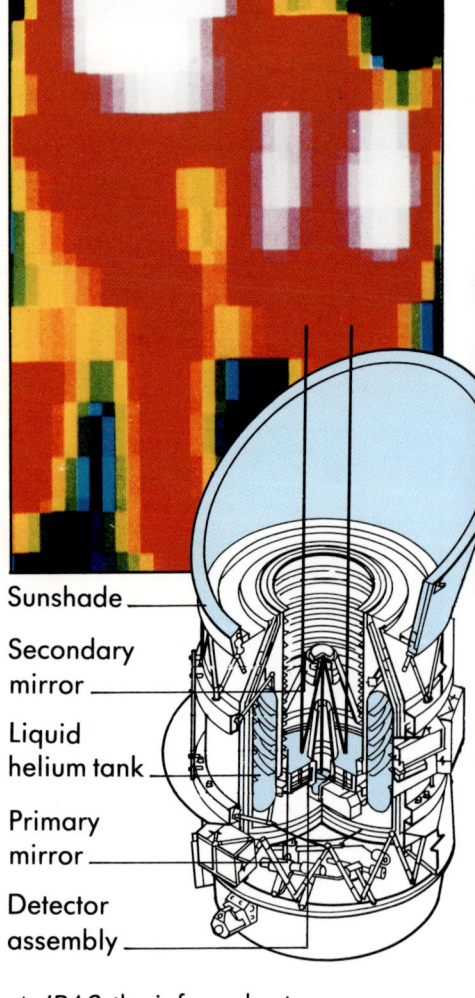

▲ *IRAS*, the infrared astronomy satellite. To make it sensitive to incoming infrared rays, the detector was cooled by liquid helium to −269°C. At the top is an *IRAS* image of the Tarantula nebula.

found evidence of the presence of those awesome bodies we know as black holes, which gobble up everything nearby, even light.

From the early 1960s numerous satellites carried instruments to detect gamma rays. The first startling gamma-ray discoveries were made in the early 1970s by US atomic scientists using spy satellites to look for gamma rays from nuclear-bomb tests on Earth. Instead, they detected bursts of gamma rays from the heavens. These lasted only a fraction of a second, but packed the power of 100,000 Suns. They became known as bursters.

Exploration of the wavelengths on either side of the visible spectrum, the ultraviolet and infrared, has been revolutionized by two satellites. One is the *International Ultraviolet Explorer* (*IUE*), still working in 1990 after 12 years in orbit. The other was the *Infrared Astronomy Satellite* (*IRAS*), which worked for 10 months in 1983.

In 1990 the Hubble Space Telescope (HST) was launched into orbit to pioneer optical astronomy from space. Amazingly, once in orbit, its light-gathering mirrors were found to be defective, and the hoped-for quantum leap in optical performance over Earth-based telescopes has not happened.

▼ The Hubble Space Telescope, launched from the Space Shuttle in April 1990, gathers light with a 2.4-m diameter mirror. Despite a design fault, it has still managed to acquire good images, including the first clear picture showing Pluto's moon Charon.

▼ An HST image showing the ring around the supernova star 1987A.

Analysing and probing

Spot facts

- By using a combination of gas chromatography (for separation) and mass spectroscopy (for analysis), chemists can identify substances in a mixture present in concentrations as low as one part per million.

- The latest echo-sounders, using sound waves vibrating up to 10 million times a second, can pick out individual fish in a shoal.

- In experiments with atom-smashers, nuclear physicists have discovered nearly 250 subatomic particles, each with an equivalent antiparticle.

- In the latest atom-smashers, colliding beams of particles create tiny fireballs that are hundreds of millions of times hotter than the Sun.

▶ A shower of subatomic particles, produced in a bubble chamber at CERN, Europe's nuclear research centre in Geneva, Switzerland. Some were created by smashing together high-speed particles in a particle accelerator. Others are cosmic rays, which reach the Earth from outer space.

Scientists and engineers use both traditional and modern methods to investigate the make-up of materials, the way these materials behave, and the processes they undergo. Chemists, for example, still mix chemicals in test tubes, but also analyse substances with advanced instruments like mass spectrographs. They also use radiation methods in analysis, while engineers use them to examine materials for hidden flaws. Nuclear physicists use the biggest and most expensive scientific equipment of all – particle accelerators, or atom-smashers – to help them probe into the very heart of the atom.

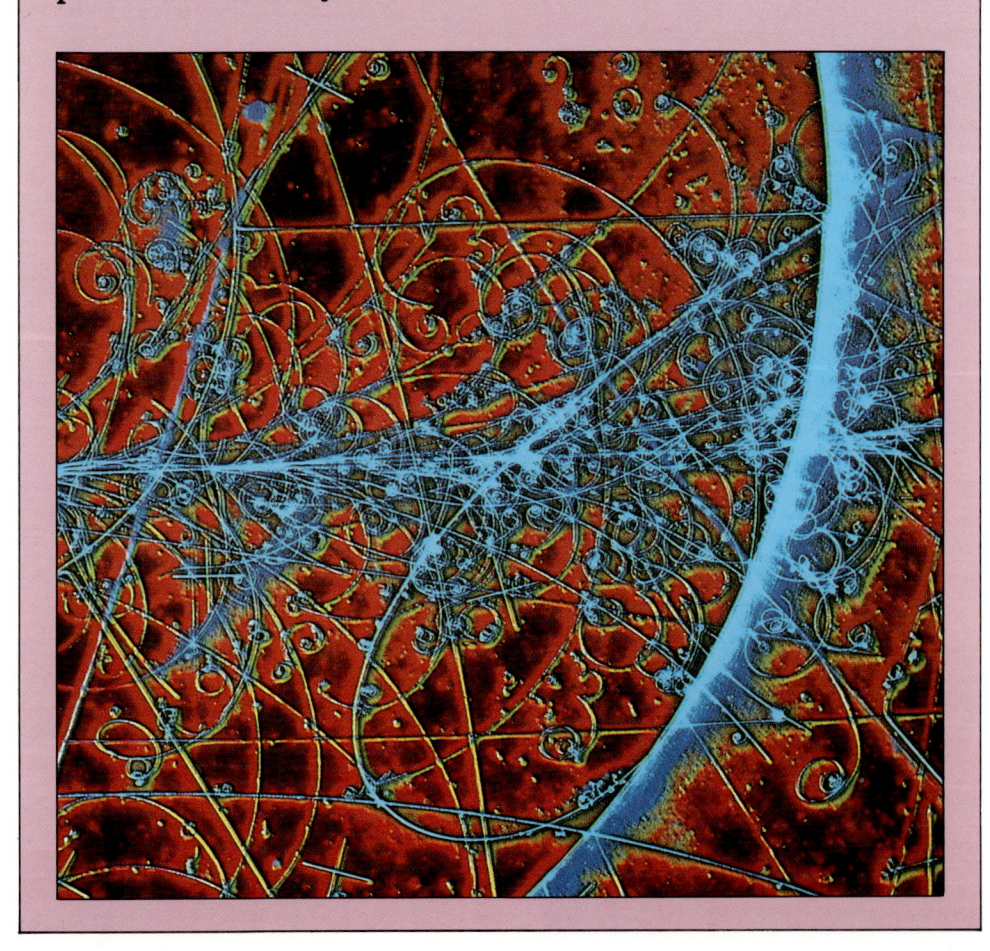

Research

Our knowledge of science and our ability to apply that science to create useful machines and devices, are the product of centuries of inspiration and perspiration by legions of dedicated scientists and inventors.

The first systematic scientific exploration, or research, into the nature of matter was carried out by the ancient alchemists. They flourished in the Middle Ages, trying to find a method of turning base metals into gold. Although they failed in their attempts, they made discoveries that laid the foundations of chemical science.

Modern science has its origins in the 1600s and 1700s, when the likes of Isaac Newton in England and Galileo in Italy established fundamental physical laws, and Robert Boyle in England and Antoine Lavoisier in France expanded chemical knowledge. All these early scientists conducted what is often called pure research – investigation for its own sake, without a particular goal in mind.

But as the chemical industry began to expand in the late 1700s, scientists began to direct their research along more practical lines. Such applied research led, for example, to the French chemist Nicholas Leblanc developing a process for converting salt into soda ash (sodium carbonate). This was much in demand for the manufacture of soap and glass. Sometimes research in one direction led by chance to another. For example, while trying to synthesize the drug quinine, the English chemist William H. Perkin in 1856 discovered the brilliant coal-tar dyes, and launched a whole new industry.

These days scientific research is usually a team effort and linked with specific industries. The prolific American inventor Thomas A. Edison set up one of the first dedicated research laboratories in New Jersey in the 1870s.

◄▲ The changing face of the chemical research laboratory, from the 1840s to the present day. The test tubes and gas burners of yesteryear have long since disappeared. In today's laboratories chemicals are often analysed automatically by electronic instruments, and the results are calculated and displayed by a computer.

Chemical analysis

The analysis of chemical substances is bread-and-butter work to the laboratory chemist. It is routinely carried out, for example, to test for the purity of food, drugs and water supplies, and to check the progress of chemical processes in industry. Chemists carry out both qualitative analysis, to identify which substances are present; and quantitative analysis, to measure the amount of those substances.

The substances chemists have to deal with are usually mixtures, and some form of separation is required before components can be identified. Physical methods of separation may be possible, which involve such processes as filtration, distillation and centrifuging. These make use of physical differences between the components, such as boiling point. Chemical separation may be possible: a chemical reagent may be added so as to cause a substance to come out of solution as a precipitate.

A much more sensitive and quicker method of separation is chromatography. This relies on the varying degrees of attraction different kinds of molecules have for an inert substance. In paper chromatography, for example, a strip of paper forms the inert substance (the stationary phase). A drop of solution of the mixture to be analysed is placed on the paper, which is then dipped in a suitable solvent. This is called the mobile phase, because it moves along the paper. It carries with it the components in the mixture. But these are attracted to differing extents by the paper, so they travel at different speeds and gradually separate out.

In gas chromatograpy, an inert gas such as

▼ An industrial chemist uses an electronic instrument called a coulometer to test the moisture content of plastic roofing material. Her colleague is examining another sample under a microscope. Electronic instruments are now standard laboratory equipment.

Emission spectroscopy

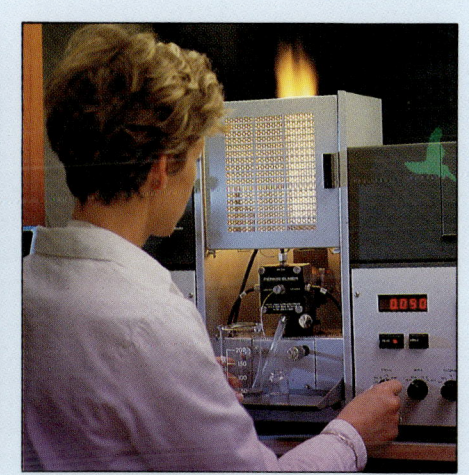

Absorption spectroscopy

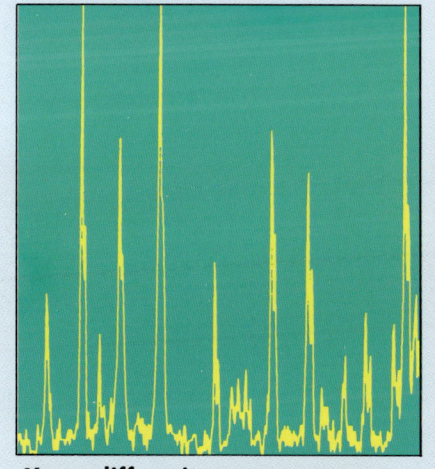

X-ray diffraction spectroscopy

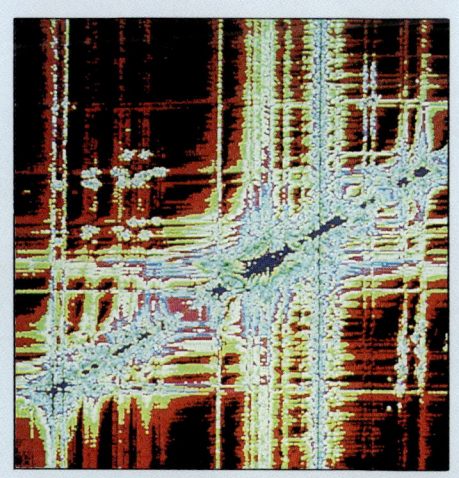

NMR spectroscopy

Spectroscopic analysis

When white light is passed through a prism, it is split up into a spectrum: a spread of different colours, or wavelengths. When chemical elements are excited, or given excess energy, their atoms afterwards lose energy by giving out light. If this light is split into a spectrum, bright lines are seen at different wavelengths. This is called a bright-line, or emission spectrum. Each element has a characteristic emission spectrum, so this method can be used in analysis.

Elements also absorb certain wavelengths. When light is passed through a sample and split into a spectrum, dark lines show up at different wavelengths. This absorption, or dark-line spectrum, is again characteristic of the elements present.

Other spectroscopic methods are used to study the make-up of crystals and molecules. X-ray diffraction spectroscopy uses the way X-rays are scattered by atoms. Nuclear magnetic resonance (NMR) spectroscopy uses radio waves in the presence of a strong magnetic field.

nitrogen is the mobile phase. It carries a vaporized sample of the mixture through a column packed with material coated with a suitable liquid (stationary phase). The components separated by this method may be identified by the time they take to pass through the column, or by mass spectroscopy. This method works by ionizing the components and then identifying them by their mass and electric charge. It is one of several spectroscopic methods now widely used. They rely on the fact that atoms and molecules display a characteristic spectrum when they absorb or give off energy when in a high-energy state.

▶ Children in a school laboratory observe the changes that have occurred in a solution after a chemical experiment.

113

Radiation methods

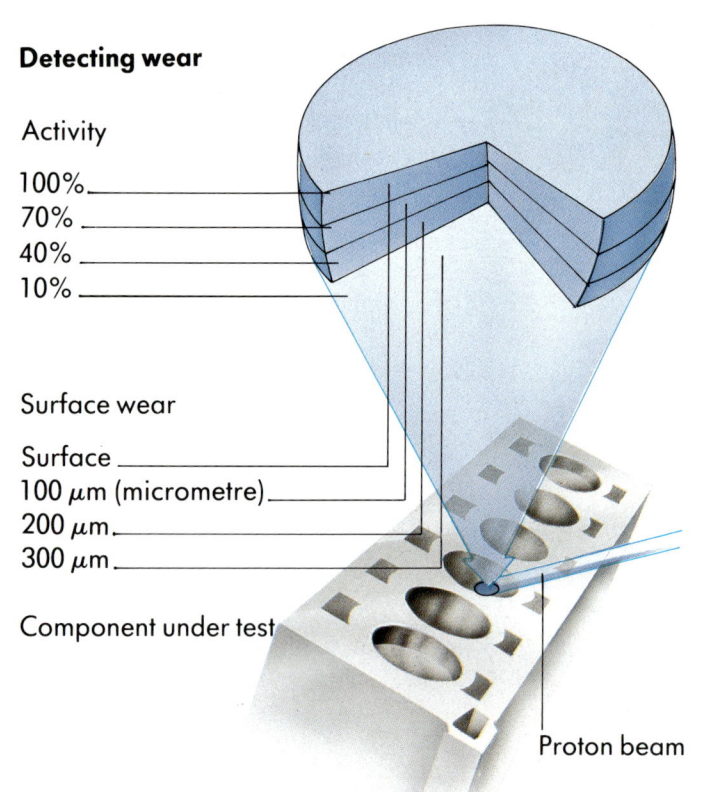

Detecting wear

Activity
100%
70%
40%
10%

Surface wear
Surface
100 μm (micrometre)
200 μm
300 μm

Component under test

Proton beam

▲▼ The amount of wear in, say, an engine cylinder can be checked by what is called thin-layer activation. The surface of the metal in the bore is covered with a radioactive layer. Later the radioactivity of the layer is measured with a radiation detector. A reduction in the level of activity will indicate that wear of the metal in the bore has taken place.

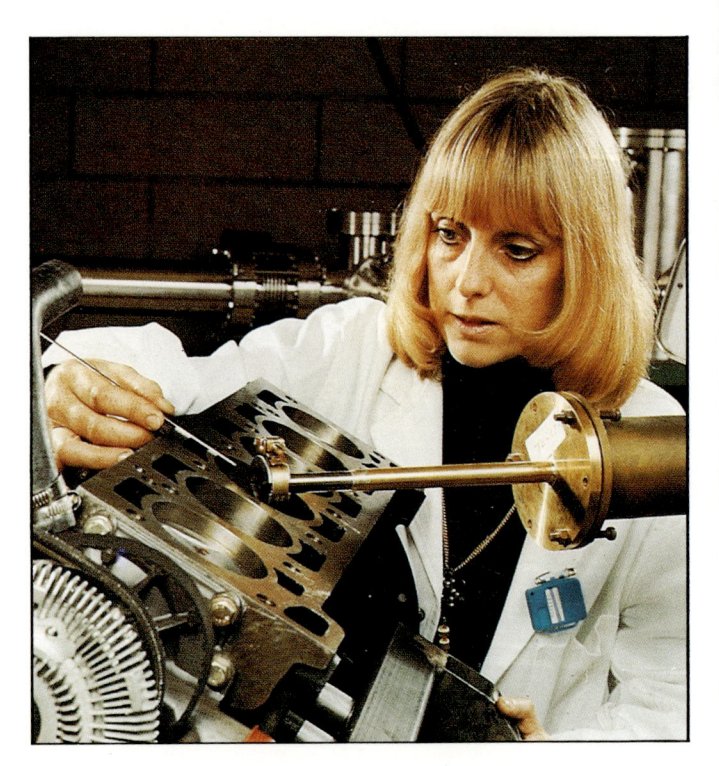

The visual inspection of materials and structures plays a vital role in many branches of industry. And over the years engineers have come up with ingenious devices, such as endoscopes, to peer into places the eye cannot see. The doctor, too, uses such devices to peer inside the body and even carry out operations there.

However, these devices give only limited information. They cannot see internal flaws in materials, or abnormalities in body tissues. The engineer and doctor therefore resort to methods employing radiation that can penetrate materials and flesh. And by means of photography or imagery generated by computers, they can produce pictures showing internal details. Study using radiation is called radiography.

The first kind of radiation used was X-rays, a short-wave electromagnetic radiation discovered by the German physicist Wilhelm Röntgen in 1895. X-rays are produced when a material (usually tungsten) is bombarded by high-energy particles (usually electrons). To make a radiograph, X-rays are passed through the subject under test on to a photographic plate. A shadow image is produced when different parts of the internal structures absorb different amounts of radiation.

In medicine the simple X-ray technique can show bones, but not delicate tissues or details of internal organs. In the 1970s a much more sensitive method came into use which can. It is called X-ray tomography, or computer-assisted tomography (CAT). It images a "slice" of the body with an X-ray source in a rotating frame. As the source rotates, sensitive detectors beneath the body measure the amount of radiation passing through. A computer analyses the readings and converts them into an image on a video screen.

Engineers now also use gamma rays for materials testing; they are even more penetrating than X-rays. Artificial radioactive isotopes, or radioisotopes, are the source of these rays. Since they are readily portable, they can be used anywhere, unlike X-ray equipment.

Radioisotopes are also useful in medical research and even for treatment. Research scientists use them as tracers, for example, to follow the blood flow through the brain. Their progress can be traced by means of radiation detectors, such as the Geiger counter.

Radioisotopes are also used to reveal how particular substances are taken up by the body. For example, iodine naturally accumulates in the thyroid gland. To check how the thyroid is working, doctors inject the patient with radioactive iodine. They can then monitor, with radiation detectors, how the gland handles the iodine, and this will tell them whether the thyroid is working normally or not.

X-ray tube

▲ X-rays are produced in a vacuum tube like this. A stream of electrons from the heated cathode is accelerated by a high voltage and strikes the tungsten anode. The atoms in the metal become highly energized, and give out their excess energy in the form of X-rays. The process generates heat, so the anode has to be cooled. This may be done by means of a liquid coolant, as here, or by rotating the anode. X-rays have wavelengths between about 10 picometres (million-millionths of a metre) and about 1 nanometre (thousand-millionths of a metre).

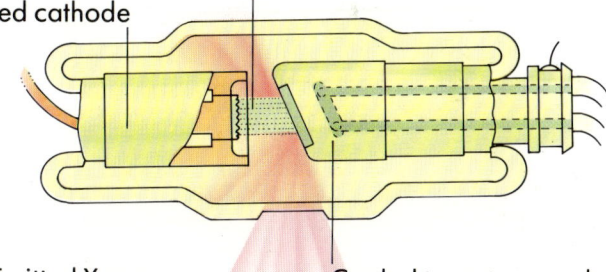

◀ Inspecting welds on a pipeline at a nuclear power station with X-ray equipment. The X-rays penetrate the metal and clearly show up any defects there might be in the welds. The inset radiograph shows a flaw.

Sound methods

Scientists, engineers and doctors not only use electromagnetic waves to examine internal structures, they also use sound waves. We normally think of sound as travelling only through air. But in fact it can travel through all kinds of materials, such as water, glass, rock and steel. Indeed it travels much faster in these materials than in air. Whereas in air the speed of sound is only about 1,200 km/h, in water it is more than four times faster, and in steel it is nearly four times faster still.

Instruments employing sound for detection are useful for probing the atmosphere, the oceans, the Earth and the human body. All these instruments work by sending out sound waves and "listening" for reflections, or echoes. They are often called echo-sounders.

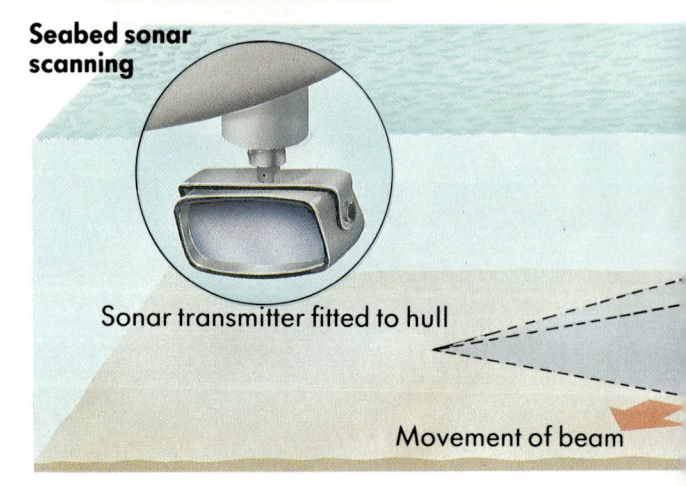

Seabed sonar scanning

Sonar transmitter fitted to hull

Movement of beam

Meteorologists may use echo-sounding in the atmosphere to measure levels of humidity, or to detect where there are temperature inversions, atmospheric conditions likely to give rise to pollution. Geologists use echo-sounding techniques to study underground rock layers. This procedure is called seismic surveying.

At sea, echo-sounders are used by ships' navigators to measure water depth and by fisherman to locate shoals of fish. They are also used as underwater radar to scan the depths for wrecks and rocks and, on naval ships, for submarines. This method of underwater echo-location is generally known as sonar (sound navigation and ranging).

The sound waves used for sonar are of very high frequency. They are too high-pitched to be heard by the human ear, and are termed ultrasonic. The ear can only hear sounds up to a frequency of about 20,000 hertz (cycles per second). But sonar uses frequencies of up to 10 megahertz (million hertz).

Ultrasonic frequencies are also used for probing inside materials. The reflections, or echoes, of internal structures are computer-processed to provide an image. This is a valuable method of non-destructive testing in engineering, useful for detecting flaws in welds, for example. In medicine a similar technique is used to image the foetus in a mother's womb.

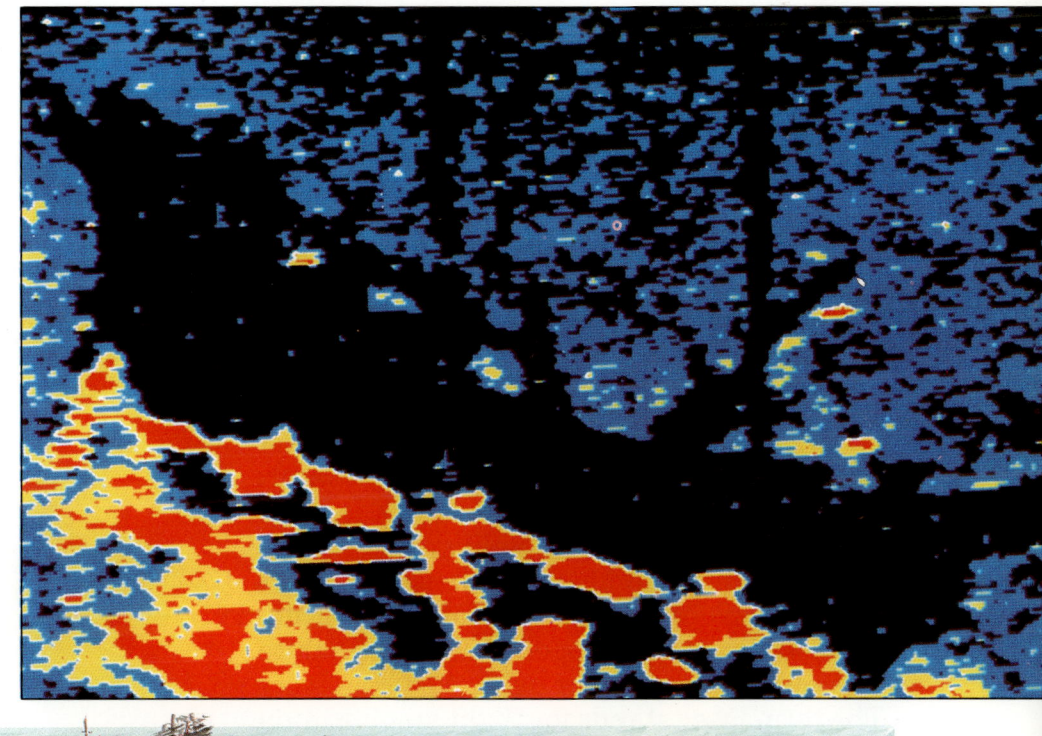

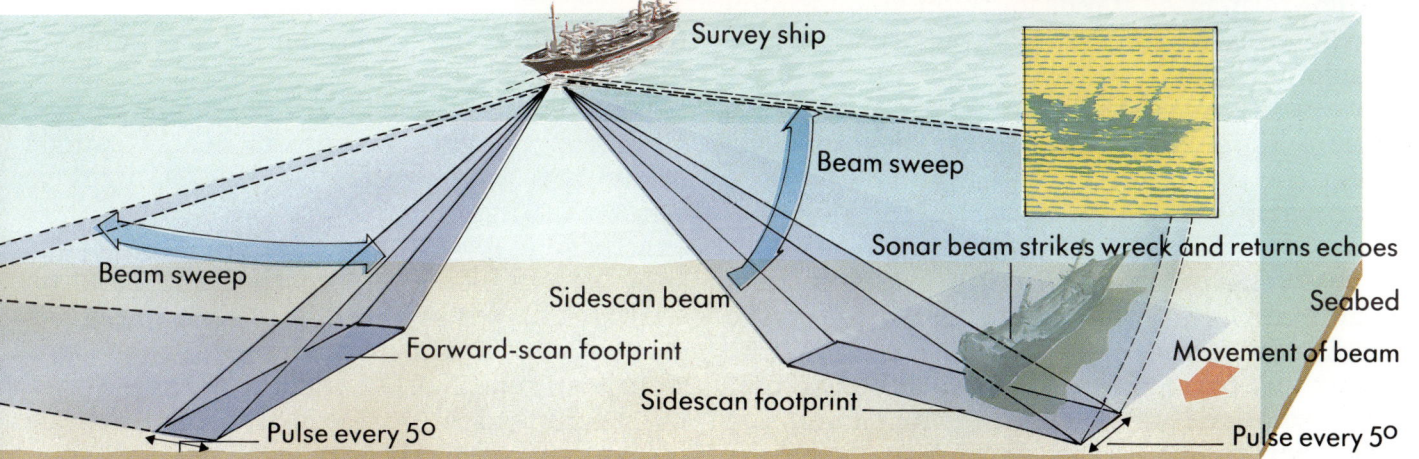

◀ Engineers checking welds in high-pressure water pipes during the construction of the Dinorwic hydroelectric power station in Wales. They are using an ultrasonic probe, linked with a computer. Flaws in the welds show up clearly on the computer screen. In the inset computer image, the flaw is the large red spot below centre.

▼▶ Ultrasound is also used in sonar methods of underwater detection. The survey ship shown in the diagram below is scanning forwards and sideways over the seabed as it travels. Echoes received back are processed by computer to form an image of anything on the seabed, such as a wreck. The image can be displayed in false colour, as here (right). It even shows the acoustic "shadow" made by the vessel.

117

Laser methods

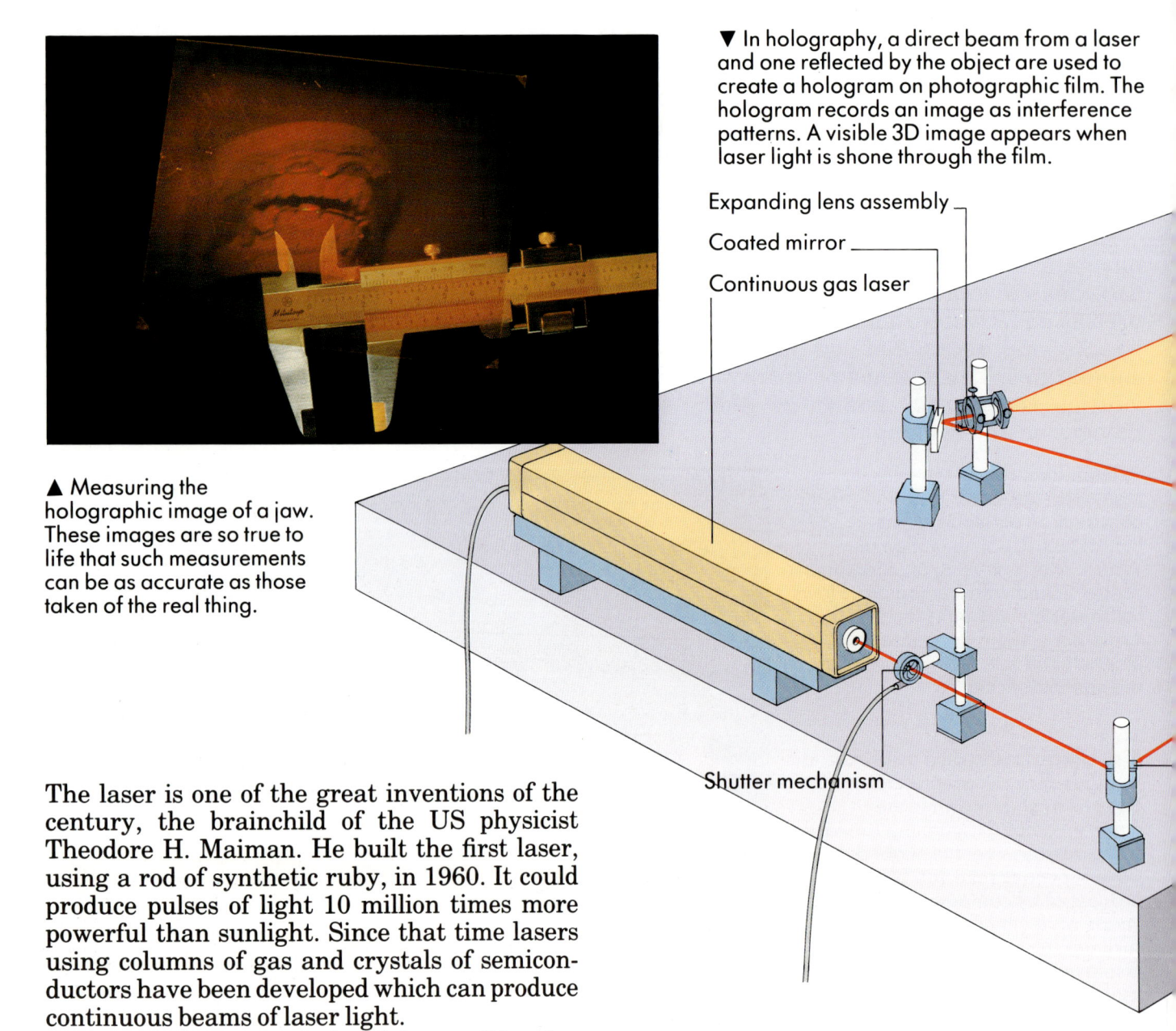

▼ In holography, a direct beam from a laser and one reflected by the object are used to create a hologram on photographic film. The hologram records an image as interference patterns. A visible 3D image appears when laser light is shone through the film.

▲ Measuring the holographic image of a jaw. These images are so true to life that such measurements can be as accurate as those taken of the real thing.

The laser is one of the great inventions of the century, the brainchild of the US physicist Theodore H. Maiman. He built the first laser, using a rod of synthetic ruby, in 1960. It could produce pulses of light 10 million times more powerful than sunlight. Since that time lasers using columns of gas and crystals of semiconductors have been developed which can produce continuous beams of laser light.

The term laser stands for "light amplification by stimulated emission of radiation", which explains how it works. Atoms in the laser medium (gas or crystal) are "excited", or forced into a high-energy state. Some spontaneously emit the extra energy as light radiation of a particular wavelength, or colour. This radiation triggers off, or stimulates, other atoms to emit light of the same wavelength, which in turn stimulates still other atoms. In this way the radiation increases rapidly in strength. The laser has parallel mirrors at each end so that the radiation is reflected to and fro, stimulating emission from more atoms each time. One of the mirrors allows the radiation to emerge as an intense parallel beam.

Laser light is special for three main reasons. One, it is very pure, being very nearly a single colour, or wavelength. Two, laser light is coherent, which means that all its waves are exactly in step with one another. This means that they reinforce one another and build up energy. The waves of ordinary light are out of step with one another, which makes them, as it were, jostle one another and lose energy. Three, laser light is highly directional, being produced in an almost perfectly parallel beam.

Making a hologram

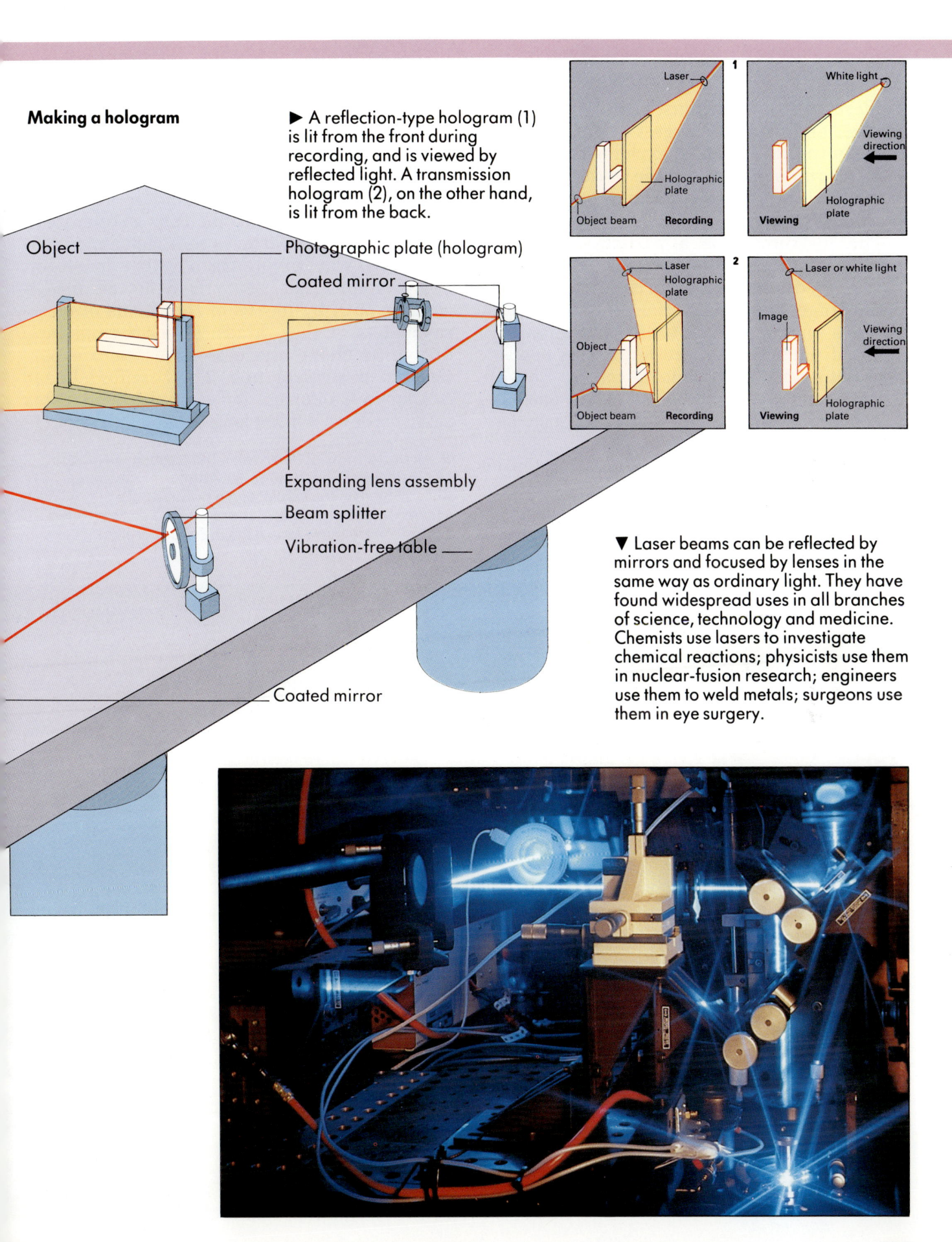

▶ A reflection-type hologram (1) is lit from the front during recording, and is viewed by reflected light. A transmission hologram (2), on the other hand, is lit from the back.

Object
Photographic plate (hologram)
Coated mirror
Expanding lens assembly
Beam splitter
Vibration-free table
Coated mirror

1
Laser
Holographic plate
Object beam **Recording**

White light
Viewing direction
Holographic plate
Viewing

2
Laser
Holographic plate
Object
Object beam **Recording**

Laser or white light
Image
Viewing direction
Holographic plate
Viewing

▼ Laser beams can be reflected by mirrors and focused by lenses in the same way as ordinary light. They have found widespread uses in all branches of science, technology and medicine. Chemists use lasers to investigate chemical reactions; physicists use them in nuclear-fusion research; engineers use them to weld metals; surgeons use them in eye surgery.

119

Atom-smashing

Underneath the fields just outside Geneva, in Switzerland, lies the largest scientific instrument in the world, which explores the smallest piece of ordinary matter in the world – the atom. It was built by CERN, the European Centre for Nuclear Research.

The instrument is a particle accelerator, popularly called an atom-smasher. It occupies a ring-shaped tunnel 27 km in circumference and on average more than 100 m underground. Inside a pipe within the tunnel two streams of charged particles are accelerated in opposite directions and then made to collide. The collision produces a shower of other particles, which are recorded by a detector, such as a bubble chamber.

This atom-smasher produces collisions between beams of negatively-charged electrons and their antiparticles, positively charged positrons. Other powerful atom-smashers of similar design, such as the Tevatron at Fermilab, near Chicago, USA, collide beams of protons and antiprotons.

These great atomic "race tracks" are accelerators known as synchrotons. They use electric fields to repel and thus accelerate a beam of charged particles in a rhythmic (synchronized) way, giving them a little "push" on every circuit. The beam is deflected into a circular path by means of thousands of electromagnets. Other kinds of atom-smashers, known as linear accelerators, accelerate particles in a straight line. The Stanford linear collider in California, for example, accelerates particles down a 3-km long tunnel.

▼ An aerial view of Fermilab's giant synchrotron, the Tevatron, near Chicago, which has been operating since 1983. The tunnels in which particles are accelerated to high energies are located underground. They measure more than 6 km in circumference.

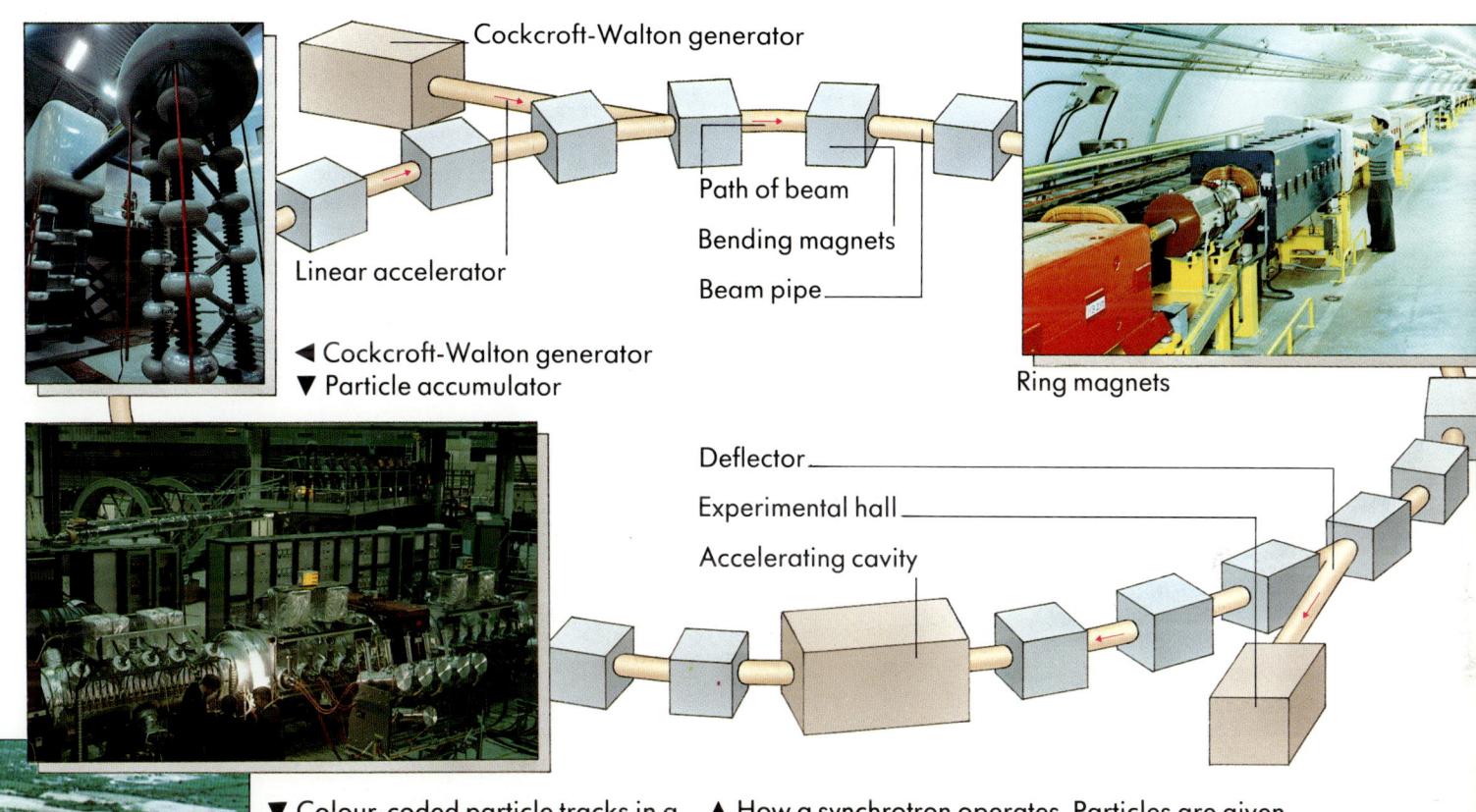

Cockcroft-Walton generator

Linear accelerator

Path of beam
Bending magnets
Beam pipe

◀ Cockcroft-Walton generator
▼ Particle accumulator

Ring magnets

Deflector
Experimental hall
Accelerating cavity

▼ Colour-coded particle tracks in a bubble chamber, showing the result of a collision between a moving proton (yellow) and a stationary one. The main subatomic particles produced are pions, both positive (red) and negative (blue).

▲ How a synchrotron operates. Particles are given their first energy boost in a Cockcroft-Walton generator, and then a linear accelerator injects them into the synchrotron. This is encased by magnets to bend the particle stream, which is given energy each lap. It is finally deflected into an experimental hall, where it collides with other particles.

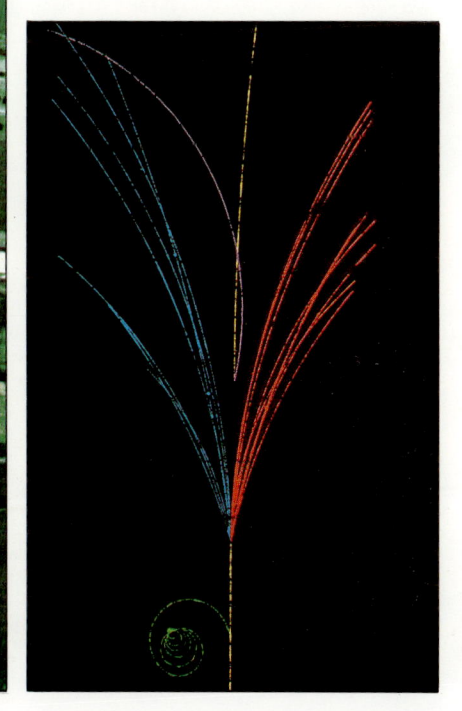

The energy generated in synchrotrons can be colossal. It is expressed in terms of electron-volts (eV), the energy of an electron when it is accelerated by an electric field of one volt. In the Tevatron, for example, colliding particles generate energies of up to 2,000,000,000,000 electron-volts. This is enough to create in collisions the ultimate subatomic particles, quarks, from which all other particles appear to be made up.

The earliest kind of accelerators, developed in the early 1930s, included the Van de Graaf and the Cockcroft-Walton generators. The Van de Graaf machine produced the high voltages needed to accelerate particles by building up electrostatic charges. The Cockcroft-Walton machine built up voltages using alternating current electricity. It is still used in modern accelerator facilities for energizing particles before they are injected into the main machine.

121

The Solar System

Spot facts

- A globe the size of the Sun could swallow more than 1 million Earths.

- The Sun contains 750 times more matter than all of the rest of the Solar System put together.

- Every second the Sun loses more than 4 million tonnes of its mass. The lost mass is converted into vast amounts of energy in its nuclear furnace.

- It can take up to a million years for the energy produced in the Sun's core to reach the surface of the Sun.

- More than 70 chemical elements have been discovered in the Sun. The element helium was found there before it was detected on Earth. Helium is named after the Greek word for the Sun.

▶ The Sun gives out light at all wavelengths. This picture was taken in ultraviolet light by the astronauts on the *Skylab* space station in 1973. It has been given false colour by computers.

Dominating our small corner of the Universe is the star we call the Sun. As it rushes headlong through space, it carries with it a collection of planets and moons, asteroids, meteoroids and comets. Together they make up the Solar System. The heat and light that pour out from the Sun's nuclear furnace make the inner planets Mercury and Venus searing hot and inhospitable. But they breathe life into the planet Earth.

The Sun was born along with the planets nearly 5,000 million years ago from a cloud of interstellar gas and dust. It is now in middle age. But it will be another 5,000 million years before the Sun begins to run out of nuclear fuel, swells up into a red giant and starts to die. When this happens, all life on Earth will perish.

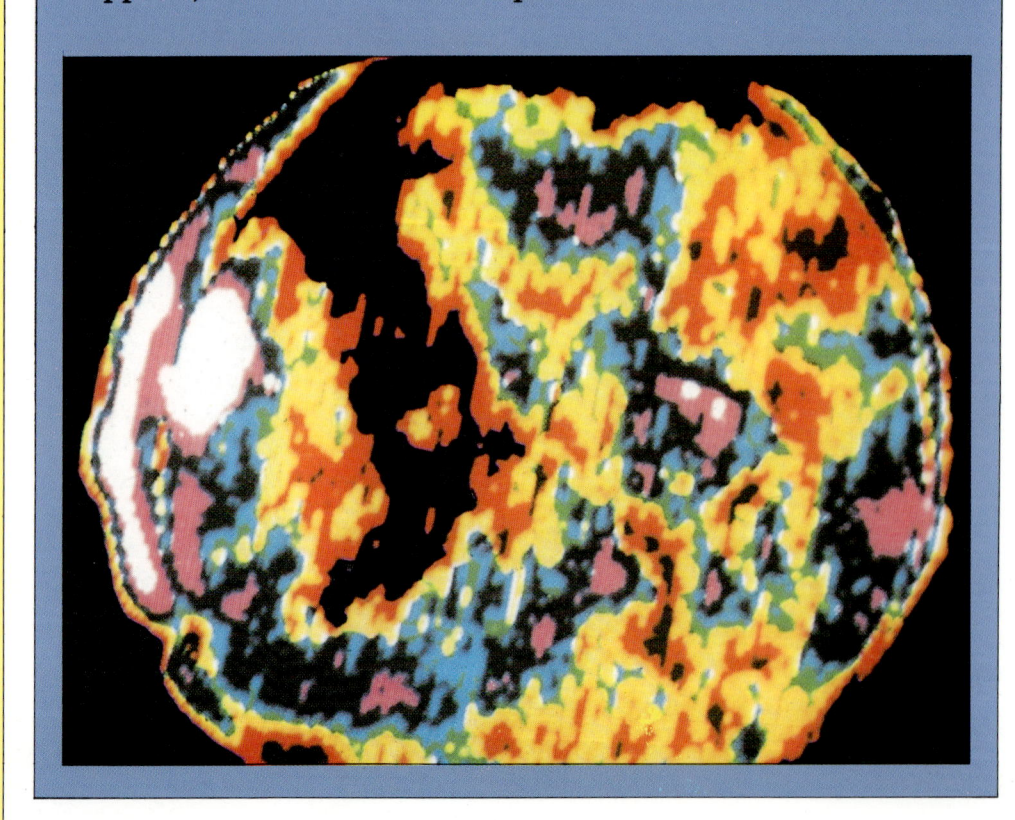

Our star, the Sun

To us on Earth the Sun is the most important star in the Universe. But as a star it is not very special. It appears bigger and brighter than the others only because it is much nearer – about 150 million km away. This compares with over 40 million million km to the next nearest star.

Like most stars, the Sun is a globe of searing hot gas. It is an averaged-sized star, with a diameter of about 1,400,000 km, or more than 100 times that of the Earth. It gives off white light, although it is classified astronomically as a yellow dwarf. As well as light, the Sun gives off other kinds of energy, from short wavelength gamma rays to long radio waves.

This energy is produced in the central core of the Sun at temperatures of up to 15 million degrees. There, nuclear reactions take place between nuclei of hydrogen. They combine, or fuse, together to form helium. During this process a little mass (m) is "lost", or rather converted into energy (E), according to Einstein's famous equation $E = mc^2$, where c is the velocity of light. The loss of only a small amount of mass releases huge amounts of energy.

The solar atmosphere

The energy produced in the Sun's core travels to the surface and then radiates into space. We call the surface the photosphere ("light sphere"). Its temperature is about 5,500°C.

Above the photosphere is a layer of gases called the chromosphere ("colour sphere"), about 10,000 km thick. It is the inner part of the Sun's atmosphere, and is named after its reddish colour. The outer atmosphere, the corona, extends for millions of kilometres until it merges into space. It can be seen well only during a total eclipse of the Sun (see next page).

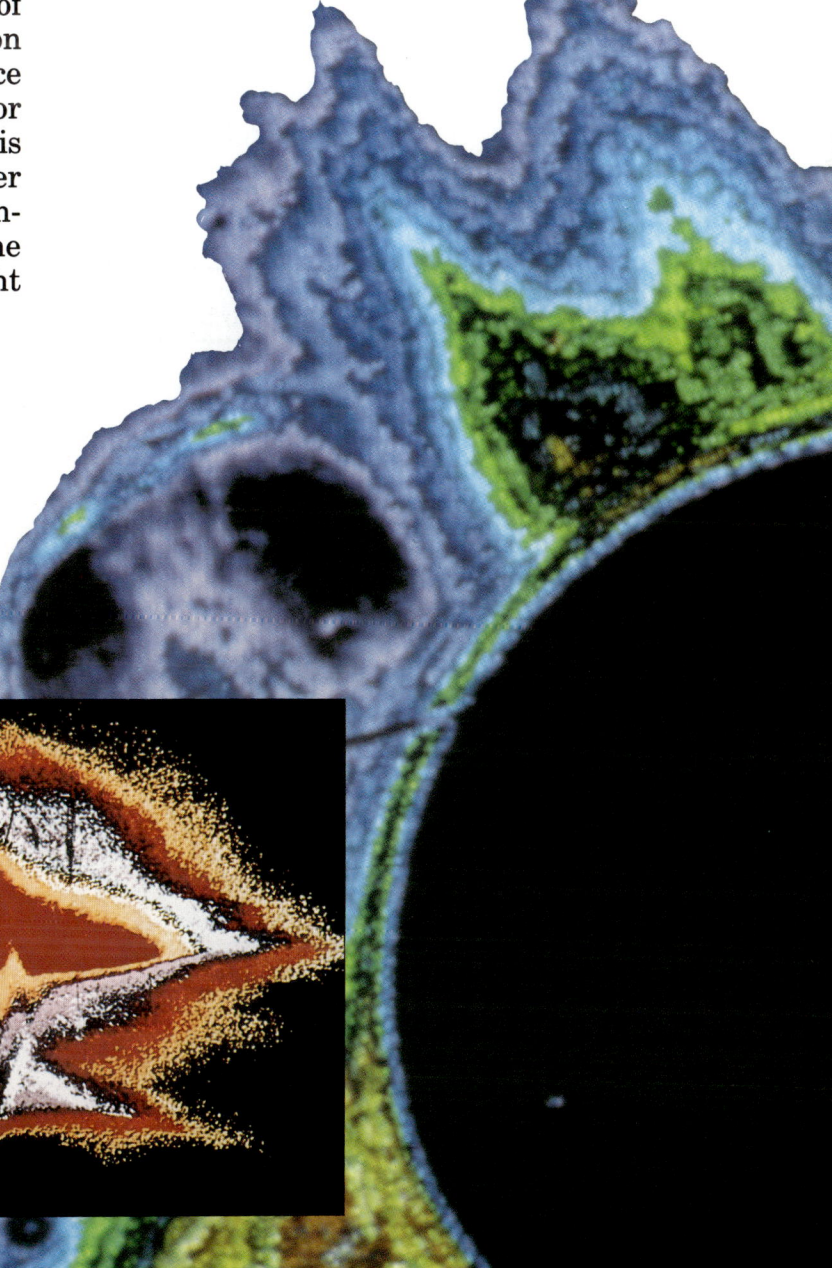

▶ This *Skylab* picture shows tongues of gas leaping thousands of kilometres above the surface of the Sun. The surface has been blotted out using an instrument called a coronagraph to create an artificial eclipse.

▼ Another *Skylab* picture shows in false colours the Sun's corona. It extends millions of kilometres out into space.

Spots and flares

The seething surface

When studied with astronomical instruments, the photosphere appears as a mass of speckles, or granules. They show where cells of hot gas are constantly welling up from below. Quite often more prominent markings appear on the surface, called sunspots. They look rather like ink blots, with a dark middle, or umbra, surrounded by a paler penumbra. Sunspots are about 2,000°C cooler than their surroundings.

Some sunspots, called pores, may be just a few hundred kilometres across and last for a day or so. Others may grow up to a width of 200,000 km or more and last for months. The number of sunspots varies from year to year according to a regular cycle. The length of this sunspot cycle is about 11 years.

Sometimes when sunspots occur, brilliant and violent eruptions called flares take place. They give off powerful streams of charged particles, such as protons and electrons. When these particles reach Earth, they give rise to brilliant displays of aurora and disrupt radio communications. Some charged particles stream out from the Sun, forming the solar wind. On reaching Earth, they become trapped in its magnetic field to form "belts" of intense radiation, called the Van Allen belts.

▲ In far northern regions of the world the night sky is often lit by shimmering curtains of coloured light. They are displays of the Northern Lights, or Aurora Borealis. Similar displays occur in far southern regions, where they are called the Southern Lights, or Aurora Australis. Aurorae happen when the solar wind blows strongly. It forces high-energy particles out of the Van Allen radiation belts. They collide with atoms in the upper atmosphere which then give off an eerie light.

Eclipses

An eclipse of the Moon occurs when the Moon moves into the Earth's shadow in space. It moves first into a region of partial darkness, called the penumbra, then into a region of almost complete darkness, called the umbra. The Moon may remain in eclipse for up to about 2½ hours. During a lunar eclipse the Moon often takes on a faint glow. This is caused by sunlight reaching it after having been bent by the Earth's atmosphere.

An eclipse of the Sun occurs when the New Moon moves in front of the Sun and blots out some or all of its light. In a partial eclipse only part of the Sun is blotted out. In an annular eclipse the Moon does not quite mask the Sun, leaving a ring (annulus) of light. In a total eclipse the Moon masks the Sun completely. A shadow, up to 250 km wide, is cast on the Earth and travels rapidly over the surface. Totality (time of total eclipse) only lasts up to about 7½ minutes.

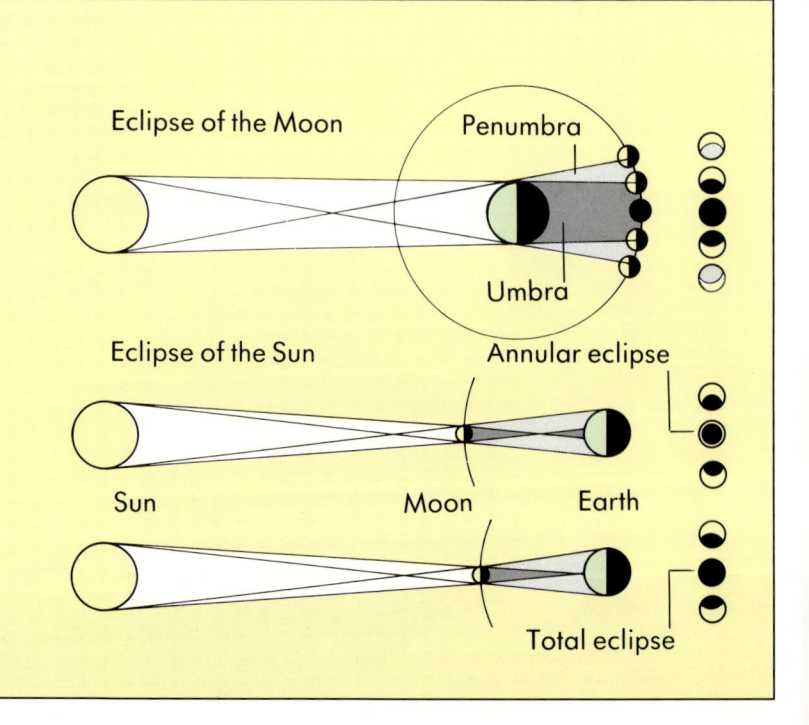

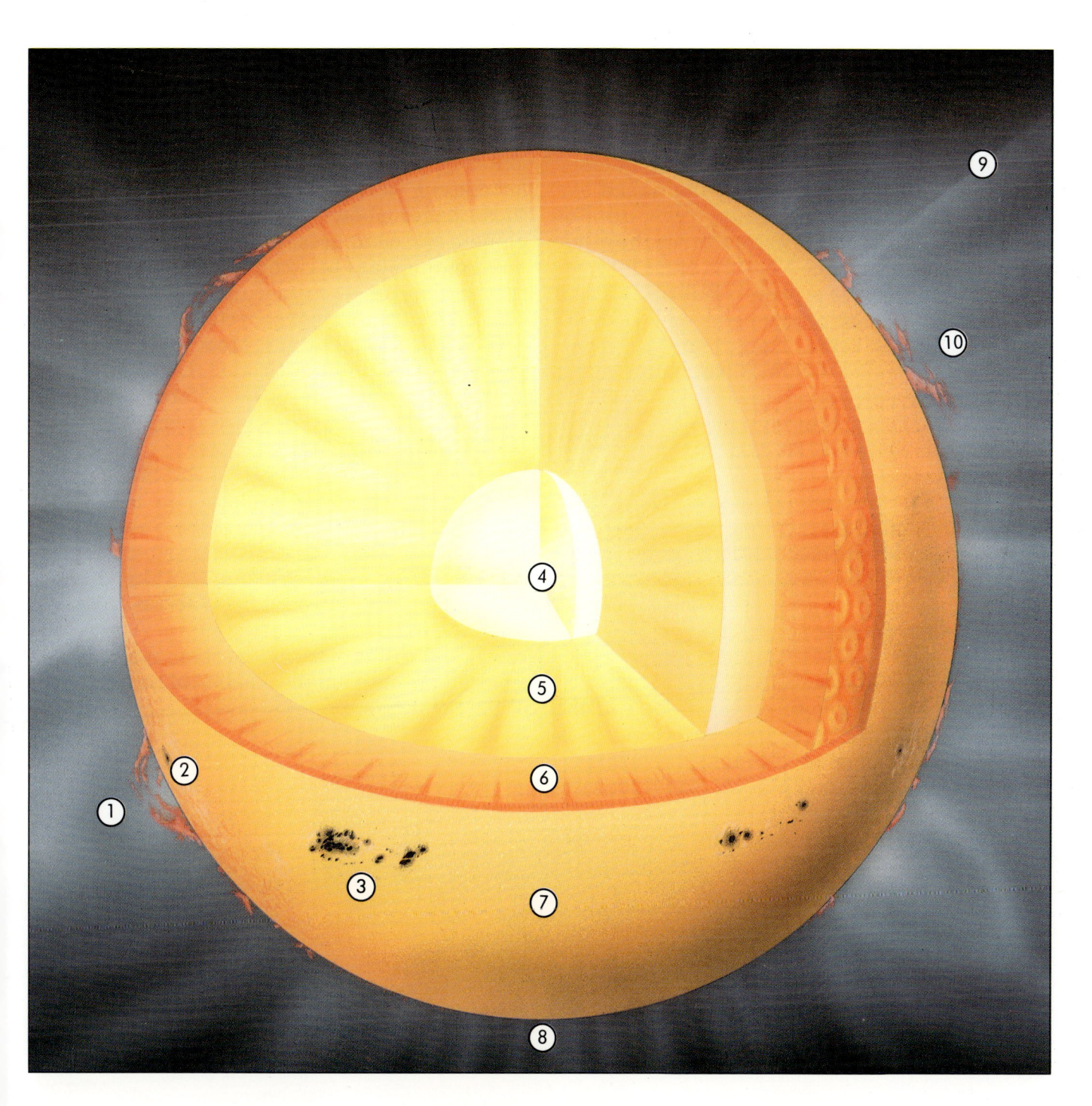

▲ In the central core of the Sun (4) the temperature is about 15 million degrees. At this temperature nuclear reactions take place which produce the energy that keeps the Sun shining. The energy travels towards the surface in two ways. On the first stage of its journey, it travels in the form of radiation, through the so-called radiative zone (5). From the top of this zone the energy also travels by convection: hot gas rises and carries the heat to the surface. This region of the Sun is called the convective zone (6). From the surface, the photosphere (7), the energy radiates into space as light, heat and other radiation. Sometimes dark patches called sunspots appear on the surface (3). Sunspots can grow very large, and may be surrounded by bright patches called faculae (2). Just above the bright surface is a thin layer of atmosphere called the chromosphere (8). Through this layer, fountains of hot gas shoot up as filaments (1) or prominences (10). Farther out, the outer atmosphere, or corona (9), billows into space.

The planets

The Sun is over 300,000 times more massive than the Earth, and the pull of its gravity is very strong. It is this pull that keeps the Earth and eight other large bodies circling in space around the Sun. These bodies are the planets.

From Earth we can see with the naked eye five planets – Mercury, Venus, Mars, Jupiter and Saturn. We need a telescope to see the other three planets – Uranus, Neptune and Pluto.

The Earth itself is a planet. But until about 450 years ago, most people believed it was the centre of the Universe. They thought that all the heavenly bodies – Sun, Moon, planets and stars – revolved around the Earth. But this could not easily explain how the planets move through the heavens. At times, for example, they appear to loop backwards.

In the 1500s a Polish priest named Nicolaus Copernicus realized what was wrong. The Sun must be the centre of everything, not the Earth. He put forward the idea of a Sun-centred, or Solar System in 1543, and in so doing gave birth to modern astronomy.

Worlds near and far
The Earth differs from the other planets in one very important respect. It boasts conditions that allow millions of different lifeforms to flourish. On all the other planets, conditions are deadly to life as we know it.

The planets divide neatly into two groups. Mercury, Venus and Mars are small rocky planets like the Earth. They are close enough to be considered neighbours. Jupiter, Saturn, Uranus and Neptune on the other hand are giant balls of gas. Like icy Pluto, they are far-distant worlds. The diagram at the foot of these pages shows the different sizes of the planets and compares them with the Sun.

All of the planets except Mercury and Venus are the centre of miniature systems of their own. They have moons circling around them. The Earth has only one Moon, but the giant planets have a multitude. Details of the numbers of moons, and other information about planetary sizes and orbits, are given in the table on the opposite page.

The remaining members of the Sun's family are much smaller. They include the asteroids – miniature planets that circle the Sun between the orbits of Mars and Jupiter – meteoroids and comets. Meteoroids that enter the Earth's atmosphere and burn up flash across the night sky as meteors, or shooting stars. A very few reach the ground as meteorites. We can consider all of these smaller bodies to be the debris of the Solar System.

Planetary statistics

Planet	Mercury	Venus	Earth	Mars	Jupiter	Saturn	Uranus	Neptune	Pluto
Diameter at equator (km)	4,878	12,104	12,756	6,794	142,800	120,000	51,800	49,500	2,284
Mean distance (million km)	59.9	108.2	149.6	227.9	778.3	1,427.0	2,869.6	4,496.7	5,900
Mean distance (Earth = 1)	0.387	0.723	1.000	1.524	5.203	9.539	19.182	30.058	39.44
Circles Sun in (d, y)	87.97d	224.7d	365.2d	686.98d	11.86y	29.46y	84.01y	164.79y	247.7y
Turns on axis in (h, d)	58.65d	243d	23.93h	24.62h	9.8h	10.2h	16.3h	16.0h	6.3d
Mass (Earth = 1)	0.056	0.815	1.000	0.107	318	95.1	14.5	17.2	0.002
Volume (Earth = 1)	0.05	0.88	1.00	0.15	1.316	755	52	44	0.005
Density (Water = 1)	5.43	5.24	5.52	3.04	1.32	0.70	1.27	1.77	2
Number of moons	0	0	1	2	16+	22+	15+	8	1

▶ The orbits of the planets in the Solar System, seen from "above". Most are nearly circular, with the Sun at the centre. Those of Mercury and Pluto are displaced off-centre, and Pluto's orbit sometimes crosses inside that of Neptune.

▼ The sizes of the planets drawn to the same scale. From left to right, they are shown in order of increasing distance from the Sun. Jupiter, Saturn, Uranus and Neptune dwarf the others, but are themselves dwarfed by the Sun.

Planets nearby

Spot facts

• Mercury travels faster in its orbit than any other planet. Its average speed is over 170,000 km/h (1½ times the speed of the Earth).

• If you were set down on Venus, you would at the same time be suffocated and roasted.

• The extinct volcano Olympus Mons on Mars is nearly three times the height of Mount Everest.

• In 1877 the Italian astronomer Giovanni Schiaparelli said he could see canali (channels) on Mars. This was translated as "canals", prompting people to think that they were artificial waterways made by intelligent beings.

▶ Part of the Martian surface, showing what could be dried-up river valleys and canyons. There is no longer any liquid water on Mars.

Mercury, Venus and Mars are close enough to Earth to be considered neighbours. But they are strikingly different from Earth in most respects. Mercury is a cratered wasteland, where noonday temperatures soar high enough to melt lead. It is a small planet, a little larger than our Moon, and is the nearest planet to the Sun. Earth's near-twin Venus is a hellish world that is as hot as Mercury, and has a dense, suffocating atmosphere. It is permanently covered in clouds. Mars, on the other hand, is icy cold, with only a whiff of an atmosphere around it. But it is the only planet in the Solar System where human beings could survive.

Mercury

Mercury orbits so close to the Sun that it always appears near the Sun in the sky. We can therefore see the planet only in the morning sky in the east just before sunrise, or in the evening sky in the west just after sunset. At its brightest Mercury looks like a bright pinkish star. Through a telescope, you can make out little detail on the planet's disc, except for a few vague markings.

Mercury is the second smallest planet, after Pluto. It is not much bigger than the Moon and, like the Moon, has no atmosphere. It looks remarkably like the Moon too, because it is covered with craters. These were formed billions of years ago when the planet was bombarded by huge meteorites.

One difference between Mercury and the Moon is that it does not have such large "seas", or maria. Its most obvious feature is the huge, ring-shaped Caloris Basin, which measures some 1,400 km across. The basin must have been carved out by the impact of a truly gigantic meteorite. The same impact caused the surface to wrinkle over vast distances, forming "waves" of mountain chains.

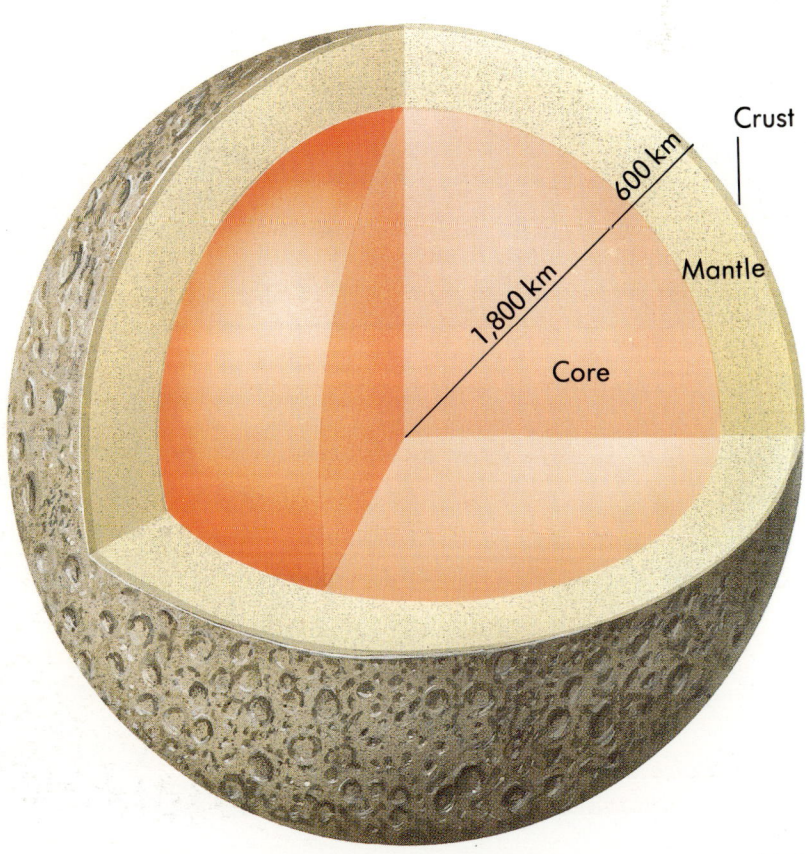

▲ This photograph of Mercury was taken by *Mariner 10* when it flew past the planet in 1974. The surface looks very similar to that of the Moon. It is peppered with craters, large and small. Some craters have bright rays radiating from them. Others have mountain peaks in their centre. Both these features are common also on the Moon. In places the surface of Mercury is quite smooth, with few craters. But there are no large "seas" like there are on the Moon.

▶ The structure of Mercury. Astronomers think that it has a large core of iron and nickel. Above the core there is probably a layer of lighter rock, which is topped by an even lighter crust.

129

Venus

Venus is the planet that comes closest to Earth (within 42 million km). In size it is almost an identical twin of the Earth. Yet the two planets are very different in other ways. For one thing, Venus rotates very slowly on its axis; its day is 243 Earth-days long. It also rotates in the opposite direction from the Earth and indeed from all the other planets. Apart from the Moon, Venus is by far the brightest object in the night sky. Sometimes we see it in the west at sunset, and call it the evening star. At other times we see it in the east at sunrise, and call it the morning star. Seen through a telescope, Venus shows phases like the Moon. Its shape changes from crescent to full and back again.

Conditions on Venus are quite different from those on Earth. The temperature at the surface is over 480°C. The atmospheric pressure is nearly 100 times that on Earth. The reason for these hellish conditions is the thick atmosphere. It is made up mainly of carbon dioxide. This is a heavy gas, which traps heat in the same way as a greenhouse.

The atmosphere of Venus is always full of clouds. They reflect sunlight brilliantly, which is what makes Venus so bright. The clouds are made of tiny droplets of sulphuric acid. The acid was probably formed from sulphur dioxide gas shot into the atmosphere when volcanoes erupted on the surface.

◀ Clouds in the atmosphere of Venus. This photograph was taken in ultraviolet light by the *Mariner 10* space probe in 1974. From the Earth, all we can see of Venus through a telescope is a poorer view of the clouds. We can, however, look beneath the clouds if we examine the planet using radar, which can pass through cloud. In a radar scan from Earth, a radio beam is transmitted and bounced off the planet. From the reflections received back, a picture of the surface can be built up.

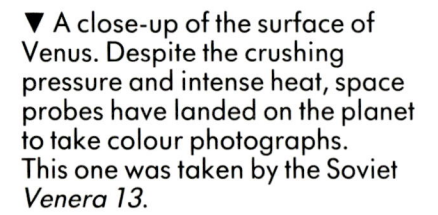

▼ A close-up of the surface of Venus. Despite the crushing pressure and intense heat, space probes have landed on the planet to take colour photographs. This one was taken by the Soviet *Venera 13*.

Beneath the thick clouds on Venus, the surface consists mainly of vast rolling plains. Here and there are isolated lowlands and highlands. The plains appear to be heavily cratered and are probably part of the ancient crust. Astronomers believe that the lowlands could once have been the beds of ancient oceans, which have evaporated.

There are two main highland regions, or continents. The one in the northern hemisphere is called Ishtar Terra. It is about the size of Australia. The other continent, the larger of the two, lies near the equator. It is called Aphrodite Terra, and is about the size of Africa. Other highland areas may be huge volcanoes.

▼ The structure of Venus. Venus probably has much the same structure as the Earth. It has a central core, a thick mantle and a thick crust. The core of iron and nickel may be partly molten. Above this lie a rocky mantle and the surface crust. The crust is more than twice as thick as that of the Earth.

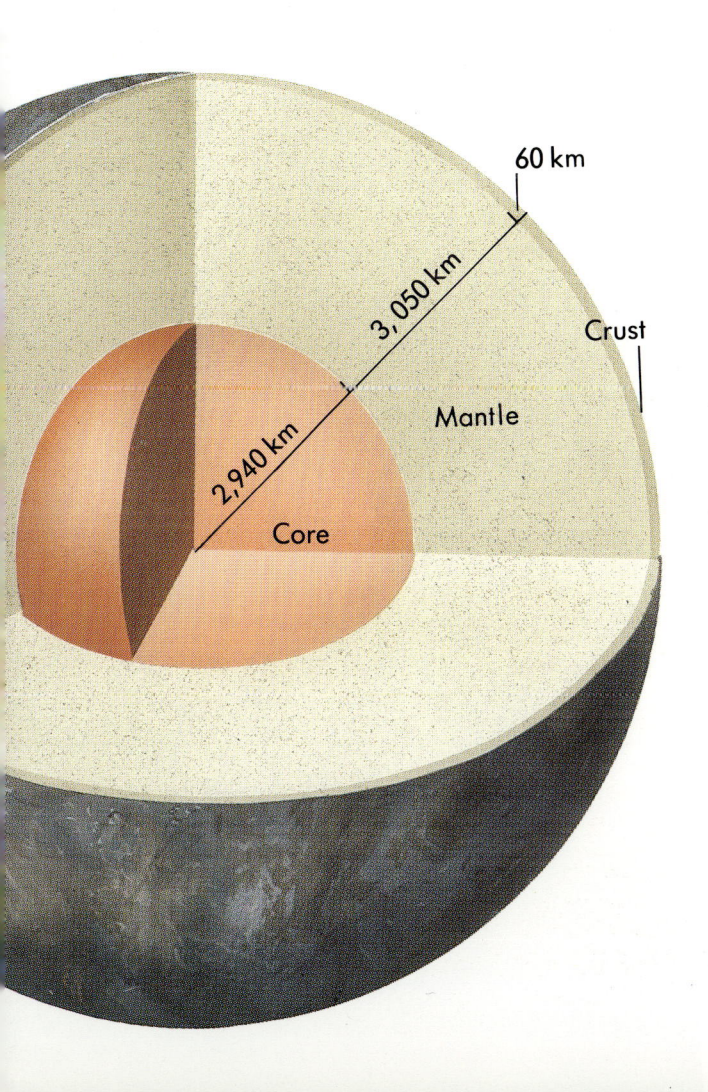

Venus's atmosphere

The atmosphere of Venus is made up mainly of carbon dioxide (96 per cent) and nitrogen. The heavy atmosphere lets in solar energy to heat up the surface, but then traps the heat.

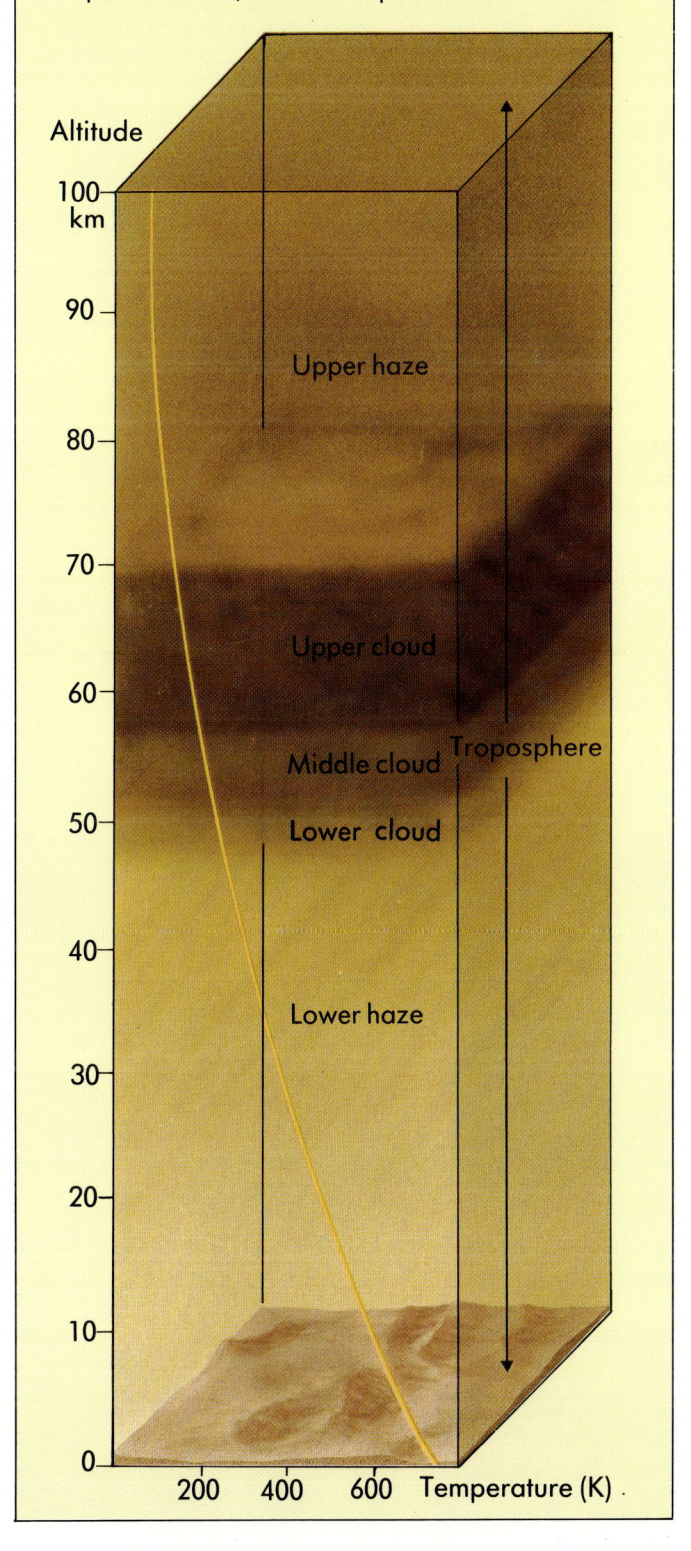

Mars

The structure of Mars

Mars is smaller than the Earth. It has a diameter of 6,794 km at the equator – just over half that of the Earth. It is also thought to have a layered structure like the Earth. At the centre is a core of iron and iron compounds. The core is surrounded by a deep mantle of silicate rock, and on top is a relatively thick crust. The crust is pitted by craters, particularly in the southern hemisphere, where it has been bombarded with rocks from outer space. The Martian surface gets its rusty red colour from iron oxides in the soil. The highest mountain is the 25-km high Olympus Mons (bottom). It is far bigger than Earth's highest peak, Mount Everest.

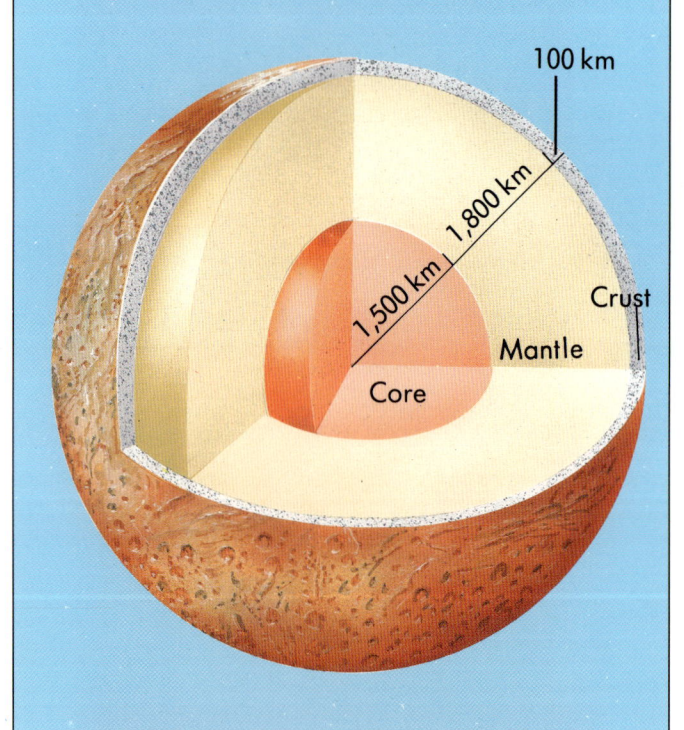

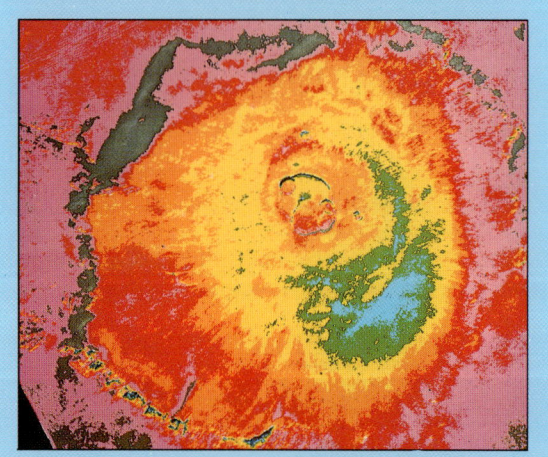

Mars comes nearer to Earth than any other planet except Venus. Close approaches occur about every 26 months at the time of opposition. At this time, the two planets orbit for a while side by side. At the closest oppositions, Mars is less than 56 million km away.

Of all the planets, Mars is the easiest to recognize in the night sky. It shines with a distinct red-orange glow. For this reason it is often called the Red Planet.

In some respects Mars is similar to Earth. A day on Mars is only 40 minutes longer than a day on Earth. Mars also has seasons. This is because its axis – like Earth's – is tilted at an angle to the plane of its orbit. But the Martian year (687 Earth-days) is almost twice as long as Earth's, so the seasons are nearly twice as long too. Mars is also similar in having ice caps at the poles. These caps vary in size from season to

▶ Ice clouds on Mars. As the Sun rises over canyons on Mars, its rays pick out puffy clouds of ice crystals. This region of Mars has the delightful name of Labyrinth of the Night (Noctis Labyrinthus).

▼ A *Viking* probe sent back this photograph as it approached Mars in 1976. Three interesting features stand out. To the left is one of Mars's four huge volcanoes, Ascraeus Mons; in the middle is the great scar of Mariner Valley; and to the right is the Argyre Basin, covered in frost.

season. They shrink in the spring and grow again in the autumn.

The wave of darkening
Mars is not covered with dense cloud like Venus. And through a telescope, we can often see various features on its surface. There are dark markings that rotate with the planet. One of the best known, near the Martian equator, is named Syrtis Major.

There are other dark markings that change with the seasons. As the ice caps shrink at the north and south poles, a so-called wave of darkening sweeps from them towards the equator. Some people once thought that this effect might be caused by the growth of vegetation, fed by water melting from the ice caps. But space probes have shown there are no plants, nor life of any kind, on Mars.

The Martian atmosphere
At the surface of Mars atmospheric pressure is less than one-hundredth of that on Earth. In the lower part of the atmosphere clouds of water ice often form. Dust is always present, giving the Martian sky a pinkish colour. The temperature (red line) changes, the higher you go above the surface.

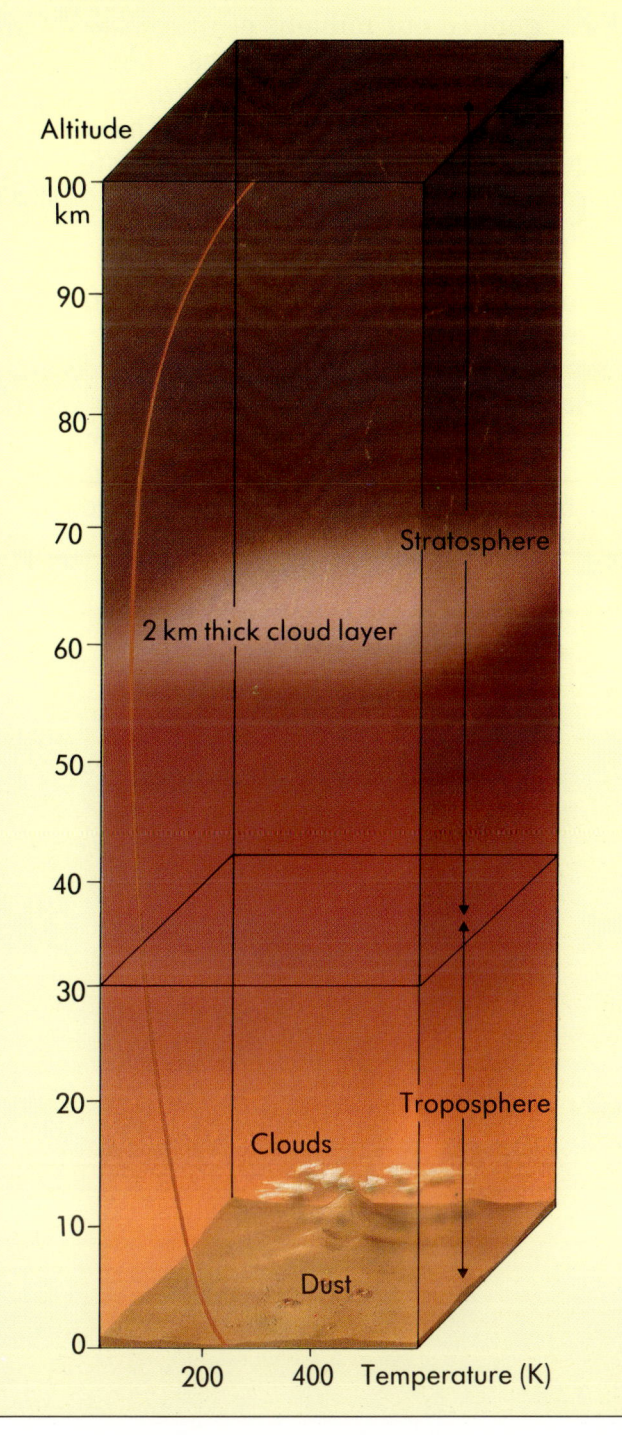

133

The Martian landscape

We can see few details of the Martian landscape through a telescope. Most of our information has come from space probes, such as *Mariner* and *Viking*. They have photographed the planet from orbit and from the surface.

The northern hemisphere consists mainly of low-lying plains with relatively few craters. The southern hemisphere has a more ancient crust, which is heavily cratered. It also has two huge basins, gouged out millions of years ago by the impacts of massive meteorites. The largest basin, Hellas, is more than 1,600 km across. It is twice as big as the Argyre Basin to the west.

One of the most interesting Martian features is a great gash in the surface that runs for some 5,000 km near the equator. In places it is more than 200 km wide and 5 km deep. It is called Mariner Valley (Valles Marineris), after the space probe that discovered it.

North-west of Mariner Valley are four huge extinct volcanoes. Three stand in a row on the Tharsis Ridge and soar to 20 km high. But even they are dwarfed by the fourth, Olympus Mons. It is 5 km higher and rises from a wide base approximately 600 km across.

Rivers on Mars

Around these and other volcanoes there are channels where molten lava once flowed. Elsewhere there are other channels that look remarkably like dried-up river beds on Earth. So did rivers once flow on Mars? Almost certainly they did, many millions of years ago.

Water probably flowed when the volcanoes were erupting. Water vapour and gases would have been given out during the eruptions. This would have created quite a dense atmosphere, allowing clouds to form and rain to fall. No rivers flow on Mars today, but there are still traces of water vapour in the atmosphere. This gives rise to the occasional cloud and early morning mist in the canyons. Astronomers believe that water is also locked in the ice caps at the two poles.

▼ This map of the landscape of Mars was produced from radar scans made from Earth. The most striking feature is the Tharsis Ridge (centred on longitude 90°). It is topped by three huge volcanoes. To the south-east is the circular depression of the Argyre Basin. Farther east is a much bigger basin called Hellas.

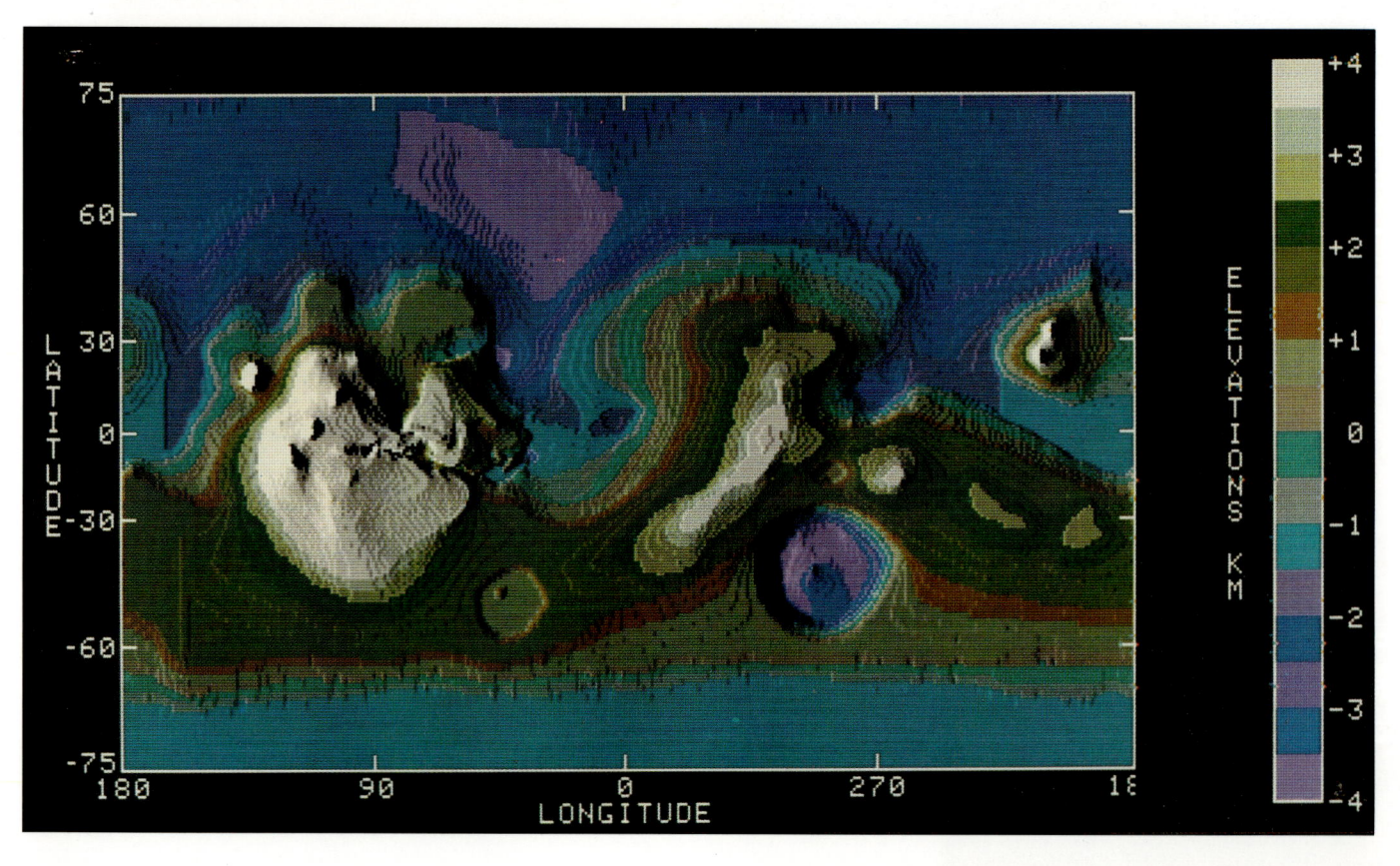

▶ Ice on Mars. Large patches of ice still linger on the rusty red ground of Mars. This picture was taken in mid-summer near the north pole. The ice patches are all that is left of the ice cap that formed the previous winter. Ice caps advance from both poles towards lower latitudes during the winter. They shrink as the ice evaporates during the following summer. The ice cap grows largest at the north pole. The ice caps are made up mainly of water ice, together with small amounts of dry ice, or solid carbon dioxide.

Vikings on Mars

In 1976 two American *Viking* space probes went into orbit around Mars. They then released craft that landed on the surface to take photographs. On the right is a view from *Viking 1* and below is a view from *Viking 2*. The pictures show a remarkably similar landscape. The ground is covered with fine rust-coloured soil, and rocks are strewn everywhere. They are pitted by the impact of soil particles blown by the wind.

Far-distant worlds

Spot facts

• Jupiter has more than twice the mass of all the other planets put together.

• Jupiter and the other three giant planets have no solid surface, but are covered by an ocean of liquid hydrogen.

• Jupiter, Saturn and Neptune give out twice as much heat as they receive from the Sun.

• Saturn has the strongest winds in the Solar System. At the equator, they blow at up to 1,800 km/h.

• Since 1979 Neptune has been the most distant planet and will remain so until 1999. In that year Pluto will move outside Neptune's orbit, and resume its role as the most distant planet.

Five frozen planets orbit in the outer reaches of the solar system, hundreds of millions of kilometres away. Four of them – Jupiter, Saturn, Uranus and Neptune – are giant gas balls of enormous dimensions. Even though these planets are so far away, we now know a lot about them. Space probes have visited them and sent back close-up photographs and all kinds of information. Only the tiny ice world of Pluto remains almost a complete mystery.

▶ *Voyager 2* took this photograph of Saturn from a distance of 43 million km. False colours have been added to show up banding in the atmosphere and the planet's very beautiful ring system.

Jupiter

Beyond the Red Planet Mars, the next planet going out from the Sun is Jupiter. Between Mars and Jupiter there is a gulf of over 500 million km. There we might perhaps expect to find another planet. But instead there is a collection of small rocky bodies, orbiting together in a broad band, or belt. They are the asteroids, or minor planets.

Apart from the Moon and Venus, Jupiter is the most brilliant object in the night sky. It shines so brightly because of its huge size and its cloudy atmosphere, which reflects sunlight very well. In Roman mythology Jupiter was king of the gods. It is an appropriate name for the largest planet, which is big enough to swallow 1,300 planets the size of the Earth. Its diameter is 11 times that of Earth.

In some respects Jupiter is more like a star than a planet like the Earth. Like a star, it is made up of gas, mainly hydrogen and helium. It has a powerful magnetic field, and gives out heat, radio waves and even X-rays. Without doubt, if Jupiter had been much bigger, it would have begun to shine as a star.

In common with the other giant planets, Jupiter is the centre of a large satellite system. At least 16 moons circle around it. The largest, Ganymede, is bigger than the planet Mercury. And like Saturn, Uranus and Neptune, Jupiter has rings around it. But they are too faint to be seen from Earth.

When you look at Jupiter through a powerful telescope, you can see that the disc is banded with light and dark, reddish stripes. The light ones are called zones; the dark ones, belts. Within the zones and belts are all kinds of contrasting patterns described as spots, ovals, streaks, wisps and plumes.

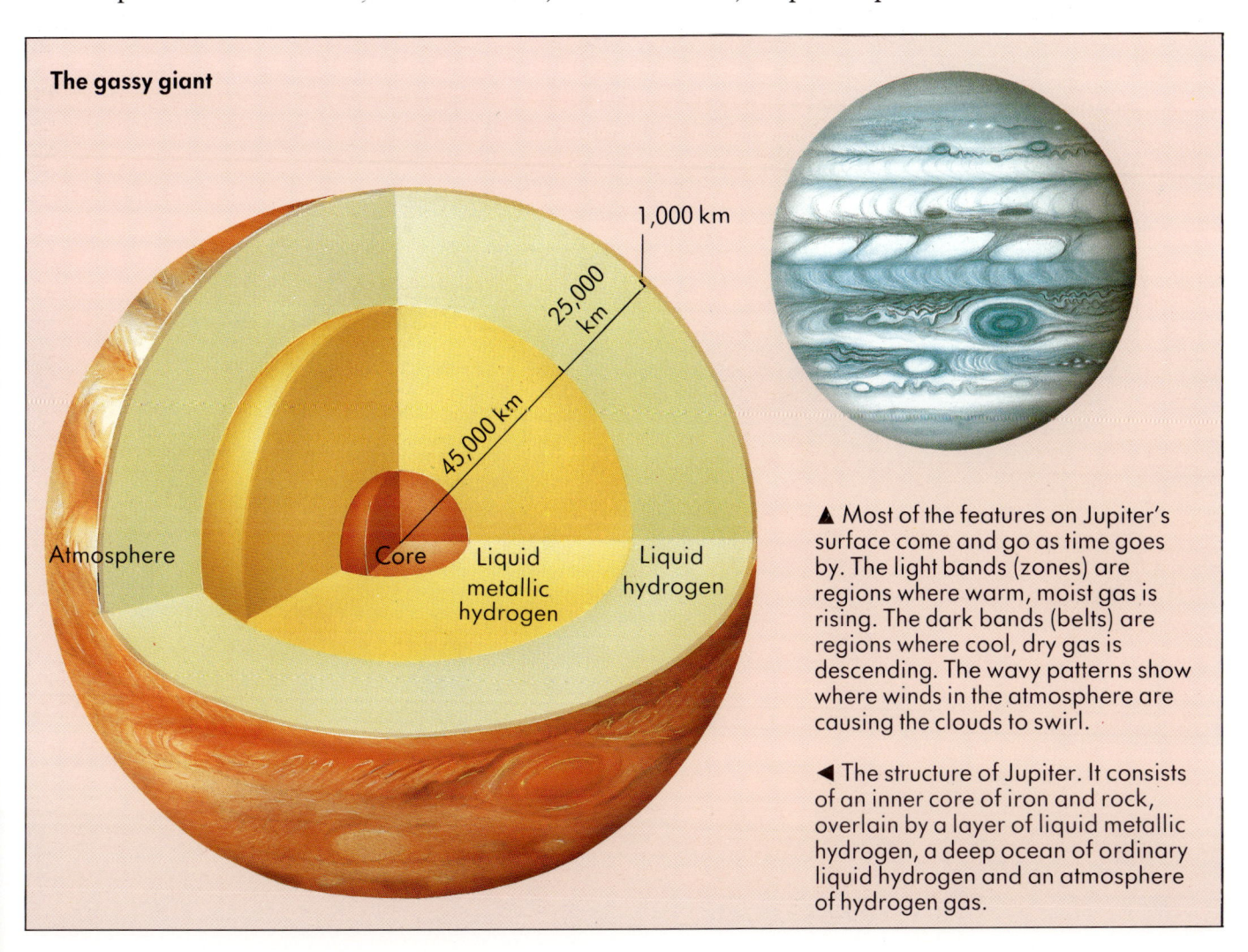

The gassy giant

▲ Most of the features on Jupiter's surface come and go as time goes by. The light bands (zones) are regions where warm, moist gas is rising. The dark bands (belts) are regions where cool, dry gas is descending. The wavy patterns show where winds in the atmosphere are causing the clouds to swirl.

◄ The structure of Jupiter. It consists of an inner core of iron and rock, overlain by a layer of liquid metallic hydrogen, a deep ocean of ordinary liquid hydrogen and an atmosphere of hydrogen gas.

Jupiter's weather

Jupiter rotates on its axis very quickly. It has the fastest rate of spin of all the planets, rotating once in less than 10 hours. This produces several noticeable effects. It causes the planet to bulge at the equator, and to be flattened at the poles. It also causes the clouds in the atmosphere to be drawn into alternate light and dark bands that run parallel with the equator (the belts and zones).

The rapid rotation sets up powerful wind belts, or jetstreams, which blow east to west. They combine with rising and descending currents to churn up the atmosphere. This churning creates the waves, eddies and other features we see on the disc. The biggest and most long-lived feature is the Great Red Spot. It is a huge storm in the atmosphere.

The white clouds in the atmosphere are made up of crystals of ammonia ice. They are higher and cooler than the reddish clouds, which are probably a compound of ammonia and hydrogen sulphide. Lower still the clouds appear bluish and are probably water-ice crystals.

Space probes have spotted other kinds of activity in the atmosphere. They have observed lightning flashes during Jupiter's night, and also displays of aurorae in the polar regions. These displays are like the Northern Lights on Earth but on a much larger scale. They occur when charged particles from Jupiter's radiation belts collide with atoms in the atmosphere. The radiation belts are regions of space in which particles have become trapped by Jupiter's powerful magnetic field.

The Great Red Spot

Astronomers first noticed Jupiter's Great Red Spot about 300 years ago. And it has been visible for most of the time since then. It has varied in size over the years and at present measures about 28,000 km long and 14,000 km across. The Red Spot lies in Jupiter's southern hemisphere, and remains at a latitude of about 20°. Astronomers once thought that it might be connected with a high point on the surface or even be a huge erupting volcano. But space probes have shown it to be a great whirling storm centre. It spins round anticlockwise every 6 days or so. Scientists think that its colour is caused by the presence of red phosphorus.

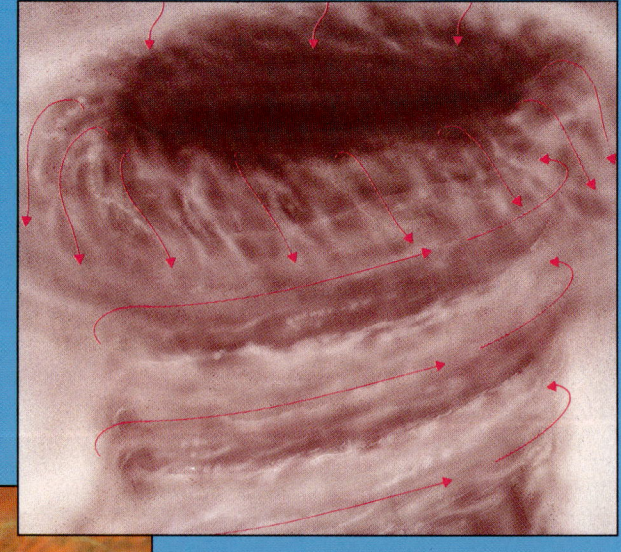

▲ The Great Red Spot appears to be a region of high pressure. The top stands about 8 km higher than the surrounding cloud layer.

◀ The *Voyager 1* space probe took this close-up picture of the Great Red Spot. It shows great swirls in the atmosphere above and three whitish spots below. They are storm centres.

▶ The Great Red Spot is also visible in another *Voyager 1* picture, which shows the whole of Jupiter's disc. The moon crossing the disc is Io, the closest of Jupiter's four large moons.

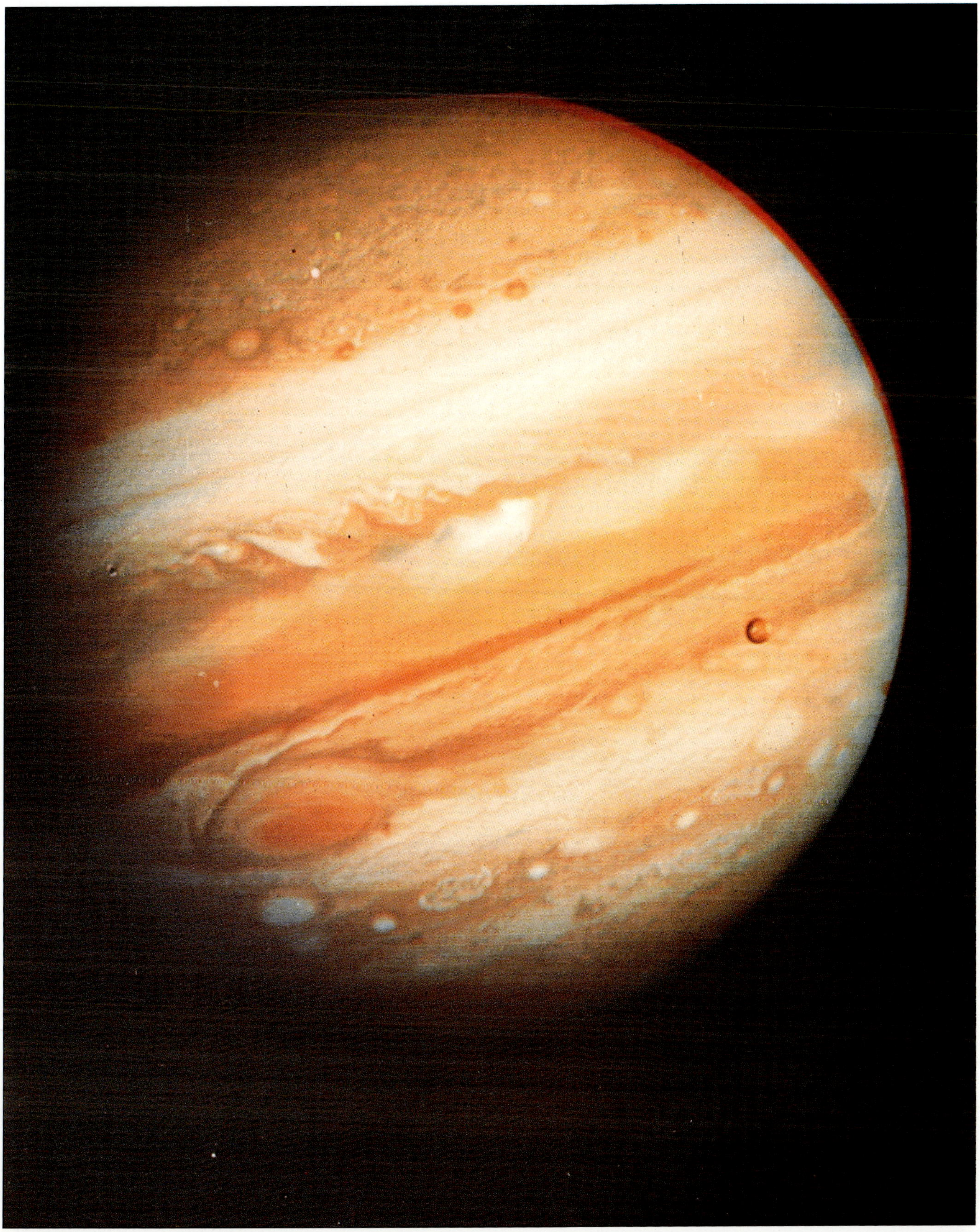

Saturn

Saturn was regarded as the outermost planet known until the late 1700s. It is nearly twice as far away from the Sun as Jupiter, at a distance of nearly 1,000 million km. It takes nearly 30 Earth-years to make one complete journey around the Sun.

Saturn is easily visible to the naked eye. But we need a telescope to see its most remarkable feature, its system of flat shining rings. From one side to the other the rings measure over 270,000 km and they are about 60,000 km wide. They reflect light well, and Saturn would be less visible and therefore appear much fainter without them. We see different aspects of them as the planet travels in its orbit.

Floating on water
After Jupiter, Saturn is the largest planet, with a diameter nearly 10 times that of Earth. Apart from its prominent rings, it resembles Jupiter in many respects. It is a gas giant, surrounded by many moons (at least 22). It rotates rapidly on its axis, bulges at the equator, and is flattened at the poles. We can see on photographs of the planet's disc parallel belts and zones, which are bands of circulating clouds. But they are not nearly as marked or as colourful as Jupiter's are. Saturn also has a powerful magnetic field and is surrounded by radiation belts similar to the Van Allen belts around the Earth.

Saturn has a much lower relative density than Jupiter, only 0.7. This is the lowest of all the planets and is less than the density of water, so if Saturn could be placed in a vast ocean of water, it would float.

▲ Saturn, like all planets, has two motions in space. It travels in orbit around the Sun, and it spins on its axis. The rings, which girdle the equator, present different aspects during the 29 or so years it takes Saturn to orbit the Sun. Sometimes we can see the rings edge-on (1, 7).

◄ Saturn, like Jupiter, probably has a small core of rock and iron with perhaps 3-10 times the mass of the Earth. Around the core is a thick layer of hydrogen in a highly compressed metallic form. Next comes a deep ocean of ordinary liquid hydrogen. The atmosphere is mainly hydrogen with a little helium.

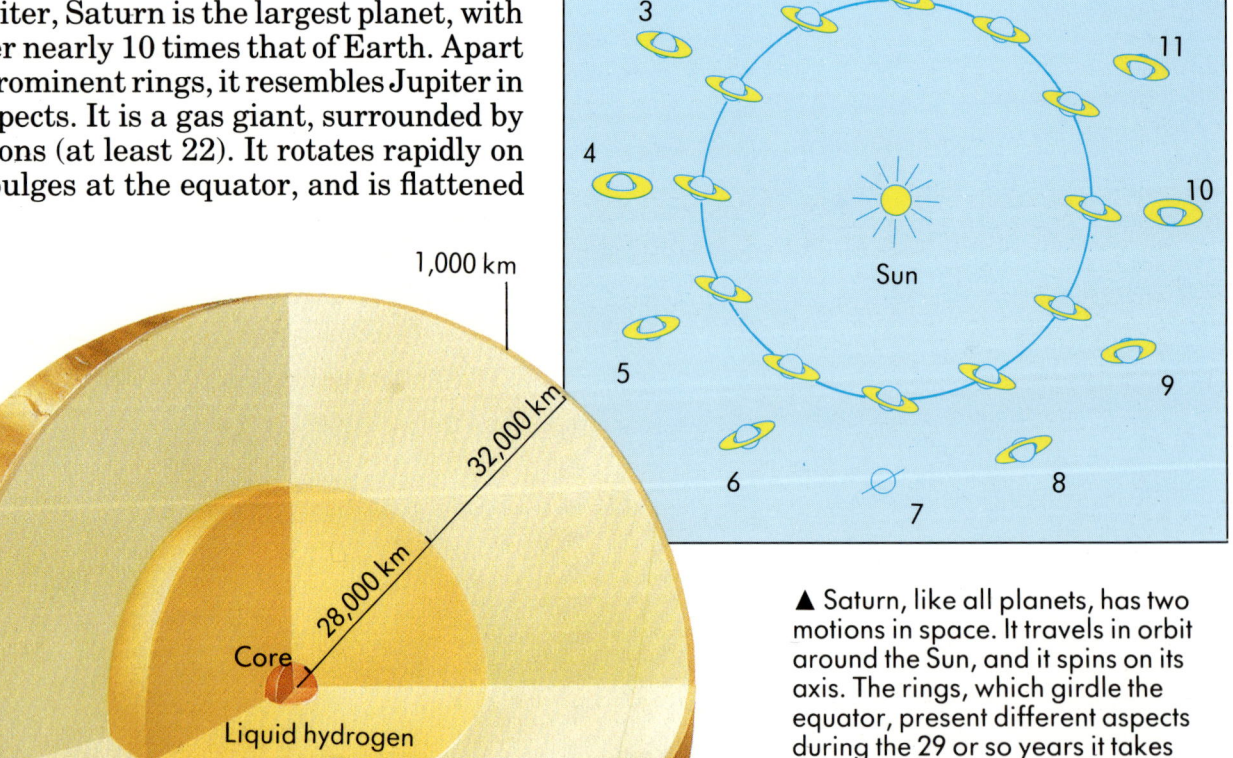

▲ Some of the wind belts in the north of Saturn. False colours have been used in this picture to bring out extra detail. The wave-like features and the oval "whirlpools" show where the atmosphere is churning up.

▼ This natural colour picture taken by *Voyager 2* shows the faint banding on Saturn's disc. The black dot below the rings is the shadow of Tethys. This is one of Saturn's 20 or more moons.

Speedy winds
Saturn is a very windy planet. Mostly the winds blow from west to east. They are strongest around the equator, where they reach the phenomenal speed of 1,800 km/h. Wind speeds fall off going northwards and southwards away from the equator.

After a latitude of about 35° north and south of the equator, a curious thing happens. The winds suddenly change direction and start blowing the opposite way, from east to west. At higher latitudes still, the winds reverse direction again, and alternate like this all the way to the poles. The winds therefore form a series of bands, or belts.

Great churning, or turbulence, occurs at the boundaries of these alternating wind belts. This is accompanied by huge storms which rage there. They appear on the disc as pale or dark spots. But none of them is as large, or lasts so long, as the Great Red Spot on Jupiter.

Rings and ringlets

The bright shining rings that girdle Saturn are one of the wonders of the solar system. Jupiter and Uranus have rings too, but they cannot be seen from Earth. Saturn's rings, however, present a magnificent sight in a telescope. They span more than two diameters of the planet, a distance of more than 270,000 km.

In a telescope we can see three main rings, called A, B and C. The outermost A-ring is separated from the B-ring by a dark gap, called the Cassini division. Both rings are bright. Inside the B-ring is the much fainter C-ring, also called the crêpe ring. The rings, made up of particles of rock and ice, are broad but thin. When they are viewed edge-on, they almost disappear from view. In places they are less than 200 m thick.

It was once thought that the rings were the remains of a close satellite of Saturn. But it is more likely that they are made up of material left over when the planet was formed.

Racing ringlets

When the *Voyager* space probes visited Saturn, the rings proved more fascinating than ever. They turn out to be made up of thousands of individual ringlets. The ringlets mark the path of particles of ice and rock as they race swiftly around the planet's equator. The particles vary in size from tiny grains to boulders more than 10 m across.

The *Voyager* probes also discovered several new rings, which we cannot see from Earth. Inside the C-ring is a very faint D-ring, which probably extends down to Saturn's cloud tops. There are also other rings outside the A-ring. The most interesting is the very narrow F-ring, which is a curiously twisted collection of several ringlets. The probes also revealed a number of tiny moons orbiting close to the edges of some of the rings. They are called shepherd moons because they appear to help to keep particles within the rings.

▲ This picture shows clearly the A- and B-rings of Saturn, separated by the dark "gap" of the Cassini division. In the B-ring are curious dark, finger-like features called spokes, which rotate with the ring.

◄ The ringlets that make up Saturn's rings. This picture has been computer-processed to show the ringlets in false colours. Mostly it shows the C-ring, with the B-ring on the left. The different colours indicate that the rings are made up of different kinds of particles.

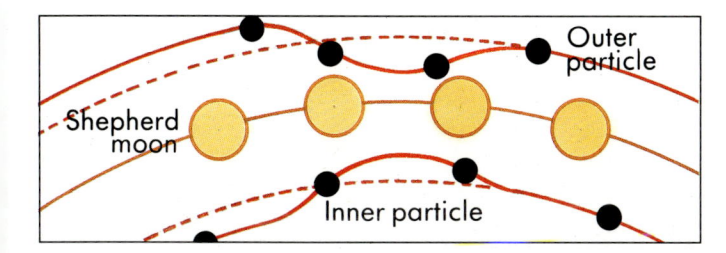

▲ Astronomers reckon that tiny shepherd moons may help to keep particles in the ringlets. As a moon moves through the ring, a faster inner particle is slowed down by the moon's gravity and falls to a lower orbit. A slower outer particle, on the other hand, is speeded up and will then move to a higher orbit.

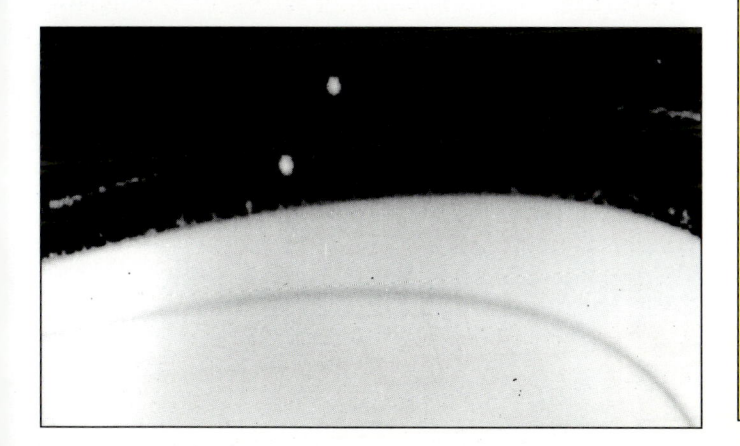

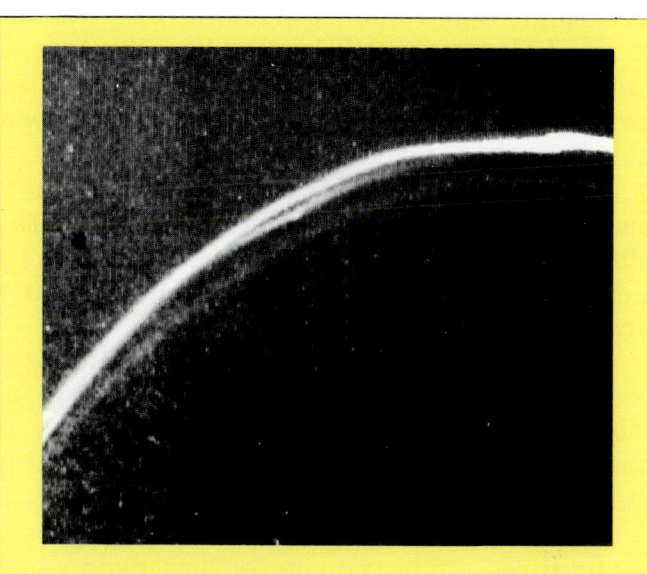

The fascinating F-ring

Outside Saturn's A-ring is the F-ring, which is less than 150 km wide. It is far too thin to be seen from Earth. It is made up of as many as 10 ringlets. Curiously, the ringlets are intertwined, giving a braided appearance (above). It is thought that the braiding effects are caused by the gravitational disturbance of a pair of nearby shepherd moons.

▶ The *Voyager 2* probe took this beautiful picture of Saturn from a distance of 3,400,000 km as it looped away from the planet to its next port of call, Uranus. The picture reveals very clearly the classic A-, B- and C- rings. Notice how transparent the inner ring is. The B-ring is denser and blots out the planet's disc. Note also the dark shadow cast on Saturn by the rings.

Outer planets

The English astronomer William Herschel discovered Uranus in 1781, after first thinking it might be a comet. The planet lies more than twice as far away from the Sun as Saturn does, at a distance of nearly 3,000 million km.

Uranus is unusual among the planets because its axis is tilted right over (at 98°) to the plane of its orbit. This means that it spins on its side as it travels through space, and not nearly upright like the other planets.

We cannot see Uranus with the naked eye because it is too far away. And even a powerful telescope shows it only as a bluish-green disc. It has this colour because its atmosphere contains methane. Methane absorbs the red wavelengths from white light, leaving blue-green. No features can be seen on the disc, even when it is viewed closely by space probes.

Like Jupiter, Uranus has faint rings around it, but they are not visible from Earth. But the *Voyager 2* probe saw them. There are at least 10 narrow rings, consisting of large chunks of dark rock. The thickest ring, the epsilon, is up to 100 km wide. The rocks in the ring appear to be kept

▲ ▶ Images of Uranus taken by *Voyager 2* in 1986. The above false-colour picture shows a bright cloud (top right) floating in the atmosphere. The natural colour picture (right) shows Uranus as a bright crescent.

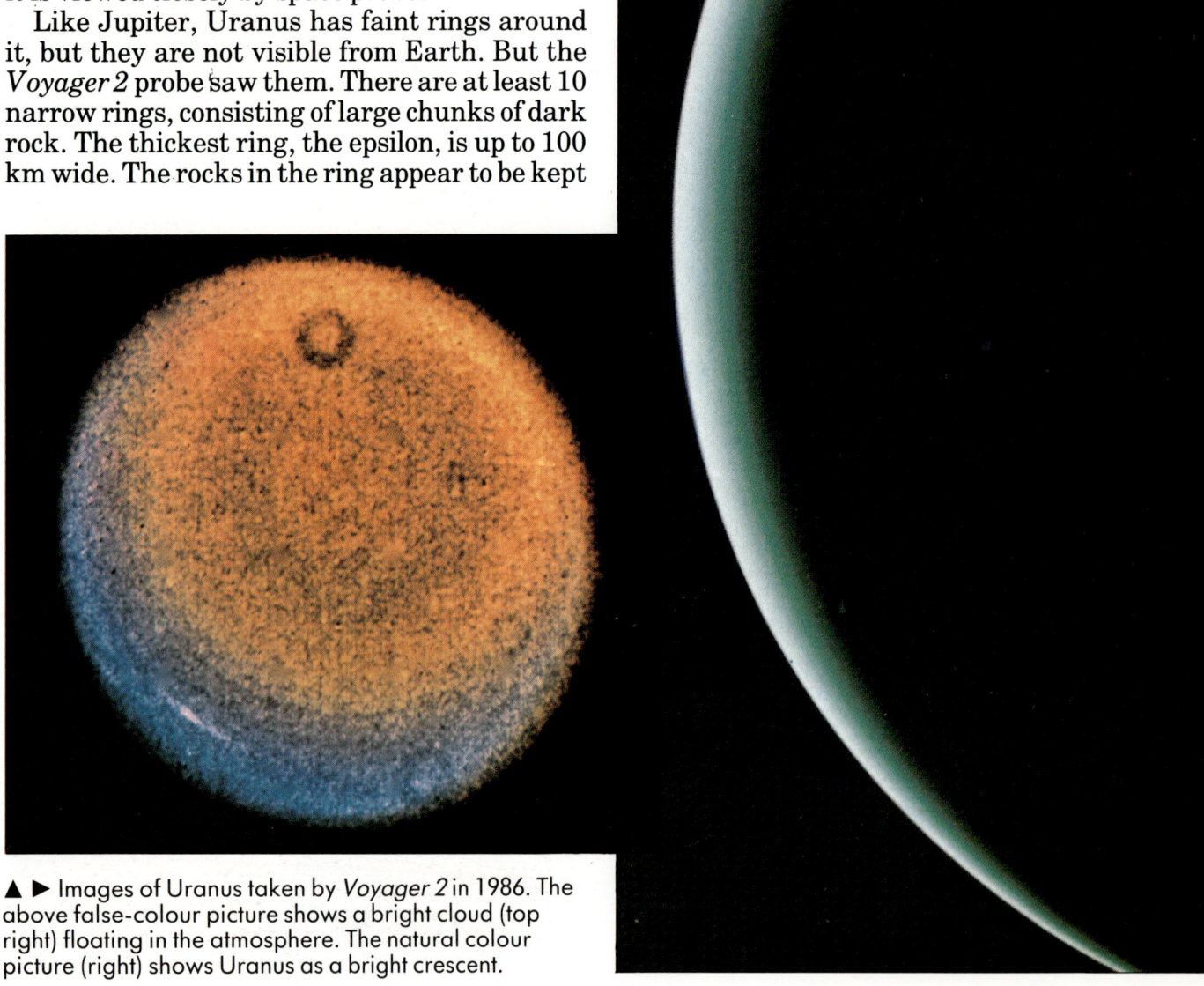

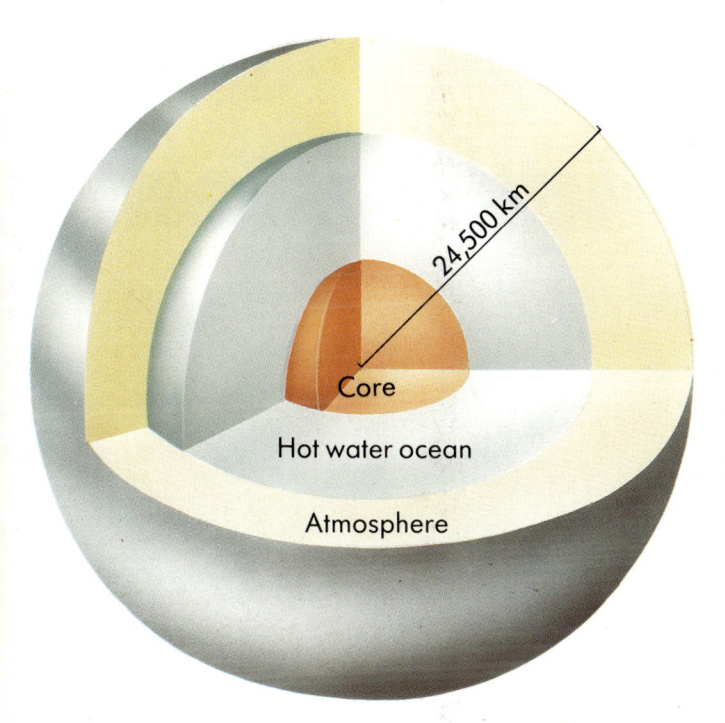

▲ Uranus has a heavy core of iron and rock about the size of the Earth. On top is an ocean of hot water and ammonia about 8,000 km deep. The atmosphere is made up of hydrogen, helium and methane.

▼ Wisps of cloud scurry across the atmosphere of Neptune in this image sent to Earth by *Voyager 2* in 1989. The red edge to the disc is a false-colour effect showing haze. The dark spot is probably a storm.

in place by a pair of tiny shepherd moons. At least 13 more moons circle the planet.

Twin Neptune

After Uranus was discovered, astronomers found that its orbit was not regular. And this made them think that there must be another planet farther out that was affecting Uranus by its gravity. In 1846 that planet was finally discovered by the German astronomer Johann Gottfried Galle. It was named Neptune.

Neptune is the fourth largest planet, and is only a little smaller than Uranus. For most of the 165 years it takes to go once round its orbit, Neptune is the second farthest planet from the Sun. But until 1999 it will be the farthest, because Pluto is currently travelling inside Neptune's orbit.

Astronomers can see few details of the planet through a telescope. But images sent back by *Voyager 2* have revealed a wealth of information. The overall colour of the planet is deep blue, because of traces of methane in the atmosphere, which consists mainly of hydrogen and helium. It has a much more active weather system than Uranus.

Pluto and Charon

The ninth planet, Pluto, was discovered as recently as 1930 by the American astronomer Clyde Tombaugh. For most of the 248 years it takes Pluto to orbit the Sun, it is the farthest planet. But at present it is travelling inside Neptune's orbit. Pluto is by far the smallest planet, smaller even than the Moon. It is probably made up mainly of rock and ice and has an atmosphere of methane. Pluto has one moon, called Charon. Charon is astonishingly large for a moon – half the size of its parent planet. The two are shown below.

Debris of the Solar System

Spot facts

- As many as 100 million meteorite particles burn up in the Earth's atmosphere every day.

- Every year the Earth gains up to 5 million tonnes in weight from meteorite particles that fall to the surface as dust.

- The massive meteorite that gouged out the Arizona Meteor Crater had the destructive power of a hydrogen bomb.

- Halley's comet has been observed on each of its returns since 87 BC.

- The nucleus of Halley's comet is about 15 km long. It is velvety black and shaped like a potato.

- The tail of the very bright "daylight comet" of 1843 stretched 330 million km across the sky.

▶ A meteorite weighing 1,000 tonnes or more skips off the atmosphere, like a flat stone skips over water. It leaves a fiery trail in its wake, clearly visible in broad daylight. If it had crashed to Earth, it might have caused great destruction.

As well as planets and their moons, the Solar System contains many much smaller bodies. Those called the asteroids circle the Sun in a broad belt, but they are too far away from Earth and too tiny to be seen with the naked eye. But other bodies make a more spectacular sight, sprouting a flaming head and a long glowing tail. We see them as comets. Smaller bits of rocky matter get trapped by Earth's gravity and plunge through the atmosphere and burn up. They leave fiery trails in their wake, which we call meteors, or shooting stars.

Asteroids and comets

Many rocky, icy chunks were left over after the formation of the planets 5,000 million years ago. The biggest collection of pieces ended up circling the Sun in a broad band between the orbits of Mars and Jupiter. These are the asteroids, or minor planets. Several thousand asteroids have been discovered, ranging in size from 1 to 1,000 km in diameter.

The asteroids are rocky bodies; the Sun long ago evaporated any ice they contained. But many lumps of matter left over since the birth of the Solar System are still in their original icy state. Most of them stay in the frozen depths of the Solar System and remain invisible. From time to time, however, some of them wander into the inner Solar System. The Sun evaporates the ice to form a gas cloud around them, and they become visible as comets.

Comets are among the most spectacular of all heavenly bodies. The brightest ones can grow a tail that stretches half-way across the sky. Some of them (in 1882 and 1910, for example) are visible in broad daylight. The appearance of most comets cannot be predicted even today. In the past people were frightened by the appearance of a comet in the sky. They thought it brought bad luck or signalled disaster.

▲This pitted lump of rock is the tiny Martian moon Phobos, which is less than 30 km across. Probably Phobos was once an asteroid, circling at the edge of the asteroid belt. Some time long ago it wandered too close to Mars and became a satellite when it was captured by the planet's gravity.

◀ Halley's comet travelling against a starry backcloth in April 1986. At that time it was getting near its closest approach to Earth, some 63 million km. The comet was best observed that year in the Southern Hemisphere. Because the comet enters our region of the Solar System only once every 76 years, it will not be seen again until the year 2061. The telescope through which this picture was taken was trained on the comet during a 1-hour exposure. This caused the backdrop of stars to make trails as the comet moved in its orbit. The comet was named after the British astronomer Edmond Halley. He was the first to recognize that the records of comets that were spotted at regular 76-year intervals, and had the same orbit, were in fact referring to the same comet. The records he studied were dated 1531, 1607 and 1682, and he deduced that the comet would reappear in 1758. It did.

Sizes and orbits

Most asteroids orbit the Sun in a band, or belt, which is about 150 million km wide. The inner edge of the belt lies about 300 million km from the Sun. The biggest asteroid, Ceres, is about 1,000 km across. Only about 200 of the others are larger than 100 km in diameter. The larger ones, such as Ceres, Pallas (600 km across) and Vesta (550 km across), appear to be ball-shaped. But the smaller ones can vary greatly in shape. Eros, for example, is a chunk of rock about 35 km by 15 km by 7 km.

Eros is also unusual in that it does not orbit in the main asteroid belt, but much closer in. Occasionally it comes within 25 million km of the Earth, as happened in 1975. Other asteroids sometimes come much closer. In March 1989 one designated 1989FC, came within 750,000 km. In cosmic terms, this was a very near miss.

The comet's tail

A comet is often described as a dirty "snowball" because it is made up of a mixture of dust and ice. We do not usually discover comets until they come within the orbit of Jupiter. Then the Sun's heat starts to evaporate the ice. This releases gas and dust to form a cloud, or coma, around the solid core, or nucleus.

As the comet gets closer to the Sun, the solar wind begins to affect it. The wind "blows" gas and dust particles away from the head of the

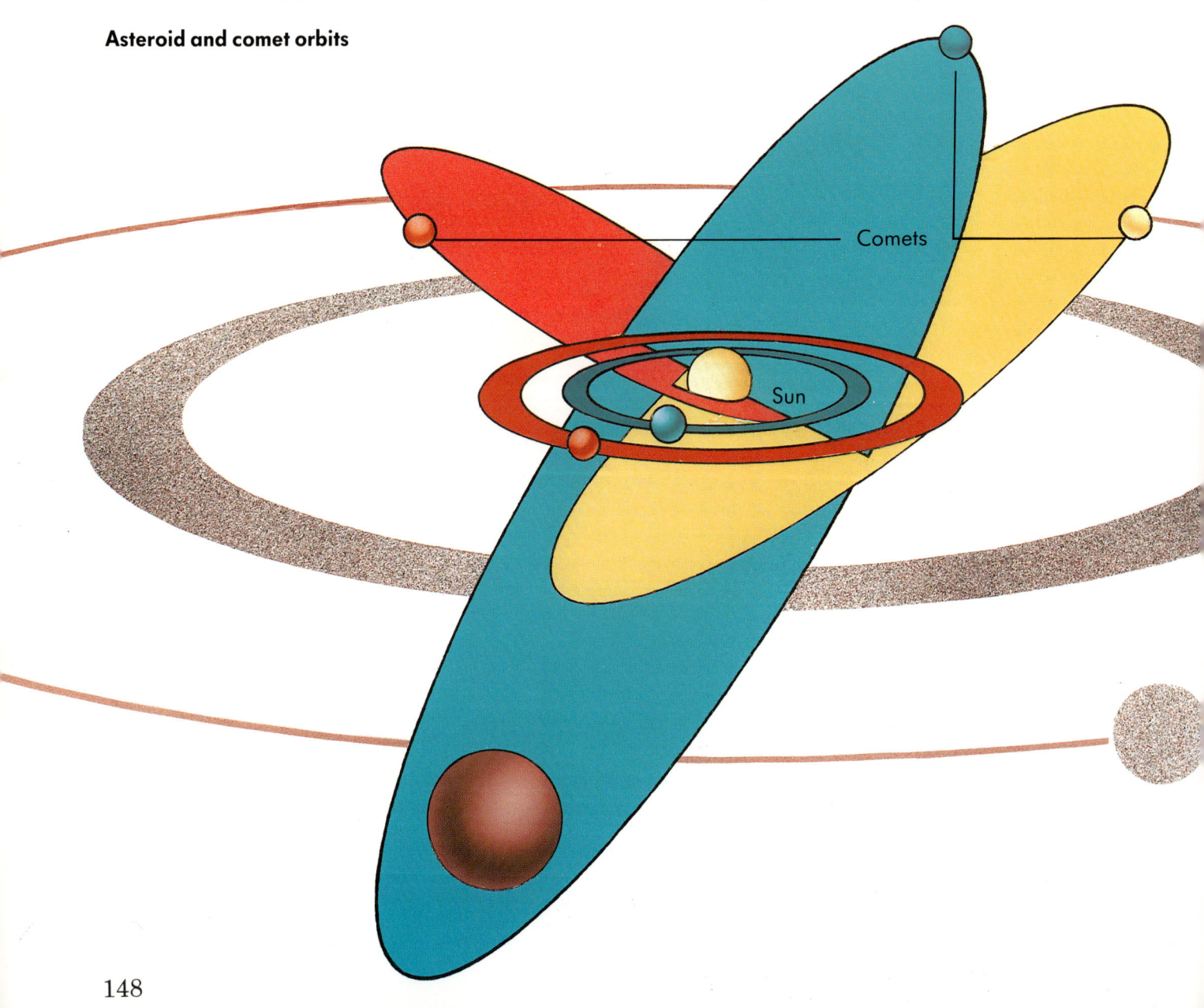

Asteroid and comet orbits

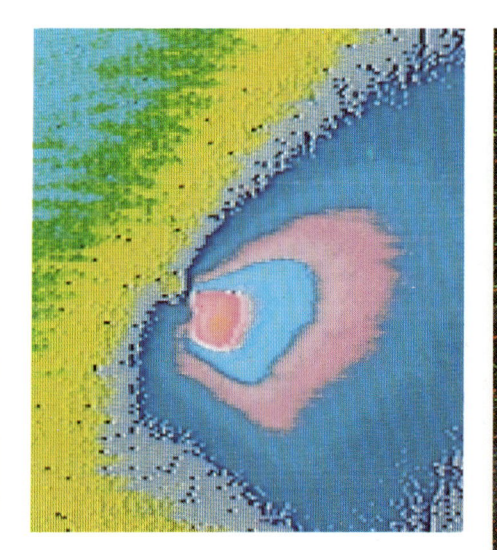

▶ Halley's comet, as photographed through a telescope in Australia in March 1986.

▲ A close-up image of the same comet, sent back by the European space probe Giotto at about the same time.

Asteroid belt

comet into a long tail. The tail therefore always points away from the Sun. So when the comet swings round the Sun and starts receding, it travels tail-first. Gradually the tail shrinks and the comet fades. By the time it reaches Jupiter's orbit, it has all but disappeared from view.

Most bright comets appear unexpectedly, blaze for several weeks, and then disappear. It is nearly impossible to predict when they will return to Earthly skies. Some comets, however, return as regular as clockwork. Their orbits are accurately known; so are their periods – the time they take to travel around their orbit. They are called periodic comets. The best-known of these is Halley's comet, with a period of about 76 years.

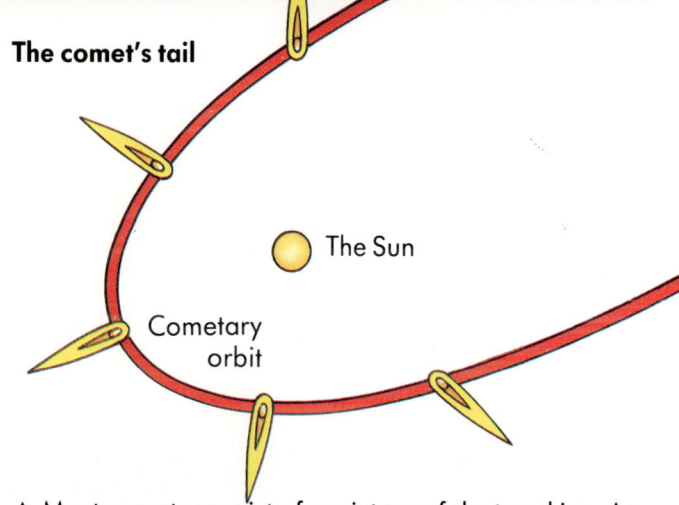

The comet's tail

The Sun

Cometary orbit

▲ Most comets consist of a mixture of dust and ice. As they get near the Sun, the ice melts to form a reflecting tail that always points away from the Sun.

149

Meteors and meteorites

Day and night the Earth is bombarded with bits of rock from outer space. They are called meteoroids, and they vary in size from microscopic specks to huge boulders perhaps hundreds of metres in diameter. Almost certainly they are pieces of asteroids that have been broken off during collisions.

Meteoroid particles travel in orbit around the Sun, just like the Earth does. When they come near the Earth, they are attracted to it by gravity. They plunge into the upper atmosphere travelling at speeds up to 70 km per second. Friction with the air heats up the particles and makes them glow white hot. In the night sky we see them as the fiery streaks we call meteors or shooting stars.

Most meteoroid particles are so small that they burn up completely. Others turn to dust and eventually settle on the ground. The biggest ones survive their fiery passage and fall to Earth as meteorites. If they are really huge they can create huge craters. Examples are the Meteor Crater in the Arizona desert in the United States, and the Henbury craters in New South Wales, Australia.

There are two main types of meteorites, stony and iron. There is also a somewhat rarer intermediate type, the stony-iron. Stony

▶ The Arizona Meteor Crater measures 1,265 m across and it is 175 m deep. Today, only small pieces remain of the massive body that gouged out the crater some 25,000 years ago.

Flattened forest

◀ Trees were felled like ninepins by the blast from an explosion that occurred near the Tunguska River in Siberia in June 1908. At first astronomers thought that the explosion was caused by the impact of a huge meteorite. But they now think a small comet was to blame. They reckon that the comet burst through the Earth's atmosphere and that its nucleus vaporized with explosive force several kilometres above the ground.

meteorites are made up mainly of silicates, like many stones on Earth. Many stony meteorites contain tiny rounded grains and are called chondrites. Some called carbonaceous chondrites are rich in carbon compounds.

Iron meteorites are made up mainly of iron and nickel, together with a little cobalt. When they are cut, polished and etched with acid, a triangular crystal pattern shows up which is unique to meteorites.

No one is certain whether the pebble-sized objects called tektites come from outer space or not. They have a different composition from meteorites, and are similar to volcanic glass. They are found mainly in four areas, in North America, Australia, Czechoslovakia and the Ivory Coast in Africa.

Missiles from outer space

There are three kinds of meteorites: iron meteorites (1), called siderites; stony ones (2), called aerolites; and stony-iron ones, called siderolites. Round grained aerolites containing traces of carbon are also known as carbonaceous chondrites. Tektites (3) are glassy pebbles.

151

Myriads of moons

Spot facts

• Earth's Moon is the fifth largest satellite in the Solar System, after Ganymede, Titan, Callisto, Triton and Io.

• The largest crater visible on the Moon is Bailly, a "walled plain" nearly 300 km across.

• A new mineral discovered in Moon rocks has been called armalcolite, after the names of the three *Apollo 11 astronauts who took part in the first Moon landing — Armstrong, Aldrin and Collins.*

• The oldest Moon rock brought back by the *Apollo astronauts is about 4,600 million years old, the same age as the Earth.*

• Jupiter's moon Io is the only body in the Solar System besides Earth where there are known to be active volcanoes.

▶ The barren, but beautiful Sea of Tranquility on the Moon. The "sea" is a vast lava plain, criss-crossed by snaking rilles (troughs) and ridges. There are only a few craters. The *Apollo 11* astronauts took this picture just before they made the first ever landing on the Moon in July 1969.

All the planets except Mercury and Venus have smaller bodies orbiting around them. These satellites, or moons, vary in size from lumps of rock a few tens of kilometres across to bodies bigger than Mercury. The giant outer planets have the most moons — Saturn has more than 20. All these tiny worlds are different from our own Moon, and from each other. Some are rugged and heavily cratered, others are as smooth as the ice that covers them. Some shine brightly, reflecting sunlight; others are dull and dark. Most are dead worlds, but at least one, Jupiter's Io, is alive with erupting volcanoes.

The Moon

Phases of the Moon

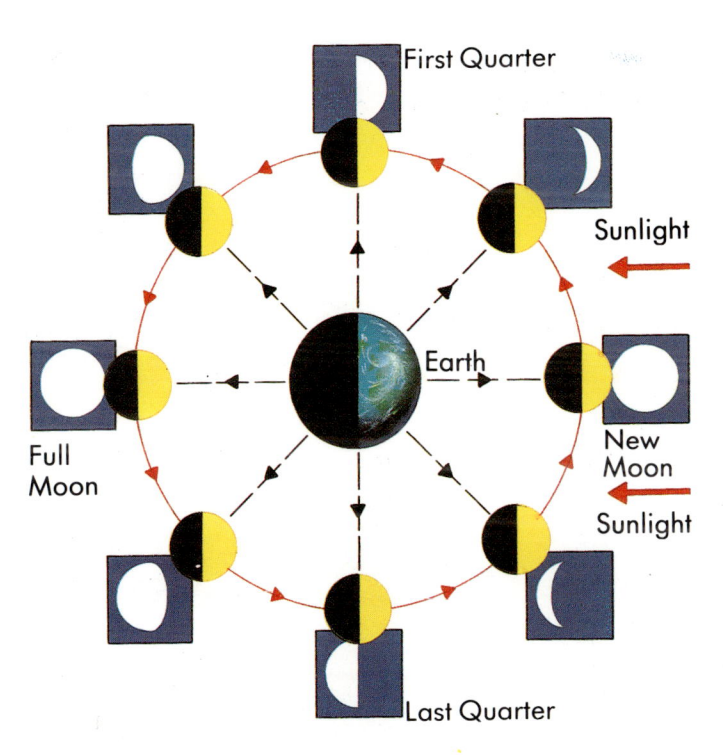

The Earth has just one natural satellite, the Moon. We know more about it than about any other moon because it is our closest neighbour. And astronauts have explored its surface.

The Moon lies at an average distance of 384,000 km from Earth, and is more than 100 times closer than the nearest planet. For a satellite, it is remarkably large compared with its parent planet. Its diameter (3,476 km) is nearly one-quarter that of the Earth.

Earth's gravity keeps the Moon circling in orbit around it, once every 27⅓ days. The Moon also spins on its axis once during this time. This results in the Moon always presenting the same face towards us. Lunar gravity is much less than the Earth's (one-sixth) because the Moon has much less mass. But it still affects the Earth by creating the rise and fall of the tides.

With its low gravity, the Moon has been unable to keep any atmosphere, so there is no air, no wind, no rain, no weather of any kind. The Moon is a dead, silent world. There is a marked contrast between day and night. In the day temperatures soar to over 100°C, but at night they drop to as low as −150°C.

▶ The Moon gives off no light of its own, but shines by reflecting sunlight. As it travels around the Earth, we see more or less of the surface lit up. It all depends on the position of the Moon in relation to the Sun. From the Earth it appears that the Moon is changing shape. We call the changing shapes of the Moon its phases. It takes 29½ days to go through all the phases. When the Moon lies between the Earth and the Sun, it presents a dark face towards us (New Moon phase). Later a crescent appears and grows in size. A week after New Moon, half the Moon is lit up (First Quarter). The visible area increases to gibbous, until two weeks after New Moon the whole face is lit up (Full Moon). Then it gradually shrinks again through gibbous, half Moon (Last Quarter), and crescent, until it disappears at the next New Moon.

▶ The structure of the Moon resembles that of a small planet. It probably has a small core, surrounded by a partly molten zone. Above is a solid mantle, covered by a thin crust.

The Moon's structure

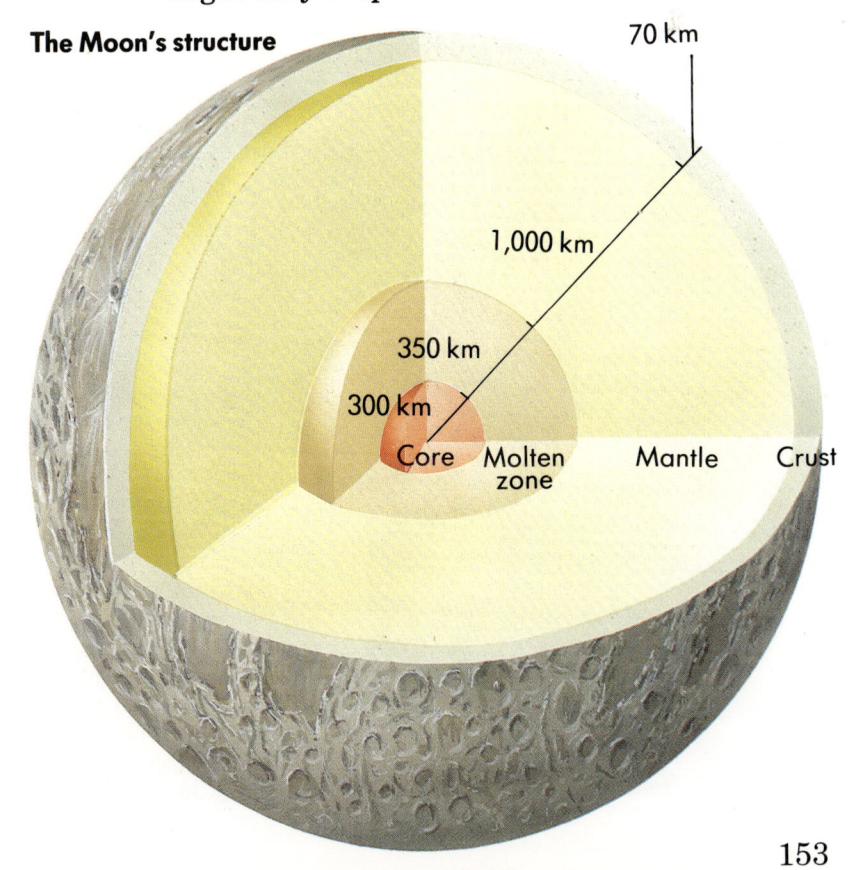

The lunar surface

These two photographs both show a full face of the Moon. But notice how different they are. The picture on the left shows the face of the Full Moon as we see it from Earth. Much of it is covered by the great lava plains we call seas. The picture above was taken from space by the *Apollo 11* astronauts. The right-hand part shows regions on the far side of the Moon which we can never see from the Earth. Notice that there are no large seas there, just rugged and cratered highlands.

Seas and highlands

Even with the naked eye we can see that the Moon's surface is made up of two main areas, dark and light. Through a telescope we can see that the dark areas are vast flat plains, while the light ones are rugged highlands.

The dark plains are vast sheets of lava. They were formed billions of years ago when massive meteorites slammed into the Moon and melted the rocks. Early astronomers thought these areas might be seas, and called them maria (Latin for "seas"). Some seas, such as the Sea of Crises, are circular and surrounded by mountains. Other seas merge together. The largest sea area is located in the north-west, where the Ocean of Storms, the Sea of Showers and the Sea of Clouds merge together.

The light-coloured highland areas of the Moon are part of the Moon's ancient crust. They are much more heavily cratered than the seas. Strangely, the far side of the Moon is almost entirely a highland region. It has only one sea of any size, the Sea of Moscow.

The highest regions on the Moon are the mountain ranges that border the seas. The Sea of Showers is ringed by the lofty Lunar Alps, Caucasus, Apennine and Carpathian ranges, in which some peaks rise to 6,000 metres.

We now know what the lunar surface is like, thanks to the 12 *Apollo* astronauts who walked on the Moon. They brought back some 385 kg of soil and rock. The dusty topsoil is made up of particles of rock smashed to pieces by meteorites. The main type of rock in the seas is a dark volcanic rock like basalt. The highlands are composed of a lighter volcanic rock. Everywhere there are breccias, rocks made up of a cemented mixture of rock chips.

▼ Man on the Moon. Harrison Schmitt examines a huge split boulder in the Taurus-Littrow Valley on the last *Apollo* mission, *Apollo 17*, in December 1972.

▶ The crater Eratosthenes, on the edge of the Sea of Showers, measures about 65 km across. Note the central mountain peaks, which are typical of large lunar craters.

Moons of the planets

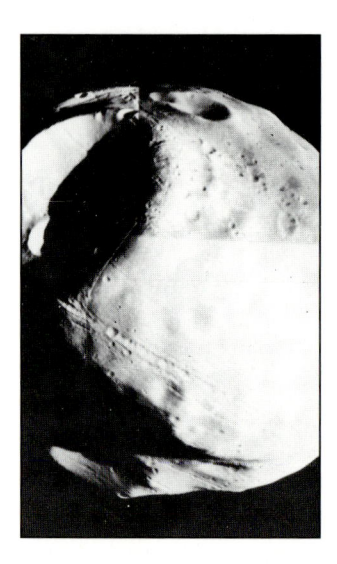

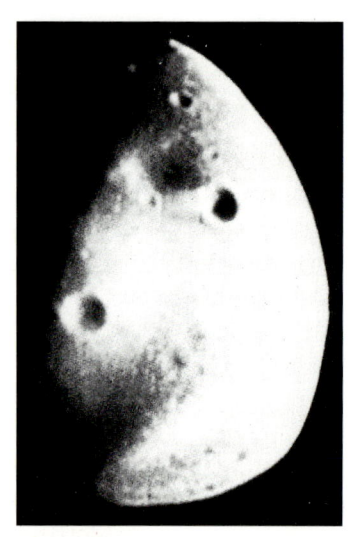

◀ The two tiny moons of Mars are Phobos (left) and Deimos. Unlike most moons, they are odd-shaped lumps of rock, and both are covered in craters. One huge crater on Phobos measures 10 km across.

Martian moons

The two moons of Mars are rocky bodies of irregular shape. They were probably asteroids which orbited close to Mars and were captured.

Even the largest moon, Phobos, is only about 28 km across. It orbits about 6,000 km above the surface and speeds around the planet in only about 7½ hours. This is much faster than Mars itself rotates (24⅔ hours). So from the Martian surface the moon would appear to move from west to east, the opposite way from usual. Deimos is about 16 km across. It circles some 20,000 km away in just over a day.

Jupiter's Galilean quartet

Apart from our own Moon, the easiest moons to spot are the four biggest moons of Jupiter. We can see them easily using good binoculars. The Italian astronomer Galileo first spotted them in the winter of 1609/10, when he trained his newly made telescope on the heavens. So they are called the Galilean satellites.

We cannot see any details of the other moons in the Solar System even through a telescope. They are too small and far away. Fortunately, space probes have visited many of the moons and sent back close-up photographs.

Most moons, like our own, orbit close to the plane of their parent planet's equator. And most travel in a nearly circular orbit in an anti-clockwise direction, when viewed from the north of the Solar System.

Active Io

Jupiter's large moon Io looks quite different from any other moon in the Solar System. It is a vivid orange-yellow, speckled with black (right). When the *Voyager* space probes flew past the moon, they took close-up pictures of active volcanoes.

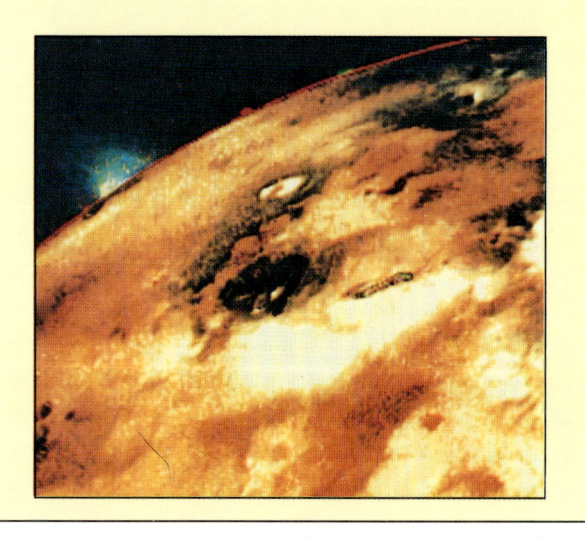

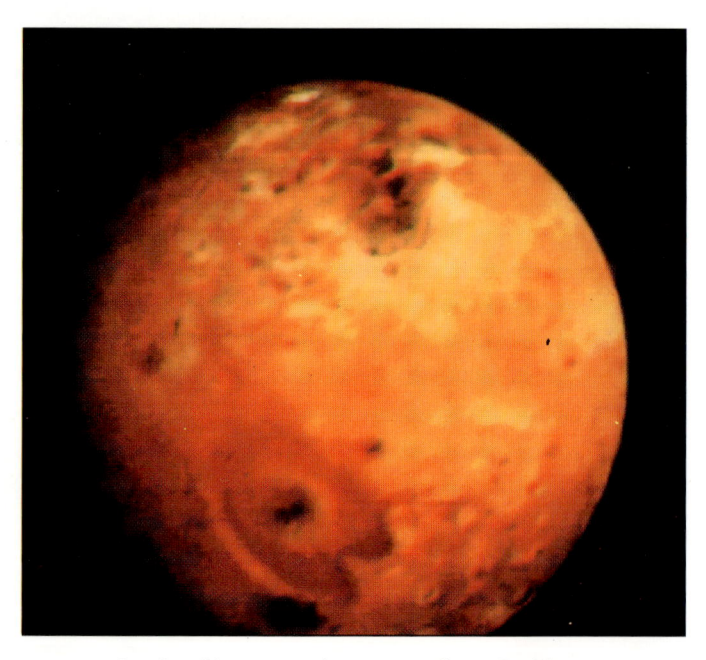

▲ Io is the third largest of Jupiter's four Galilean moons. But it is the most lively. Its orange-red surface is pitted with active volcanoes, which eject matter to heights of 250 km or more. The orange colour is thought to be caused by the presence of sulphur.

156

▶ On Jupiter's moon Ganymede, dark regions alternate with paler grooved areas. The bright craters show where meteorites have fallen to the surface and exposed fresh patches of white ice.

These four large moons are Io, Europa, Ganymede and Callisto, in order of distance from the planet. With a diameter of 5,276 km, Ganymede is the biggest moon in the Solar System, bigger even than the planet Mercury. Callisto is of similar size. Both are made up of a mixture of rock and ice.

Europa and Io are about the size of the Moon and are made up mainly of rock. But they could not look more different. Europa has a very smooth icy surface, criss-crossed with a network of dark lines. These are probably fractures in the ice that have become filled with darker material from underneath. Io is a vivid orange-yellow and is volcanically active. All the other 12 or so Jovian moons are very much smaller. Four of them orbit closer to the planet than Io. Two other groups of four moons orbit very much farther out. The four most distant moons have retrograde (backwards) and highly eccentric orbits. Astronomers reckon that they could well be captured asteroids.

▲ Callisto has an ancient crust, completely covered in craters.

▶ Europa has an icy crust, which is extremely smooth. There are few signs of any craters.

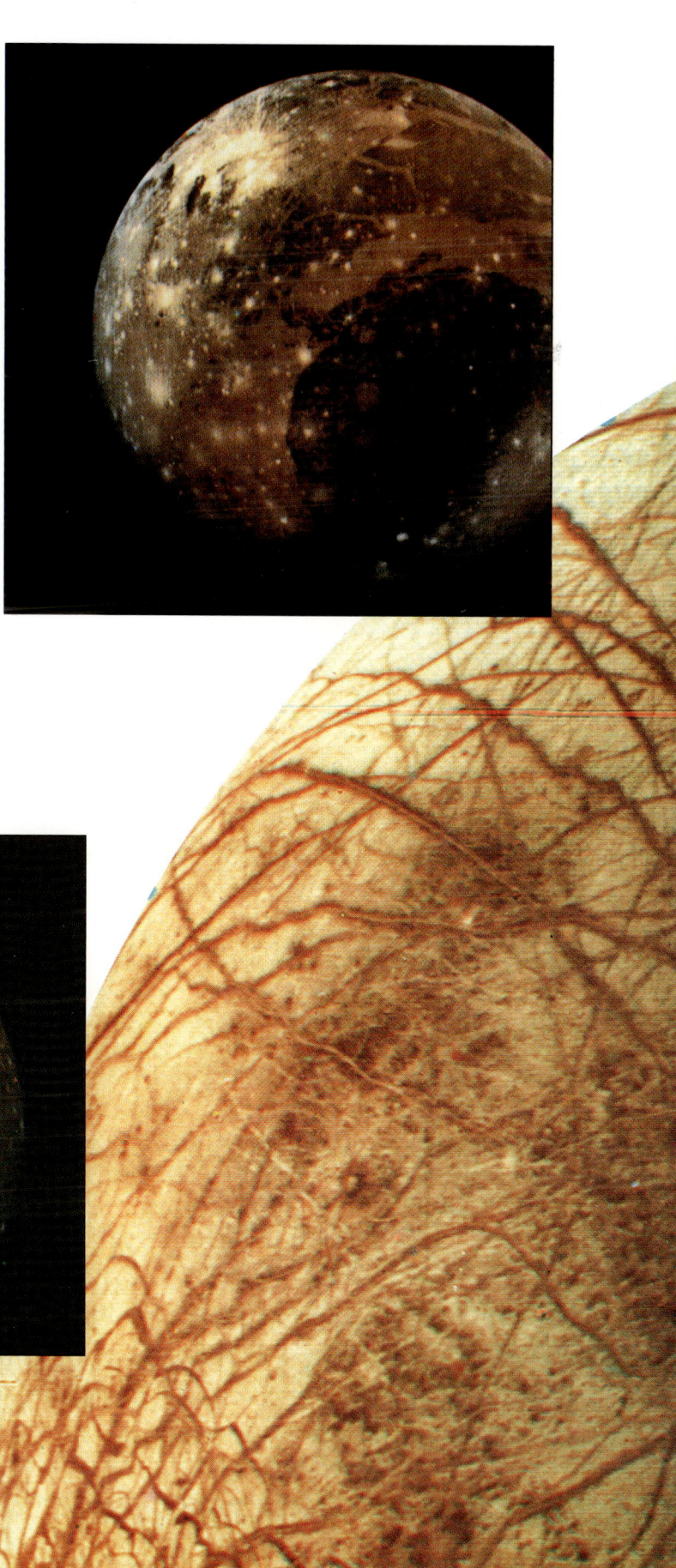

New discoveries

◀ Enceladus is the eighth farthest moon from its parent planet Saturn. Much of its surface is smooth, and it reflects light well. There are also many craters and deep grooves, which may be cracks in the crust.

▼ Titan is Saturn's largest moon by far, with a diameter of 5,150 km. It is unique among moons in having a thick atmosphere, of nitrogen and methane. Clouds in the atmosphere completely hide the surface. The orange colour of Titan is thought to be caused by a "smog" of hydrocarbon chemicals.

Saturn's satellites
Only five of Saturn's 22 or more moons have a diameter greater than 1,000 km. In order of distance from the planet they are Tethys, Dione, Rhea, Titan and Iapetus. By far the largest is Titan. After Jupiter's Ganymede, it is the largest moon in the Solar System, with a diameter of 5,150 km. It is particularly interesting because it has an atmosphere which is denser than Earth's.

Most of the moons are made up of rock and ice and are heavily cratered. Mimas has an enormous crater, 130 km across – a third of its diameter! The outermost moon, tiny Phoebe, is probably a captured asteroid.

The moons of Uranus
We can see only five of the 15 or so moons of Uranus through a telescope. Going out from the planet, they are Miranda, Ariel, Umbriel, Titania and Oberon. They seem to be made up of rock and ice in about equal proportions. All are covered in craters, some of which show up bright where fresh ice has been thrown out.

Miranda is by far the most interesting moon. Its surface is a patchwork of totally different kinds of regions, with sharp boundaries between them. Rolling cratered plains suddenly give way to curious grooved regions. In the past Miranda may have collided with an asteroid and been smashed to pieces. Then gravity pulled the pieces back together again to re-form the moon.

▲ Ariel, the fourth largest moon of Uranus, with a diameter of about 1,160 km. Its surface is scarred by long, deep faults. The bright areas show where ice has been exposed by bombarding meteorites.

▼ Pink "snow" covers most of the southern hemisphere of Neptune's moon Triton. The snow is a mixture of frozen methane and nitrogen.

Neptune's new moons

Viewed from Earth, Neptune appears to have only two moons. The largest and the one closest to the planet is Triton, which is only a little smaller than our own Moon. The other is Nereid, only about 170 km across.

In 1989 *Voyager 2* discovered a further six moons. One is rather bigger than Nereid and circles close to the planet. Four others are located inside Neptune's faint ring system.

Voyager's close-up view of Triton showed a fascinating world tinged with pink. It has a faint atmosphere of nitrogen. It is also the coldest known place in the Solar System, at a temperature of some −240°C.

Pluto's Charon

The American astronomer James Christie discovered Pluto's icy moon Charon in 1978. Its diameter (about 1,190 km) is no less than half that of Pluto! It orbits close to the planet, at a distance of only about 20,000 km. It makes one revolution about every six Earth-days, the same time it takes Pluto to spin on its axis.

Measuring the heavens

Spot facts

- The Sun lies some 150 million km away. Its light takes 8½ minutes to reach us, and 5½ hours to reach the farthest planet, Pluto.

- It would take the Space Shuttle, travelling at its usual speed in orbit, 100,000 years to reach the nearest star, Proxima Centauri, about 4¼ light-years away.

- Light from the Large Magellanic Cloud, the nearest galaxy, takes 170,000 years to reach the Earth.

- The farthest object in the heavens which we can see with the naked eye is the Andromeda galaxy. Its light takes more than 2 million years to reach the Earth.

▶ The night sky is ablaze with stars and shining clouds of gas and dust. This picture shows a starry vista in the constellation of Orion. Orion is one of the easiest star patterns to recognize.

People began stargazing thousands of years ago, a practice that grew into the science of astronomy. Perhaps surprisingly, the heavens have changed little over this period. Early astronomers saw the same kind of constellations, or star patterns, as we do today. They thought that the stars were fixed to the inside of a great celestial sphere that spun around the Earth. They had no idea of how big the Universe was. Only in this century have astronomers gained an idea of the true scale of the Universe. It is bigger than anybody can imagine.

Scale of the Universe

Simply by looking up at the night sky we can see that the Universe of stars and space is vast. But just how big is it? How far away are the stars and the galaxies we can see with our eyes and through telescopes?

Nobody had any real idea of the distances to the stars until 1838. In that year the German astronomer Friedrich Bessel used a method called parallax to measure the distance to a star in the constellation Cygnus. The distance turned out to be 105 million million km!

Light travels extremely quickly, at the fastest speed we know. Yet over such a vast distance, it takes the light from that star 11 years to reach us. Astronomers say that the star is 11 light-years away.

Other stars are up to tens of thousands of light-years away. The galaxies are even more remote. But the most remote objects of all appear to be the quasars. They lie up to 13,000 million light-years away at the edge of the observable Universe.

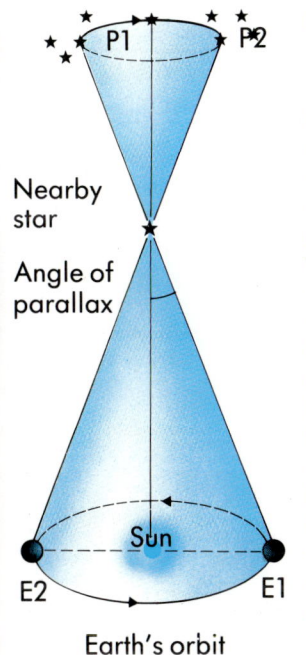

The infinite Universe

The Solar System

The nearby stars

The Galaxy

The Local Group

The local supercluster

The observable universe

How far away?

◄ We can measure the distance to nearby stars by using the method of parallax. This works on the principle that a nearby object appears to change its position (P1/P2) when seen from different viewpoints. We can view a nearby star from opposite sides of the Earth's orbit (E1/E2). From each side, the star appears in a different position against the background of distant stars. From the amount the star appears to shift, its distance can be calculated. The distance from Earth at which the parallax of a star is an angle of 1 second (equal to 1/3600 of a degree) is called a parsec.

► We can gain an idea of the vastness of the Universe by tracing how the Solar System fits into it, and looking at larger and larger areas of space. The Sun is part of a galaxy, which is part of a cluster of galaxies, which is part of a supercluster. Many superclusters make up the whole of the Universe.

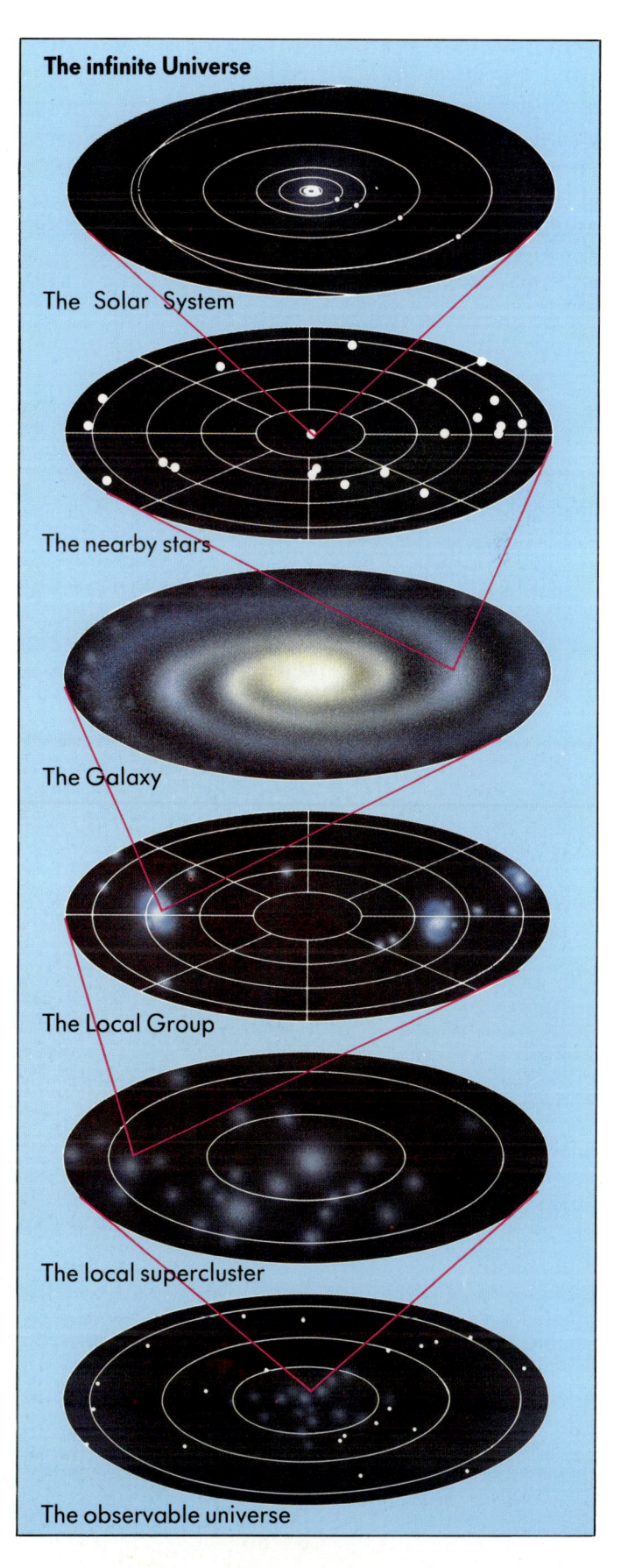

The celestial sphere

When we look at the night sky for any length of time, the stars appear to wheel round the sky. It is as if they were fixed to the inside of a great globe, or sphere, that rotates around the Earth. Ancient astronomers believed there was such a sphere. But we now know that it does not exist, and that it is the Earth rather than the stars that rotates. Nevertheless, astronomers find the idea of a celestial sphere extremely useful for pinpointing the positions of the stars.

The Earth revolves on its axis once every 24 hours, relative to the Sun. This is the basis of our ordinary, solar time. But during the same period the Earth moves slightly along its orbit. And relative to the stars it takes only 23 hours 56 minutes to revolve once. This period is known as the sidereal day.

Astronomers use this period as the basis of star time, or sidereal time. Stars rise and set at the same sidereal time each day, and reach the

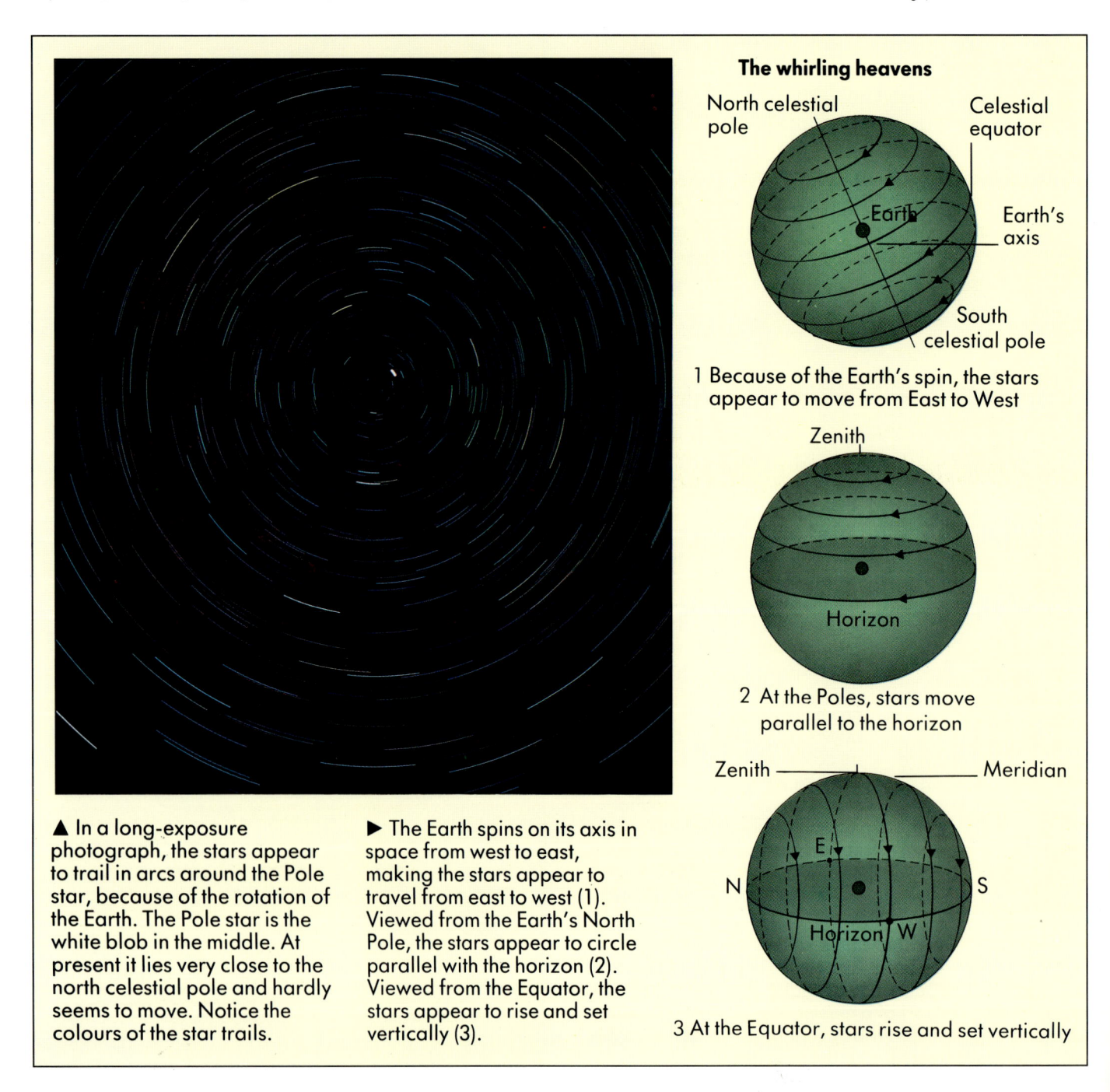

The whirling heavens

1 Because of the Earth's spin, the stars appear to move from East to West

2 At the Poles, stars move parallel to the horizon

3 At the Equator, stars rise and set vertically

▲ In a long-exposure photograph, the stars appear to trail in arcs around the Pole star, because of the rotation of the Earth. The Pole star is the white blob in the middle. At present it lies very close to the north celestial pole and hardly seems to move. Notice the colours of the star trails.

▶ The Earth spins on its axis in space from west to east, making the stars appear to travel from east to west (1). Viewed from the Earth's North Pole, the stars appear to circle parallel with the horizon (2). Viewed from the Equator, the stars appear to rise and set vertically (3).

same points in the sky at the same sidereal time. So sidereal time provides a means of locating the stars on the celestial sphere.

The revolving celestial sphere
The celestial sphere rotates about the same axis as the Earth. The north and south celestial poles are located where the Earth's axis meets the sphere, vertically above the Earth's poles. The celestial equator is the circle where the plane of the Earth's Equator meets the sphere. The stars appear to travel in circles parallel with the celestial equator. During the year the Sun appears to travel around the sphere in a circle called the ecliptic. Its path crosses the celestial equator at two points – the equinoxes – the spring (vernal) equinox on about 21 March and the autumnal equinox on about 23 September. On these dates throughout the world, day and night are of equal length.

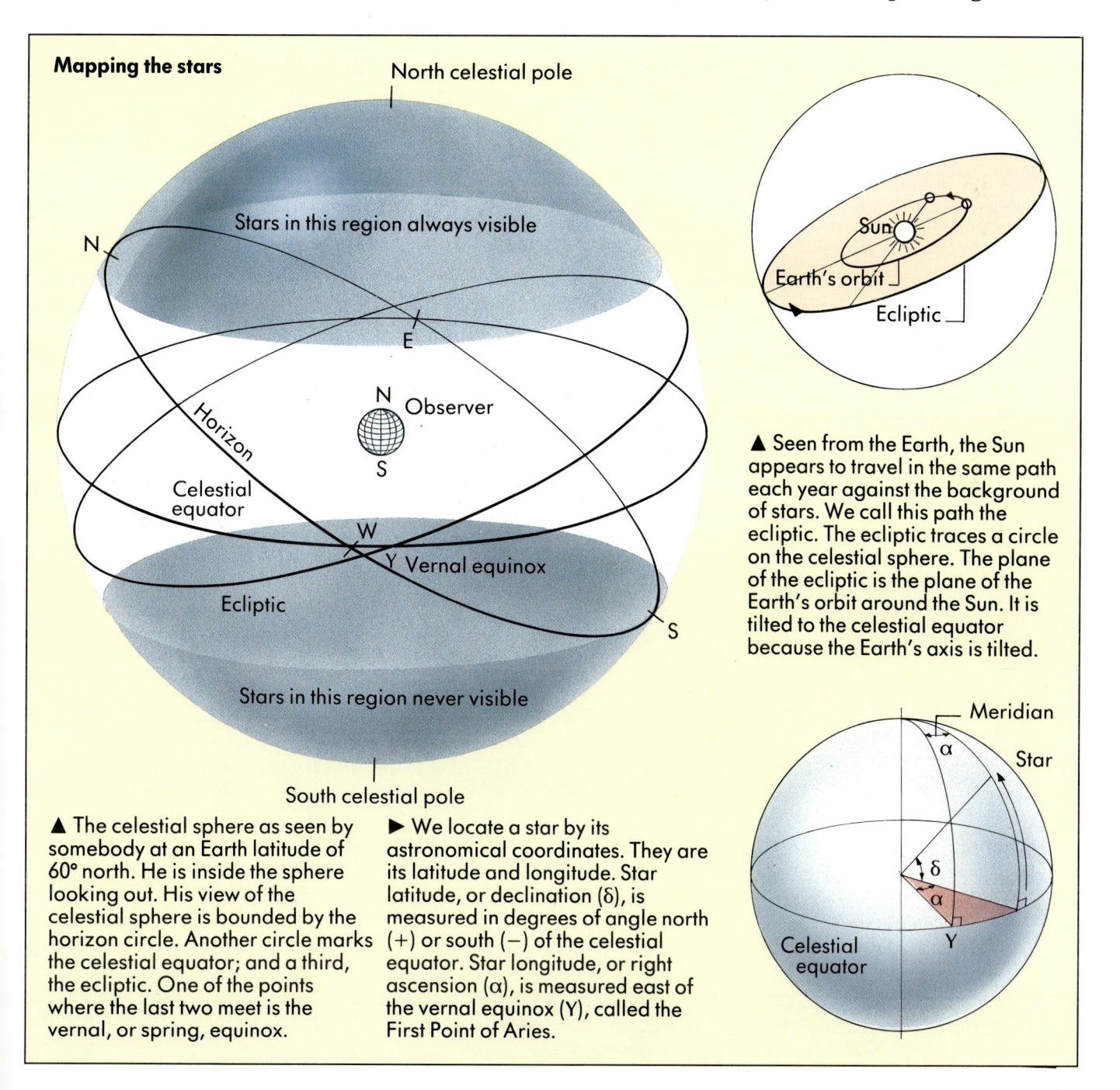

Mapping the stars

▲ The celestial sphere as seen by somebody at an Earth latitude of 60° north. He is inside the sphere looking out. His view of the celestial sphere is bounded by the horizon circle. Another circle marks the celestial equator; and a third, the ecliptic. One of the points where the last two meet is the vernal, or spring, equinox.

▶ We locate a star by its astronomical coordinates. They are its latitude and longitude. Star latitude, or declination (δ), is measured in degrees of angle north (+) or south (−) of the celestial equator. Star longitude, or right ascension (α), is measured east of the vernal equinox (Υ), called the First Point of Aries.

▲ Seen from the Earth, the Sun appears to travel in the same path each year against the background of stars. We call this path the ecliptic. The ecliptic traces a circle on the celestial sphere. The plane of the ecliptic is the plane of the Earth's orbit around the Sun. It is tilted to the celestial equator because the Earth's axis is tilted.

Constellations

Some of the brightest stars appear to make patterns in the sky, which we call the constellations. The patterns remain much the same year after year, and change only very slowly. But usually the stars in a constellation are not close to each other in space. They appear that way because they lie in the same direction.

The earliest stargazers in ancient Babylon named the northern constellations they could see after figures they thought the star patterns looked like. The names passed to ancient Egypt, Greece and Rome. We still use the Latin names for them. Some are animals: Ursa Major, the Great Bear; Cygnus, the Swan; and Leo, the Lion. Some are mythological people, such as Hercules, Orion, and Andromeda. Others are everyday objects, such as Crater, the Cup; and Libra, the Scales.

Some individual stars within a constellation have names of their own. For example, the brightest star in the constellation Canis Major, the Great Dog, is called Sirius. But in general astronomers identify a star in a constellation by a Greek letter, according to its brightness or position. The brightest is usually labelled alpha (α), the next brightest beta (β), and so on.

We measure the brightness of a star on a system pioneered by the ancient Greeks, who grouped stars they could see into six categories of brightness. The brightest were described as 1st magnitude and the dimmest star as 6th magnitude. For exceptionally bright stars the scale is extended backwards to give negative values. Sirius, for example, has a magnitude of −1.45. For stars too dim to see without a telescope, the scale is extended forwards beyond 6.

◀ This illustration from an early book on astronomy shows constellations of the northern hemisphere. The artist has used a vivid imagination to create suitable figures to match the pattern of bright stars. The Great Bear (Ursa Major) stands out clearly, as it does in the heavens. A circle marks the ecliptic, and the constellations it goes through are known as the constellations of the zodiac. These play an important role in astrology.

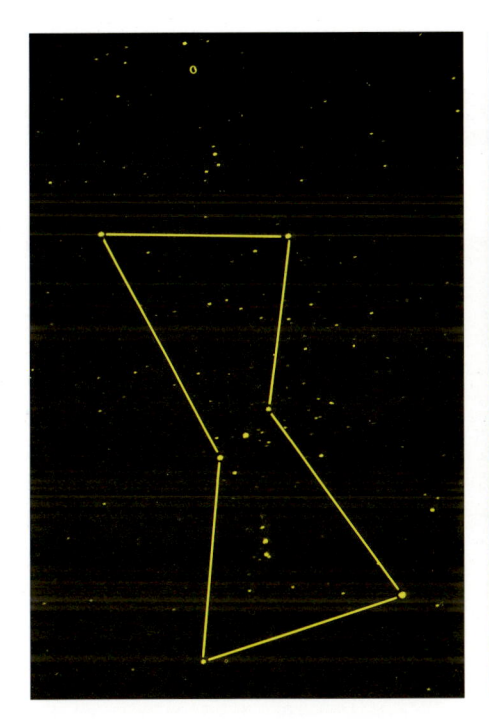

▲ The constellation of Orion, the Mighty Hunter, sits on the celestial equator and is easy to spot. The three stars across the middle form Orion's Belt.

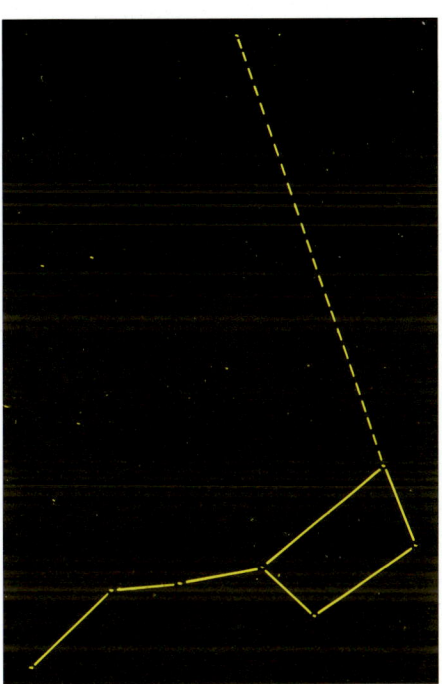

▲ The Plough is part of the constellation of the Great Bear (Ursa Major). The two stars at the end of the "ploughshare", pointing towards the Pole star, are the Pointers.

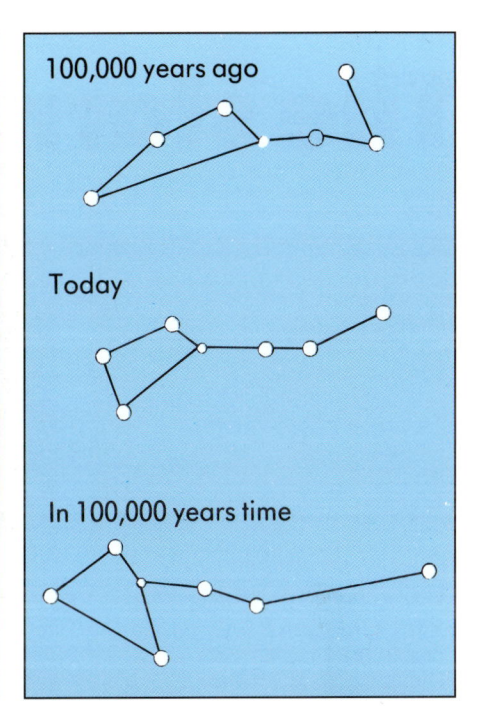

▲ The seven main stars that make up the Plough lie at different distances in space and move at different speeds. That is why the shape of the constellation gradually changes.

▲ An astrologer's star signs from the island of Bali. Astrologers believe that people's lives are affected by the positions of the planets among the constellations.

The main constellations

Latin name	Common name	Latin name	Common name
Andromeda		Dorado	Swordfish
Aquarius	Water-Bearer	Draco	Dragon
Aquila	Eagle	Eridanus	
Aries	Ram	Gemini	Twins
Auriga	Charioteer	Hercules	
Boötes	Herdsman	Hydra	Water Serpent
Camelopardus	Giraffe	Leo	Lion
Cancer	Crab	Libra	Scales
Canes Venatici	Hunting Dogs	Lyra	Lyre
Canis Major	Great Dog	Ophiuchus	Serpent-Bearer
Canis Minor	Little Dog	Orion	
Capricornus	Sea Goat	Pegasus	Flying Horse
Carina	Keel	Perseus	
Cassiopeia		Pisces	Fishes
Centaurus	Centaur	Puppis	Poop
Cepheus		Sagitta	Arrow
Cetus	Sea Monster	Sagittarius	Archer
Columba	Dove	Scorpius	Scorpion
Coma Berenices	Berenice's Hair	Serpens	Serpent
Corona Australis	Southern Crown	Sextans	Sextant
Corona Borealis	Northern Crown	Taurus	Bull
Corvus	Crow	Triangulum	Triangle
Crater	Cup	Ursa Major	Great Bear
Crux	Southern Cross	Ursa Minor	Little Bear
Cygnus	Swan	Vela	Sails
Delphinus	Dolphin	Virgo	Virgin

165

The starry heavens

Spot facts

• More than 5,500 stars can be seen with the naked eye.

• One of the most massive stars known is Plaskett's star (HD 47129), in the constellation Monoceros, the Unicorn. It is a binary system, in which each star has more than 50 times the mass of the Sun.

• The most powerful star is the variable Eta Carinae, in the constellation Carina, the Keel. In the mid-1800s it became 4 million times brighter than the Sun.

• Barnard's star moves across the celestial sphere faster than any other star. Its movement (proper motion) is a little over 10 seconds of arc (1/360 of a degree) per year.

▶ Stars mass together in their thousands in the far southern constellation of Vela, the Sails. In certain places, patches of gas and dust show up as wispy clouds.

The velvety black heavens are studded with stars of every description. There are massive red supergiants hundreds of times bigger than the Sun, and blazing hot blue stars that shine 10,000 times more brightly. There are stars that wink as regularly as clockwork, stars with companions, and stars that cluster together in their thousands. Astronomers today can tell us all manner of things about these stars – their size, mass, colour, temperature, speed and distance. All this, and more, they find out from faint smudges of starlight.

Varieties of stars

All the stars we see in the night sky belong to the Milky Way. This is the star system, or galaxy, to which our Sun belongs. Like the Sun, the stars are great globes of searing hot gas, which pour out energy as light, heat and other forms of radiation. They are enormous distances away, and even in telescopes appear only as tiny points of light.

The light from a star is very feeble by the time it arrives at the Earth. Yet it can be made to reveal many of the star's secrets. Starlight can be gathered by a telescope and fed to an instrument called a spectroscope. The spectroscope splits the light into a spectrum in which a number of dark lines appear.

An enormous amount of information can be gained from the spectrum, such as the chemical composition of the star. The star's temperature can be found by observing which colours in the spectrum are brightest. The coolest stars are red, the hottest blue-white. The blue-white giant star Spica has a surface temperature of about 25,000°C, which is over four times the Sun's surface temperature. Stars of similar temperature also display similar spectra, in which certain lines are prominent.

▶ Stars of all sizes, colours and ages crowd into this photograph of part of the constellation of Sagittarius, the Archer.

▼ The face of a giant. This is a picture of one of the biggest stars we can see in the night sky. It is a red supergiant called Betelgeuse in the constellation Orion.

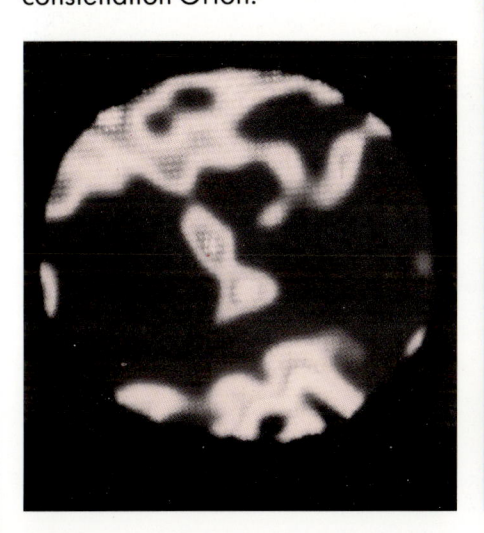

Stellar spectrum

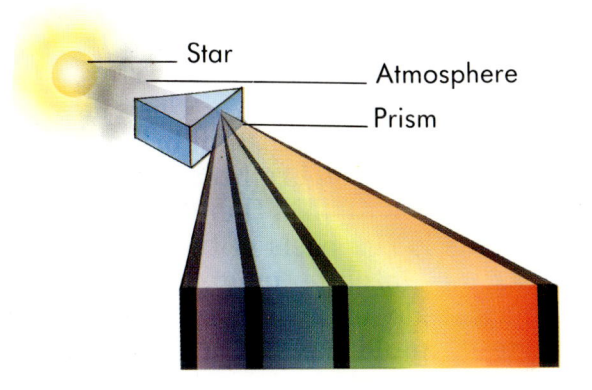

Dark-line spectrum

The hot surface of a star gives out light of all wavelengths. If this light were passed through a prism, a complete spectrum, or rainbow, of colour would be produced. But before the star's light reaches us, it has to pass through the star's outer atmosphere of cool gases. The various chemical elements in the atmosphere absorb certain wavelengths (or colours) from the light. As a result, sets of dark lines appear in the star's spectrum. From the positions of the lines, we can tell which elements caused them to appear. The lines in the Sun's spectrum are known as Fraunhofer lines after the German optician Josef von Fraunhofer, who first studied them.

Brightness, size and speed

True brightness

We describe how bright a star is in the sky by its magnitude. It is an apparent magnitude – because it is a measure of the star's brightness as it appears to us. It depends on how far away the star is as well as on the star's true brightness. For this reason, a close dim star may look brighter than a distant bright star.

To compare the true brightness of stars, we would have to look at all of them from the same distance. This is the basis of the scale of true brightness, or absolute magnitude. The absolute magnitude of a star is the apparent magnitude it would have at a distance of 10 parsecs, or about 33 light-years away.

Sirius, the brightest star to our eyes, has an apparent magnitude of -1.45. But as stars go it is really not particularly bright, having an absolute magnitude of only $+1.41$. Deneb, the brightest star in Cygnus, has an apparent magnitude of $+1.25$, but an absolute magnitude of -7.3. It is one of the brightest of all the stars. On the absolute scale the Sun is fairly dim, with a magnitude of only 4.8.

We can calculate a star's absolute magnitude from its apparent magnitude when we know how far away it is. This is because we know how brightness changes with distance.

Star size

The absolute magnitude is a measure of a star's luminosity, or how luminous it is. This depends on how hot the surface is and on the surface area. If we know the temperature of a star and its luminosity (absolute magnitude), we can work out its surface area and diameter.

We find that stars vary enormously in size. The Sun has a diameter of some 1,400,000 km. As stars go, it is fairly small, and is classed as a dwarf. Many stars have a similar size. But some are very much larger, and others very much smaller. Supergiant stars can be 500 times larger in diameter, whereas white dwarfs can be 100 times smaller.

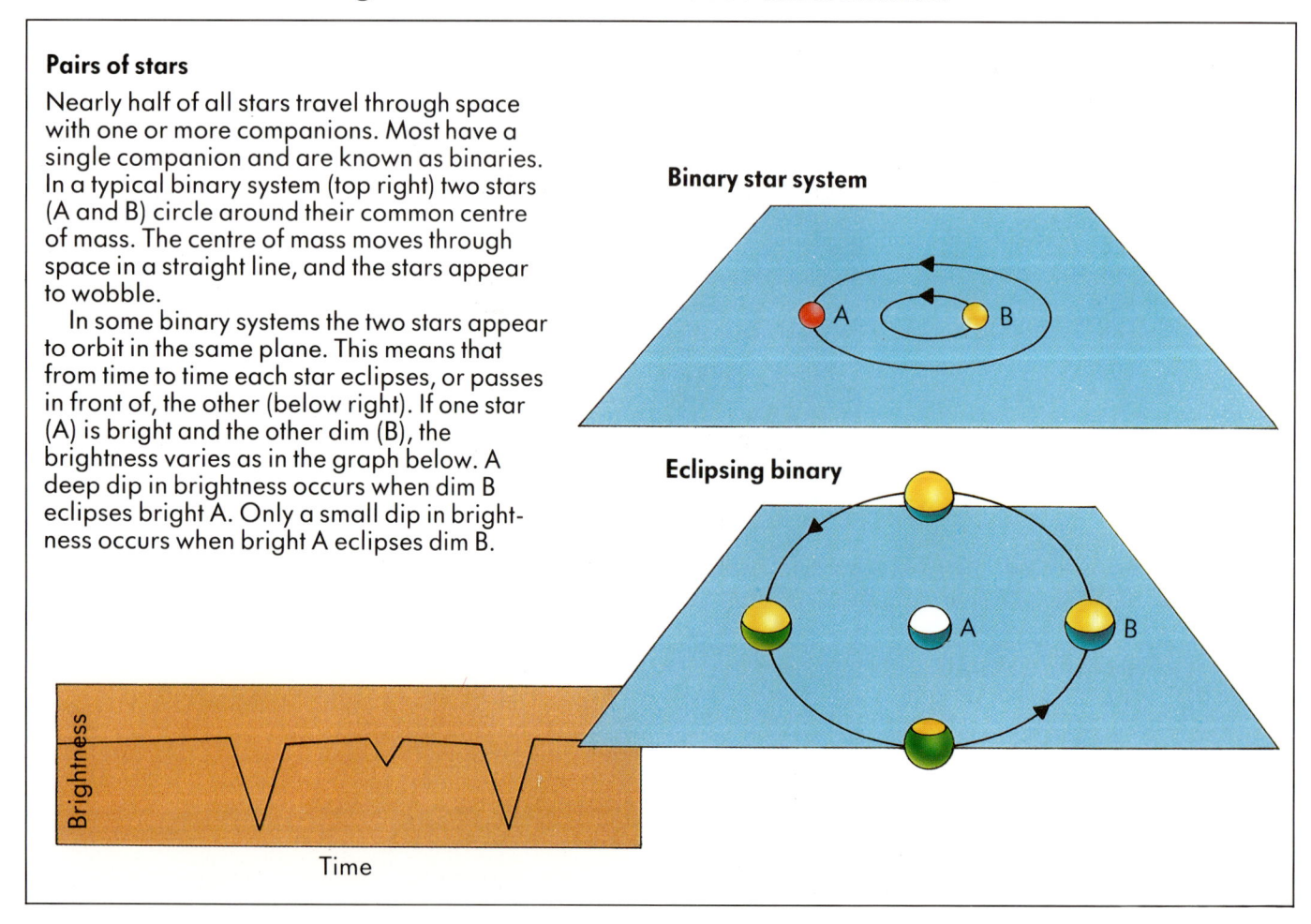

Pairs of stars

Nearly half of all stars travel through space with one or more companions. Most have a single companion and are known as binaries. In a typical binary system (top right) two stars (A and B) circle around their common centre of mass. The centre of mass moves through space in a straight line, and the stars appear to wobble.

In some binary systems the two stars appear to orbit in the same plane. This means that from time to time each star eclipses, or passes in front of, the other (below right). If one star (A) is bright and the other dim (B), the brightness varies as in the graph below. A deep dip in brightness occurs when dim B eclipses bright A. Only a small dip in brightness occurs when bright A eclipses dim B.

Binary star system

Eclipsing binary

▲ The double star cluster in Perseus, called the Sword Handle, is an open cluster containing about 350 stars.

▶ This open star cluster in Crux is known as the Jewel Box because its stars flash different colours and sparkle like jewels.

Finding the mass of a star is difficult. But when a star is one of a pair in a binary system its mass can be found. It can be calculated from the distance the two stars are apart and the time it takes them to revolve around each other. Many stars have a similar mass to the Sun. Some have only one-tenth the Sun's mass, and others are 10 times more massive.

Stars on the move

The stars appear to be fixed in the sky, but they are travelling rapidly through space. But most are so far away that their movement cannot be detected, even after hundreds of years.

In general stars move towards us or away from us at an angle. So in time they should show a movement across our line of sight. We can detect this movement, called proper motion, for a few nearby stars, but it is only slight.

We can detect a star's movement towards or away from us by examining its spectrum. We call this movement radial motion. When a star is moving towards us, the lines in its spectrum are displaced towards the blue end. When a star is moving away, the lines are displaced towards the red end. The amount of blue or red shift is a measure of how fast the star is moving.

Globular clusters

Some of the most spectacular of all heavenly bodies are globular clusters. They consist of hundreds of thousands, even millions, of stars packed closely together in space in the shape of a globe. The brightest in the Northern Hemisphere is M13 (above), in Hercules. It is made up of about half a million stars. As in most globular clusters, the stars are very old. M13 is one of about 200 globular clusters found in the great halo that surrounds the centre of our Galaxy.

The H-R diagram

As we have seen, two important features of a star are its temperature and its true brightness. Temperature is represented by the spectral class, and brightness by the absolute magnitude, or luminosity.

During the early part of this century, two astronomers were investigating how the two are related. They were the Dane Ejnar Hertzsprung and the American Henry Russell.

In 1914 they published diagrams in which they plotted luminosity against spectral class for a number of stars. This type of diagram helped astronomers gain a new insight into the way stars are formed and how they gradually change during their lifetimes.

The illustration shows the Hertzsprung-Russell (H-R) diagram for some of the brightest and some of the dimmest stars. Diagonal lines across the diagram indicate the size of the stars compared with the Sun.

The most striking feature of the H-R diagram is that most stars lie along a diagonal band. We call this band the main sequence. Stars on the main sequence shine steadily. They include, about half-way down, our own Sun. Stars at the top of the band are very hot, with a surface temperature approaching 30,000°C. They are also very bright – up to 10,000 times brighter than the Sun. In contrast, the dim red stars at the bottom of the band are only about one ten-thousandth as bright as the Sun.

Giants and dwarfs

A number of stars lie off the main sequence. In the upper right are the cool but highly luminous group known as red giants because of their enormous size. The stars above them are even bigger and brighter supergiants. In contrast, below the main sequence are a group of hot, but tiny, white dwarf stars. These represent a late stage in the life of stars similar to the Sun.

▶ On the Hertzsprung-Russell diagram, the stars are bigger and brighter towards the top of the diagram. Going from right to left they get hotter. The inset diagrams show the very great differences in the sizes of stars. A supergiant has a diameter several hundred times that of the Sun. On the other hand the Sun is about 100 times bigger than a white dwarf. A neutron star is a thousand times smaller still, only about 20 km across. A black hole may be even smaller.

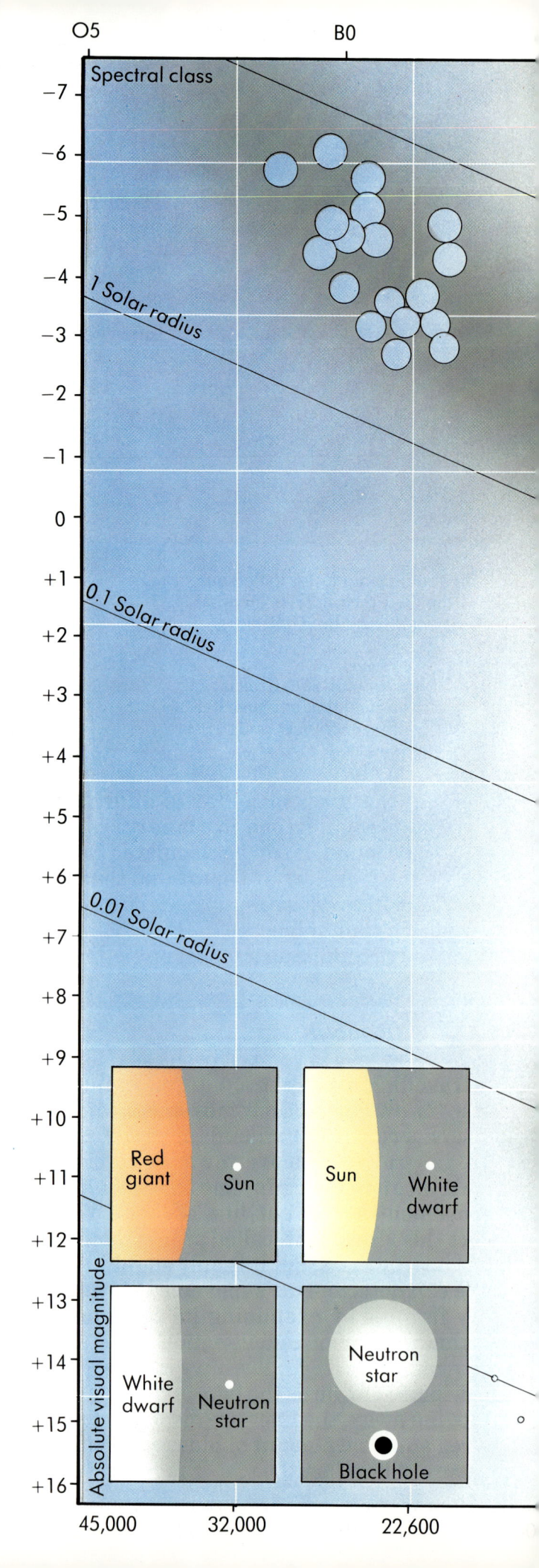

170

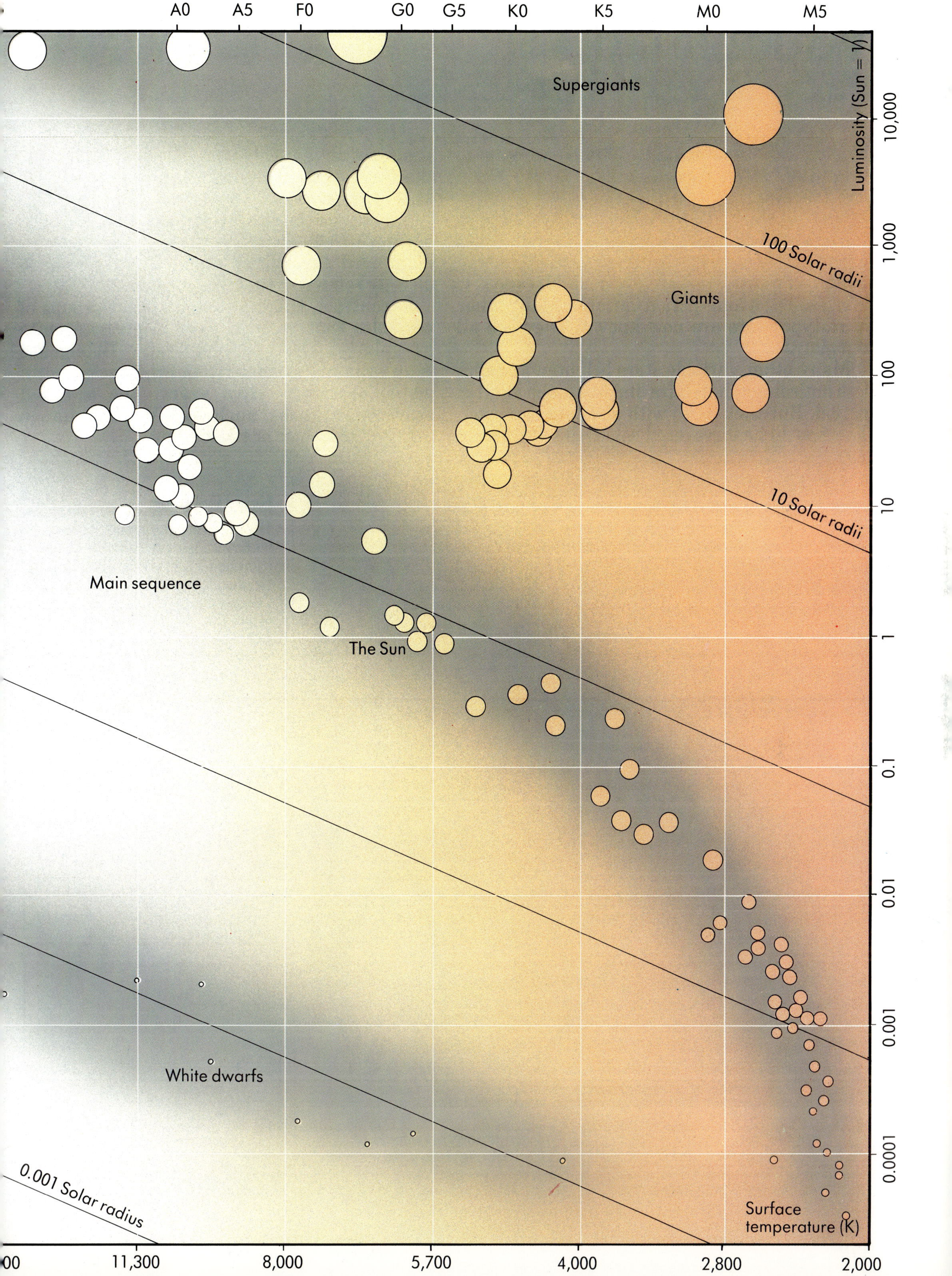
A0
A5
F0
G0
G5
K0
K5
M0
M5
Supergiants
Luminosity (Sun = 1)
10,000
1,000
100
10
1
0.1
0.01
0.001
0.0001
100 Solar radii
Giants
10 Solar radii
Main sequence
The Sun
White dwarfs
0.001 Solar radius
Surface temperature (K)
11,300
8,000
5,700
4,000
2,800
2,000

Variable stars

Most stars shine steadily. But some vary in brightness from time to time. We call them variable stars.

One of the first variables discovered was Mira, in the constellation Cetus. It changes in brightness between about magnitude 3 and magnitude 10 over a period of about 331 days. It is classed as a long-period variable. Many other Mira-type stars are now known. They are all red giants.

Many other variable stars brighten and dim less noticeably than Mira and in a much shorter period. But they do so with absolute precision. The first to be discovered was Delta Cephei, in the constellation Cepheus. It gave its name to the class of short-period variables, which are known as Cepheids.

Typical, or classical, Cepheids have a period of from 1 to 60 days. They are young, massive giant and supergiant stars. Their period is directly related to their magnitude.

In contrast, there are many other groups of variables that vary in their brightness quite irregularly. The supergiant star Betelgeuse is an example. It changes in brightness over a period of roughly five years. The so-called eruptive variables, such as UV Ceti, change in brightness once or twice a day.

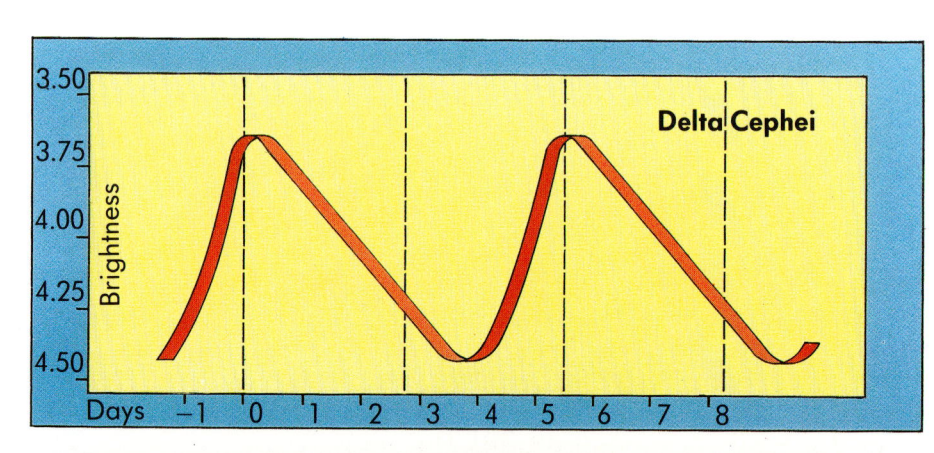

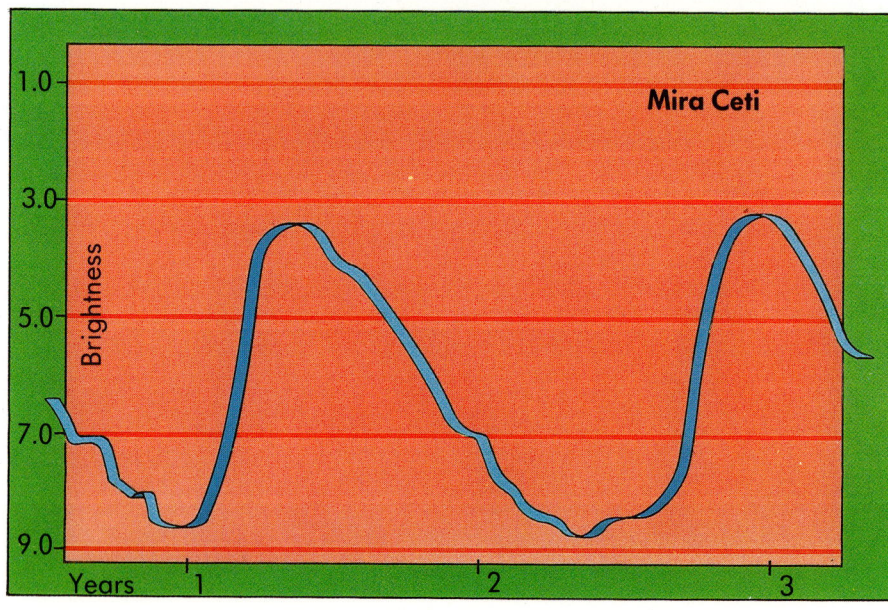

Noted stargazer

The American astronomer Henrietta Leavitt became famous for her work with Cepheid stars at Harvard Observatory. In 1912 she made the discovery of the period-luminosity law for these stars. She found that the period of a Cepheid is directly related to its absolute brightness, or luminosity. The longer its period, the greater is its luminosity. This law enables us to measure the distance to a Cepheid, where one is found.

▲ The brightness of Cepheid stars varies in a regular way. Delta Cephei itself (top graph) varies in brightness between magnitudes 3.5 and 4.3 in a period of precisely 5.37 days.

▲ The variation in brightness of Mira Ceti (above) is irregular over a much longer period. It varies between about magnitudes 3 and 10 about every 11 months.

Bright nebulae

The space between the stars is not completely empty. It contains minute traces of gases and tiny grains of dust. In places this interstellar matter clumps together to form denser clouds, which we call nebulae.

Some nebulae contain very hot stars and shine brilliantly. Energy absorbed from the stars causes the gases present to emit light. We call this type emission nebulae. We can see one with the naked eye in the constellation Orion. It is called the Great Nebula in Orion, or M42.

When a star lies outside a gas and dust cloud, the cloud may reflect starlight. We then see it as a reflection nebula.

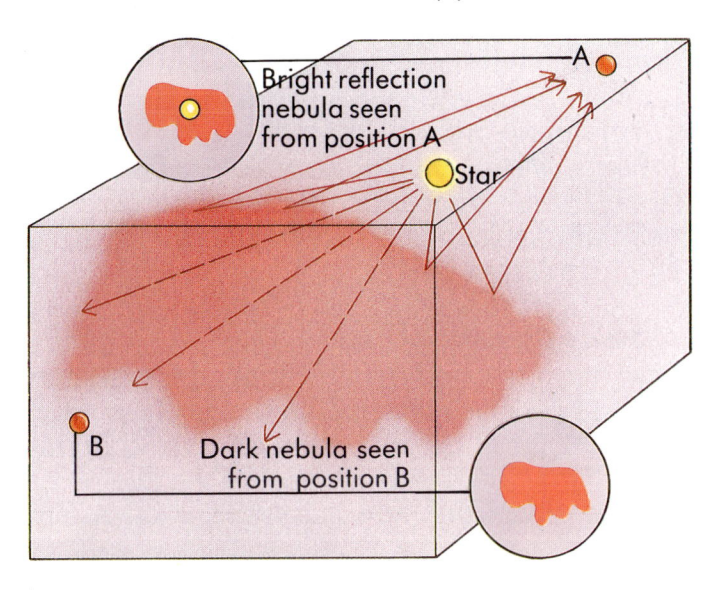

▼ A photograph of part of a gigantic gas and dust cloud that lies approximately 700 light-years away near the star Rho Ophiuchi (top left). The dark regions show where the cloud is thickest. The blue area is a reflection nebula, and the pink colour is due to the presence of the gas hydrogen.

▼ When a star lies in front of a cloud of dust, the dust reflects light and we see a bright reflection nebula (A). If the cloud lies between us and the star, the star's light is blocked and we see a dark nebula (B).

Between the stars

Dark nebulae

There are many dust clouds in the heavens that are not lit up by nearby stars. We can see them as dark patches against a starry background. They blot out the light of the stars behind them. We call them dark nebulae. Two of the best-known ones, the Coal Sack and the Horsehead, are pictured on these pages.

Some dark nebulae, called globules, are round, small and much denser than ordinary nebulae. Astronomers believe that they will one day turn into stars.

A dusty disc

Dust does not only occur clumped together in clouds. It is also scattered haphazardly in space between the stars. It has the effect of dimming their light. There is much dust in the disc of our Galaxy, the Milky Way. It masks our view of the centre of the Galaxy.

Interstellar dust appears to be made up of specks of carbon or of silicates (like many Earth rocks). Often the specks are coated with ice. Traces of other substances are also found in between the stars, including water, alcohol and glycine. Glycine is an amino acid, one of the building blocks of life.

▼ The constellation of the Southern Cross stands out in this view of part of the Milky Way in the Southern Hemisphere. Near the two brightest stars of the Cross is what appears to be a dark "hole". In fact it is a dark nebula, called the Coal Sack.

Molecules between the stars

Astronomers have discovered more than 50 different chemicals in the gas and dust clouds between the stars. Some of their molecules show up in the spectra of the light the clouds give out. Others make themselves known by the radio waves they emit. It is interesting that there are several carbon compounds among the molecules, because such compounds are the basis of life on Earth.

Name	Formula
Cyanogen	CN
Hydroxyl	OH
Ammonia	NH_3
Water	H_2O
Methanal (formaldehyde)	H_2CO
Carbon monoxide	CO
Hydrogen	H_2
Hydrogen cyanide	HCN
Methanol (methyl alcohol)	CH_3OH
Methanoic acid (formic acid)	HCO_2H
Silicon monoxide	SiO
Ethanal (acetaldehyde)	CH_3CHO
Hydrogen sulphide	H_2S
Methoxymethane (dimethyl ether)	CH_3OCH_3
Ethanol (ethyl alcohol)	CH_3CH_2OH
Sulphur dioxide	SO_2
Ethyl cyanide	CH_3CH_2CN
Nitric oxide	NO

▶ Of all the dark nebulae, none is better known or better named than the Horsehead, in Orion. The dark cloud of dust in the shape of a horse's head stands out vividly against a bright emission nebula.

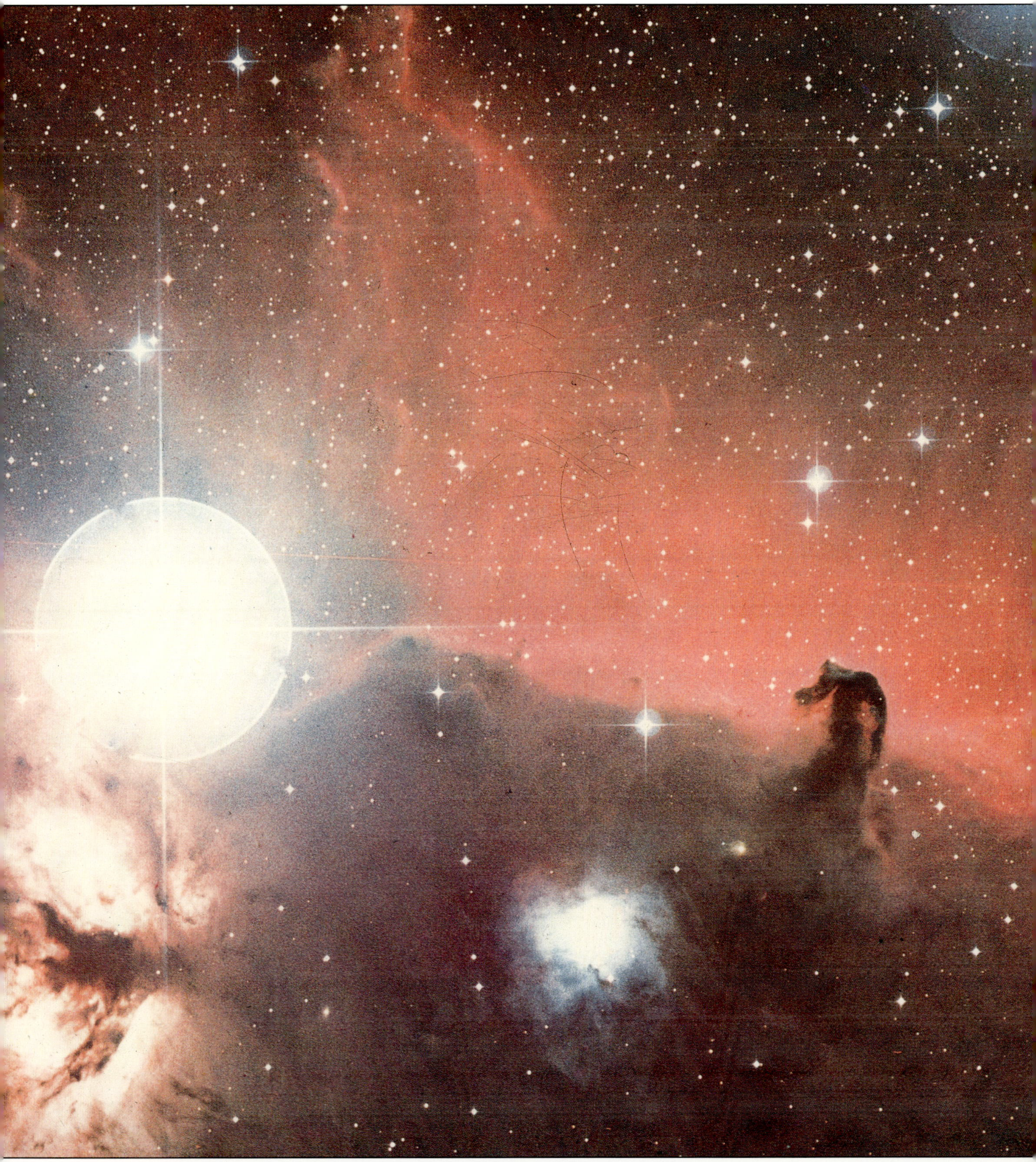

Birth and death of stars

Spot facts

- In about 5,000 million years time the Sun will swell up into a red giant. It will certainly expand beyond the orbit of Mercury and maybe even beyond that of Venus.

- The first white dwarf to be recognized (in 1915) was the faint companion of the Dog Star, Sirius, brightest star in the heavens. Properly termed Sirius B, it is often called the Pup. It is about the size of Earth, but is 350,000 times more massive.

- In 1982 astronomers discovered the first millisecond pulsar, PSR 1937 + 21. It flashes on and off every 1.56 milliseconds (thousandths of a second) as it spins 642 times per second.

▶ The Helix nebula in the constellation Aquarius. It is a puff of gas blown long ago by the star you see in the centre. It is a beautiful example of a planetary nebula.

Stars are born in the great clouds of gas and dust that are scattered throughout the galaxies. When their nuclear furnaces light up, they start to shine. Some glow feebly; others blaze like great celestial beacons. The bigger and brighter a star is, the shorter is its lifetime, and the more spectacular is its death. Stars like the Sun meet quite a peaceful end, shrinking into a super-dense white dwarf as small as a planet. But more massive stars go out with a bang. They blast themselves apart in a mighty supernova explosion and briefly shine nearly as bright as a galaxy.

A star is born

Nobody knows quite what triggers a cloud of gas and dust to turn into a star. As the cloud collapses, energy is released, which causes it to heat up. The centre of the cloud reaches a temperature of 10 million degrees or more.

Most of the gas in interstellar clouds is hydrogen. And at such high temperatures, the hydrogen atoms start to combine, or fuse together. This fusion reaction produces enormous amounts of energy as light, heat and other radiation. When this happens, the collapsing cloud starts to shine as a star.

The outward "pressure" of the radiation coming from the core of the new star acts against the matter that is collapsing under gravity. Eventually the two balance each other, and the collapse ceases. The star settles down and begins to shine steadily. It takes a star the size of the Sun about 50 million years to reach this state.

A Sun-sized star shines steadily for about 10,000 million years, until the hydrogen fuel in its core is used up. The star then begins to collapse again under gravity. The heat triggers off hydrogen fusion in the gassy shell surrounding the core. The shell heats up, causing the star to expand and brighten. But the core continues to shrink and get hotter.

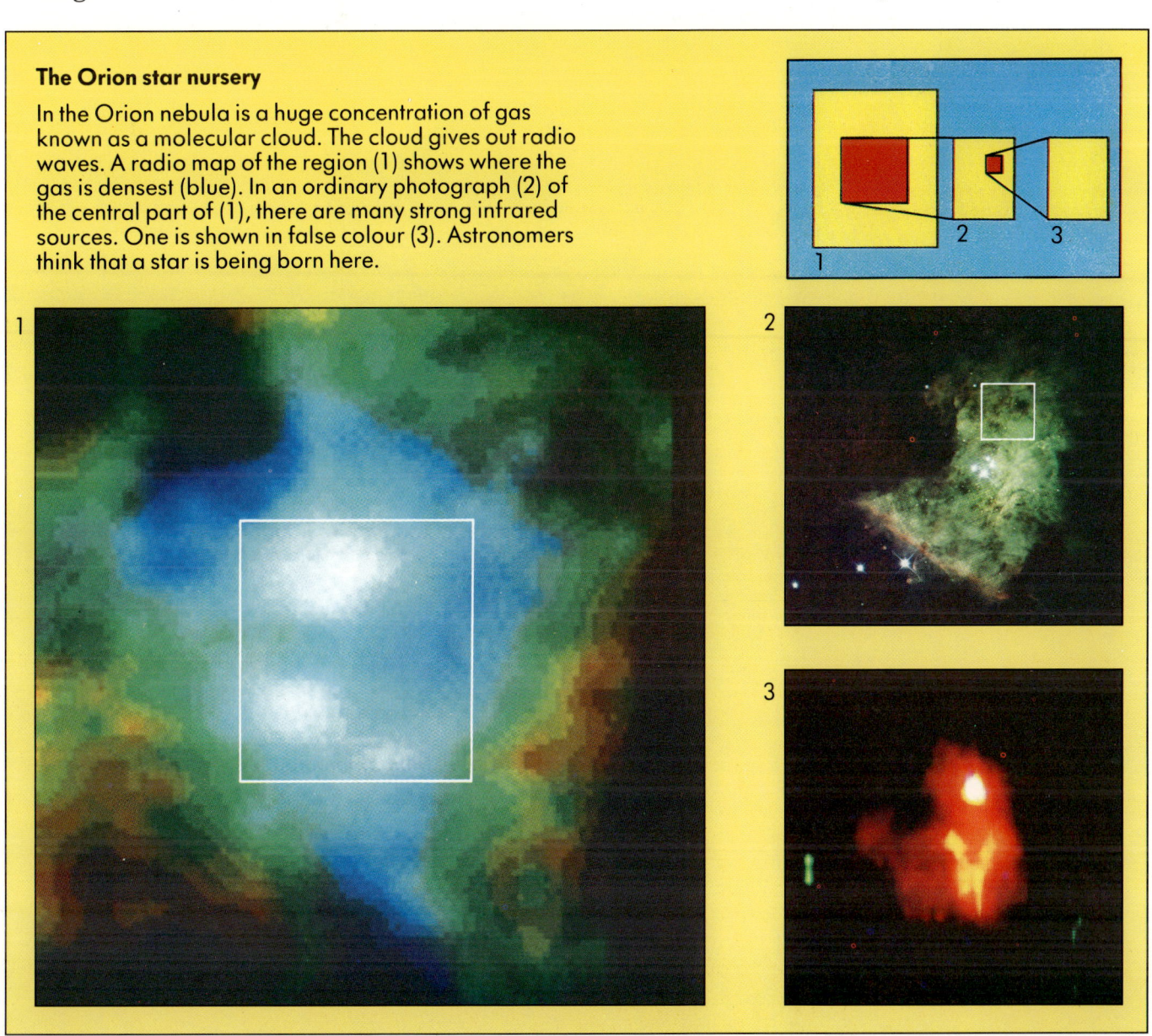

The Orion star nursery

In the Orion nebula is a huge concentration of gas known as a molecular cloud. The cloud gives out radio waves. A radio map of the region (1) shows where the gas is densest (blue). In an ordinary photograph (2) of the central part of (1), there are many strong infrared sources. One is shown in false colour (3). Astronomers think that a star is being born here.

Stellar life cycles

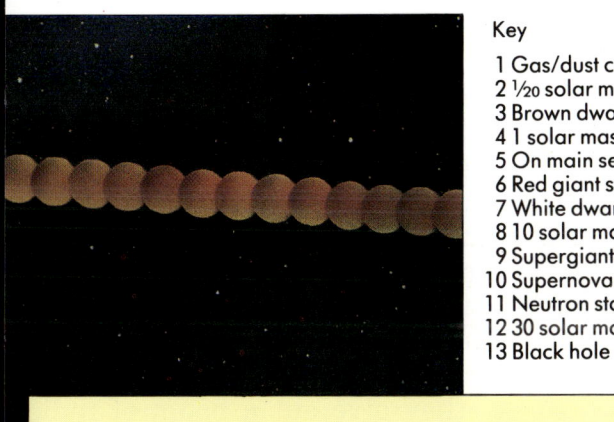

Key
1 Gas/dust cloud
2 ¹⁄₂₀ solar mass star
3 Brown dwarf star
4 1 solar mass star
5 On main sequence
6 Red giant star
7 White dwarf star
8 10 solar mass star
9 Supergiant star
10 Supernova
11 Neutron star
12 30 solar mass star
13 Black hole

When the temperature in the shrinking core of a star reaches 100 million degrees, another fusion reaction is able to take place. This reaction changes the nuclei of helium atoms into carbon nuclei. It is known as the triple-alpha reaction, because it combines three helium nuclei, otherwise known as alpha particles. The triple-alpha fusion process provides the energy to keep the expanded star shining, as a red giant.

A Sun-sized star expands up to 100 times while becoming a red giant. But in time all the helium in the core is used up, and the star again begins to shrink. Eventually gravity crushes the matter in the star into a small planet-sized body of immense density. This body is called a white dwarf.

Stars with a smaller mass than the Sun have a longer life. Those with a larger mass have a shorter life. Some massive stars may live for only a few million years. They burn up their fuel rapidly, then swell up into a supergiant. Finally they become a brilliant supernova and blast themselves apart.

◀ The way a star lives and dies depends on how massive it is. A brown dwarf (2) never shines brightly. A star such as the Sun spends much of its life on the main sequence (5). The Sun's main-sequence lifetime is about 10,000 million years. More massive stars (8, 12) have a much briefer life, and blast themselves apart spectacularly as a supernova (10).

▼ A planetary nebula. Near the end of its red-giant stage, a star may blow out some of its matter into space. Through a telescope the matter looks much like a planet and so early astronomers called it a planetary nebula. This one has shed its matter in lobes, rather like a butterfly's wings.

Gas from stars

Red giant

Shell

Core star

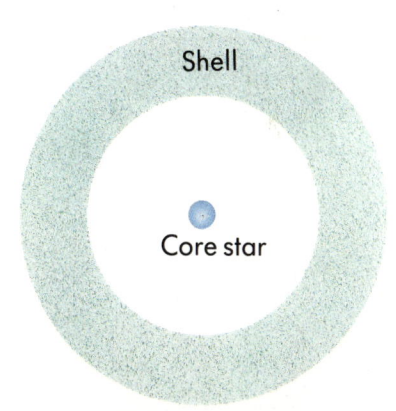

▲ ▶ An old red giant star (above right) often blows out gas, which forms an enveloping shell around it. The shell expands as time goes by. It absorbs energy from the central star and glows as a planetary nebula.

▶ The Ring nebula is one of the most beautiful of the ring-shaped planetary nebulae. Looking like a colourful smoke ring, it lies in the constellation Lyra, and measures about 0.6 light-years in diameter.

Violent death

Stars with several times the mass of the Sun meet their end in the most spectacular way. They grow to an enormous size as a supergiant and then blast themselves apart. We call these exploding stars supernovae.

When a star becomes a supernova, its brightness increases many millions of times. For a while it may even shine more brightly than a whole galaxy. In our Galaxy three supernovae have been seen during the last 1,000 years. Chinese astronomers spotted one in AD 1054, Tycho Brahe studied one in 1572, and Johannes Kepler saw one in 1604. Supernovae are so brilliant that we can see them in other galaxies.

What happens to a star after it has exploded as a supernova depends on how massive it is. One with a mass of more than seven times the Sun's mass becomes a neutron star. Stars more massive still become black holes.

A neutron star is formed when the central core of an exploding star collapses under gravity. As it gets smaller and smaller, the protons and electrons in its atoms are crushed together and form neutrons. The star turns into a "sea" of neutrons that is incredibly dense: a teaspoonful would weigh 100 million tonnes! Astronomers think that the bodies known as pulsars are neutron stars that rotate rapidly. As they do so, they give out a beam of radio waves. This sweeps round in space like the beam from a lighthouse. We receive radio pulses when the beam flashes in our direction. One of the first pulsars to be discovered was the Crab pulsar, in the Crab nebula.

Brighter than a billion Suns

Near the end of their life, heavy stars blow themselves apart as a supernova. They may first blaze more brightly than a billion Suns. On 23 February 1987 a star erupted into a supernova in the Large Magellanic Cloud (left), the galaxy nearest to Earth. It became the brightest supernova seen for nearly 400 years (below left).

A supernova blasts vast amounts of gas and dust into the surrounding space, which forms an expanding cloud. The gas cloud we call the Crab nebula (below) is what remains of a supernova first seen in the year 1054.

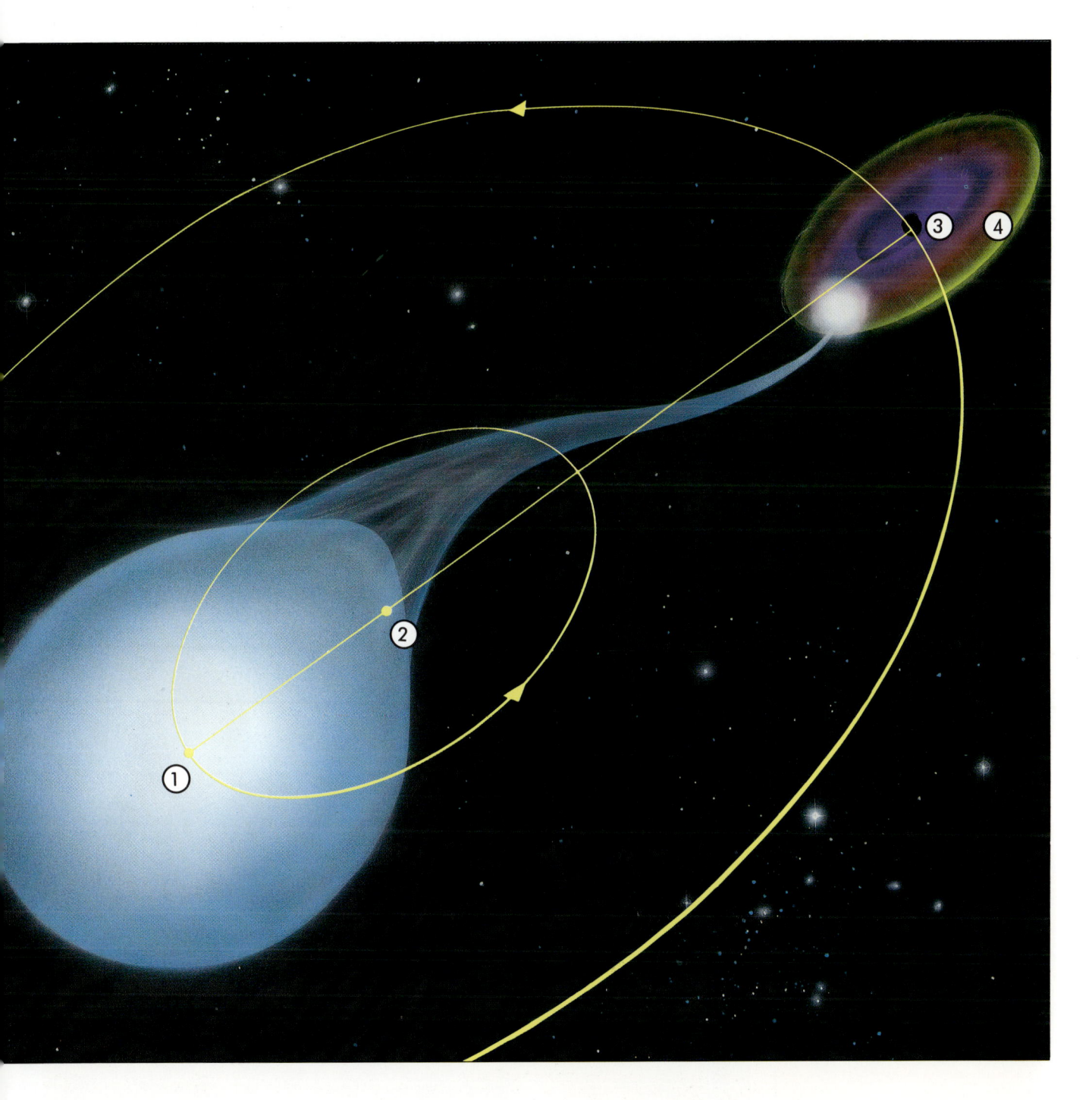

Very heavy stars do not stop collapsing even when they shrink to the neutron-star stage. Their gravity is so great that the collapse continues. The matter which they contain is eventually crushed into a point, known as a singularity. In the region around this point gravity is so intense that nothing, not even light, is able to escape from it. That is why astronomers call such a region a black hole.

▲ Astronomers believe that there may be a black hole in a binary star system in the constellation Cygnus. They think that the system is made up of a supergiant star (1) and a black hole (3), rotating around the centre of mass (2). The star system is known as Cygnus X–1 because of the powerful X-rays it gives out. In such a system, X-rays are produced as matter, torn from the supergiant. These spin into a spiral disc of matter (4) racing furiously around the black hole.

181

Galaxies galore

Spot facts

- The halo that surrounds our Galaxy may itself be surrounded by a "dark halo" of invisible matter. It could extend more than 200,000 light-years from the Galaxy's centre.

- The Sun orbits around the centre of the Galaxy at a speed of about 900,000 km/h. It is now travelling on its 23rd orbit.

- A belt of fast-moving gas connects the Galaxy with its two nearest neighbours in space, the Large and Small Magellanic Clouds.

- The Andromeda galaxy (M31) is the most distant object we can see in the heavens with the naked eye. We see it today as it was when our earliest ancestors were living on Earth 2.2 million years ago. That is how long its light has taken to reach us.

▶ A beautiful spiral galaxy in the southern constellation Antlia, the Air Pump, which has wide-open spiral arms. It is identified as NGC 2997 (number 2997 in the *New General Catalogue* of nebulae and star clusters).

Stars are not scattered about haphazardly in space. They gather together into great spinning star islands, or galaxies. All the stars we see in the sky belong to our home galaxy, which we call the Milky Way, or just the Galaxy. It is one of perhaps 100 billion galaxies in the Universe. Each one contains billions of stars. Most galaxies lead relatively peaceful lives, giving out a steady output of light. Some, however, are noticeably more active. They pour out up to a million times more energy than normal, particularly as radio waves.

The Milky Way

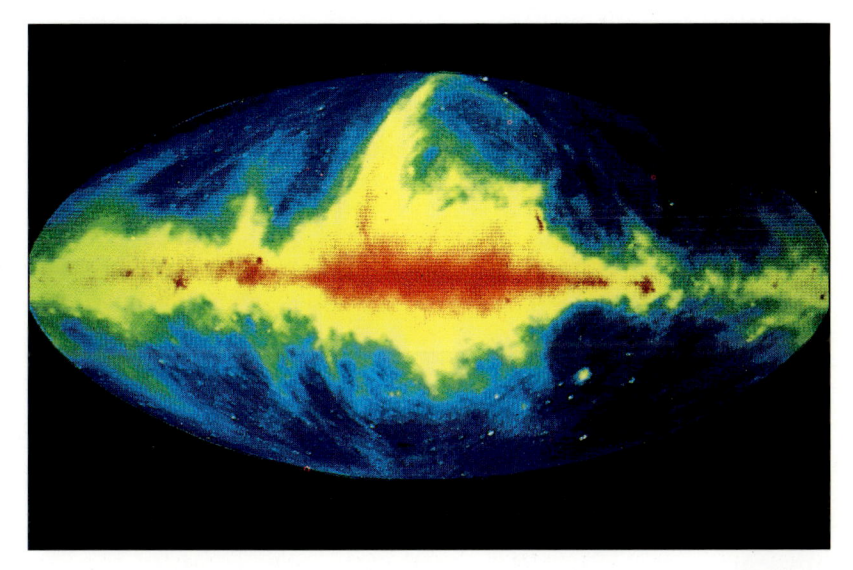

Dimensions of the Galaxy

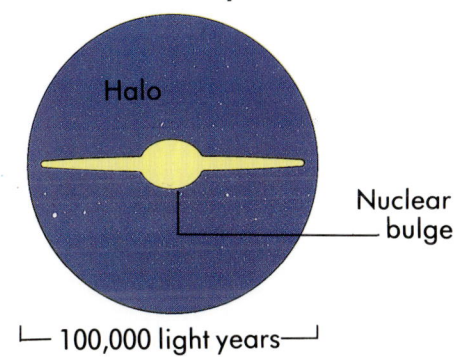

◀ This map of our Galaxy was prepared using radio waves rather than light waves. Red shows regions with strong signals.

On a moonless night, you can see a fuzzy band of light arcing across the heavens. We call it the Milky Way. In the Northern Hemisphere it passes through the easily recognized constellations Cassiopeia and Cygnus. In the Southern Hemisphere it passes through the unmistakable Scorpius and Crux, the Southern Cross. The brightest part lies in Sagittarius.

When viewed through a telescope, the Milky Way turns out to be a region containing millions of faint stars seemingly packed close together. This is because when we look at the Milky Way, we are seeing a cross-section of our own Galaxy. The stars are really far apart. They just appear to be close together because of the way we view them from the Earth.

The Galaxy takes the form of a flattish disc with a bulge (nucleus) in the middle. The 100,000 million stars it contains are spread out on the disc. In practice, the stars in the disc group together on arms that spiral out from the centre. The whole Galaxy rotates around the centre, but not at a uniform speed. Stars at the centre travel faster than those farther out. The Sun (30,000 light-years from the centre) takes 225 million Earth-years to make one rotation. This period of time is called a cosmic year.

▶ If we could view our Galaxy from far out in space, it would look something like this, the famous Andromeda galaxy. The Andromeda galaxy is one of the few we can see with the naked eye. It is much bigger than our Galaxy, but has a similar spiral shape.

Spirals, ellipticals and irregulars

Our Galaxy, the Milky Way, is one of many millions of spiral galaxies in the Universe. The others are of much the same shape and contain much the same mix of stars, clusters, gas and dust. Some are smaller than the Milky Way; others are larger. One of our galactic neighbours, the Andromeda galaxy, contains three times as many stars as our Galaxy.

Close to Andromeda in the sky are two much smaller galaxies. They are elliptical in shape, and have no spiral arms. Ellipticals are made up mainly of old stars, unlike spirals, which contain many young stars.

The galaxies nearest the Earth, however, have no distinct shape, and are classed as irregulars. They are called the Large and Small Magellanic Clouds (LMC and SMC). They can both be seen with the naked eye in far southern skies. The LMC measures 30,000 light-years across, a third of the size of our Galaxy. It appears to be twice as big as the SMC.

Classifier of galaxies
The American astronomer Edwin Hubble pioneered the study of galaxies. He is pictured here at the controls of the 100-inch reflector at Mount Wilson Observatory in California. In 1923 he set to work to devise a method of classifying the galaxies. He came up with a system that grouped galaxies according to their shape: as ellipticals, spirals and barred spirals.

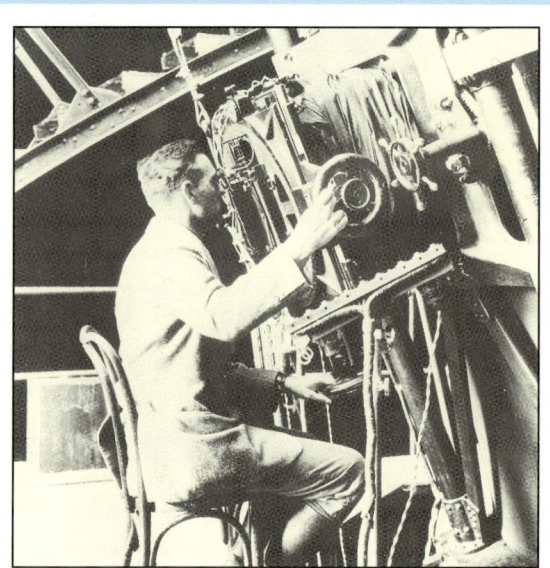

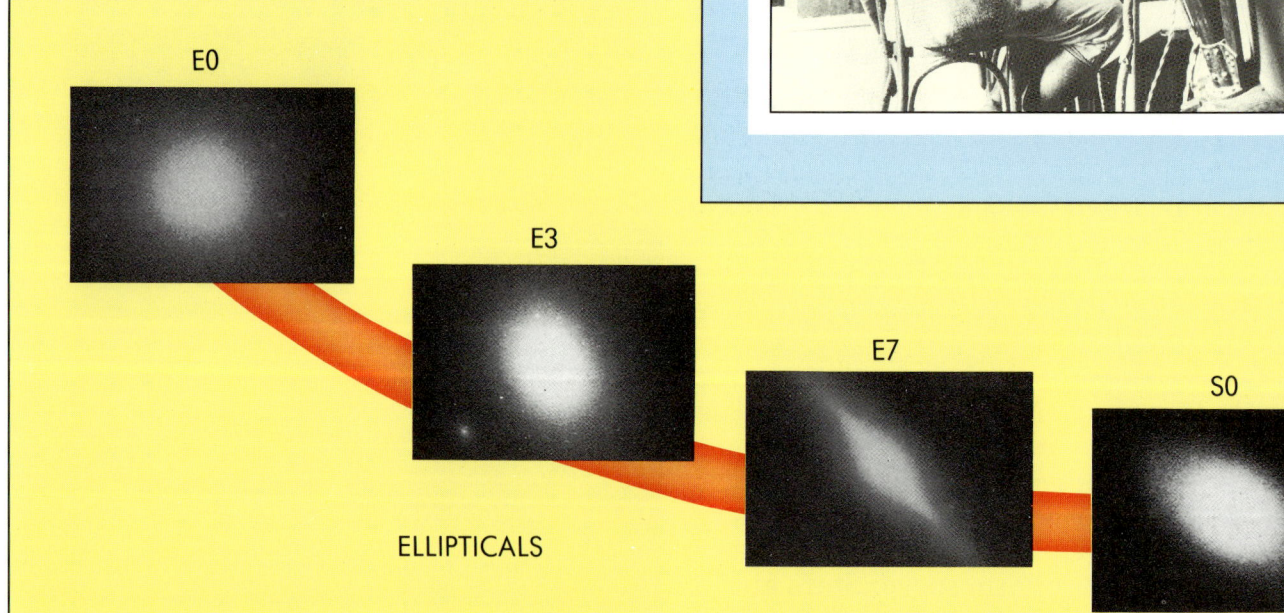

ELLIPTICALS: E0, E3, E7, S0

Portrait of the galaxies
The pictures show examples of the types of galaxies Hubble included in his historic classification. He thought that the galaxies evolved, in the sequence shown, from ellipticals into spirals. But astronomers no longer believe this. They are not certain how the different types of galaxies are related.
Elliptical galaxies (E) are described by a number from 0 to 7 which indicates how flattened they are.

Spiral galaxies (S) have a central bulge, or nucleus, from which a number of arms curve out. They are classed as a, b or c, depending on how far open the arms happen to be. S0 galaxies are similar to spirals, but have no arms.
Barred-spiral galaxies (SB), on the other hand, have spiral arms that come out of the ends of a line of stars (bar) through the nucleus. They are classed as a, b or c.

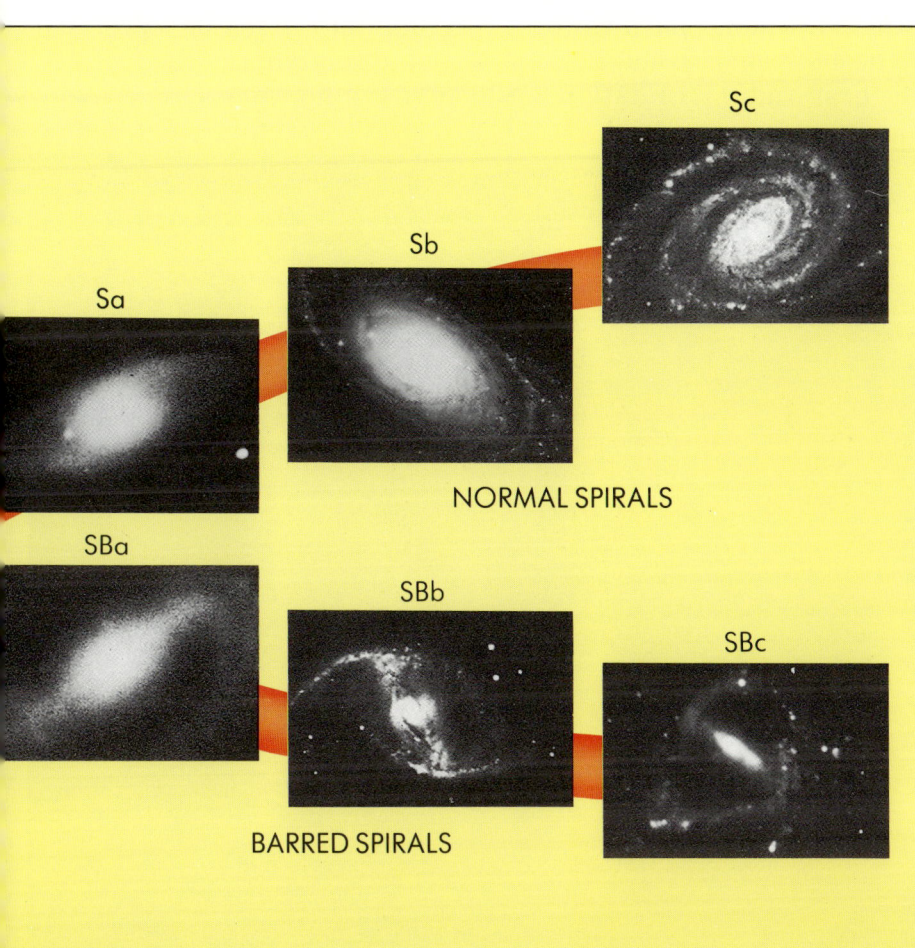

▲ The galaxy M33 is number 33 in a catalogue of "nebulae" drawn up by French astronomer Charles Messier in 1774. It is a spiral galaxy, class Sc, which has wide-open arms that spiral from the nucleus. The stars in the spiral arms (blue) are young. The nucleus is choked with dust (orange), lit by old yellow stars.

▼ The Small Magellanic Cloud is our second nearest neighbouring galaxy. It is a milky white patch, visible only in the Southern Hemisphere.

Sa Sb Sc
NORMAL SPIRALS

SBa SBb SBc
BARRED SPIRALS

185

Active galaxies

There are some galaxies that do not fit neatly into any class. Many are spiral galaxies with unusual features. Some may be colliding or have collided at some time in the past. For example, astronomers reckon that the Cartwheel galaxy was a spiral whose centre was knocked out in a collision hundreds of millions of years ago.

Seyfert galaxies

Another type of spiral galaxy is notable for having a very bright nucleus and very faint spiral arms. Such galaxies were first studied in the 1940s by the American astronomer Carl Seyfert, and are now termed Seyfert galaxies. Because their centres shine so brightly, they can sometimes be mistaken for stars.

Seyfert galaxies are one type of active galaxy, one that has an unusually powerful energy source at its centre. Other active galaxies are very strong emitters of radio waves, and are termed radio galaxies.

The study of radio galaxies started when astronomers began "tuning into" the heavens with radio telescopes. One of the first radio sources found was located in the southern constellation Centaurus and named Centaurus A. In 1949 Australian astronomers identified this source with the bright galaxy NGC 5128.

Twin lobes

In light photographs Centaurus A looks like an ordinary elliptical galaxy, as do most radio galaxies. But as radio wavelengths they can outshine an ordinary galaxy by up to a million times. Strangely, the radio waves do not come from the centre of the galaxy. They come from twin lobes, regions of space on each side of the galaxy. In the case of Centaurus A, the lobes extend over 2,500,000 light-years. Yet the galaxy is only 30,000 light-years across!

Images of the radio galaxy M87 indicate how the radio waves are emitted. They show a jet, or stream of matter shooting out of the nucleus. The jet is a high-speed beam of electrons. As they pass through the magnetic field surrounding the galaxy, they are forced into a spiral path around the magnetic lines of force. This constant change of direction triggers off the radiation we detect as radio waves.

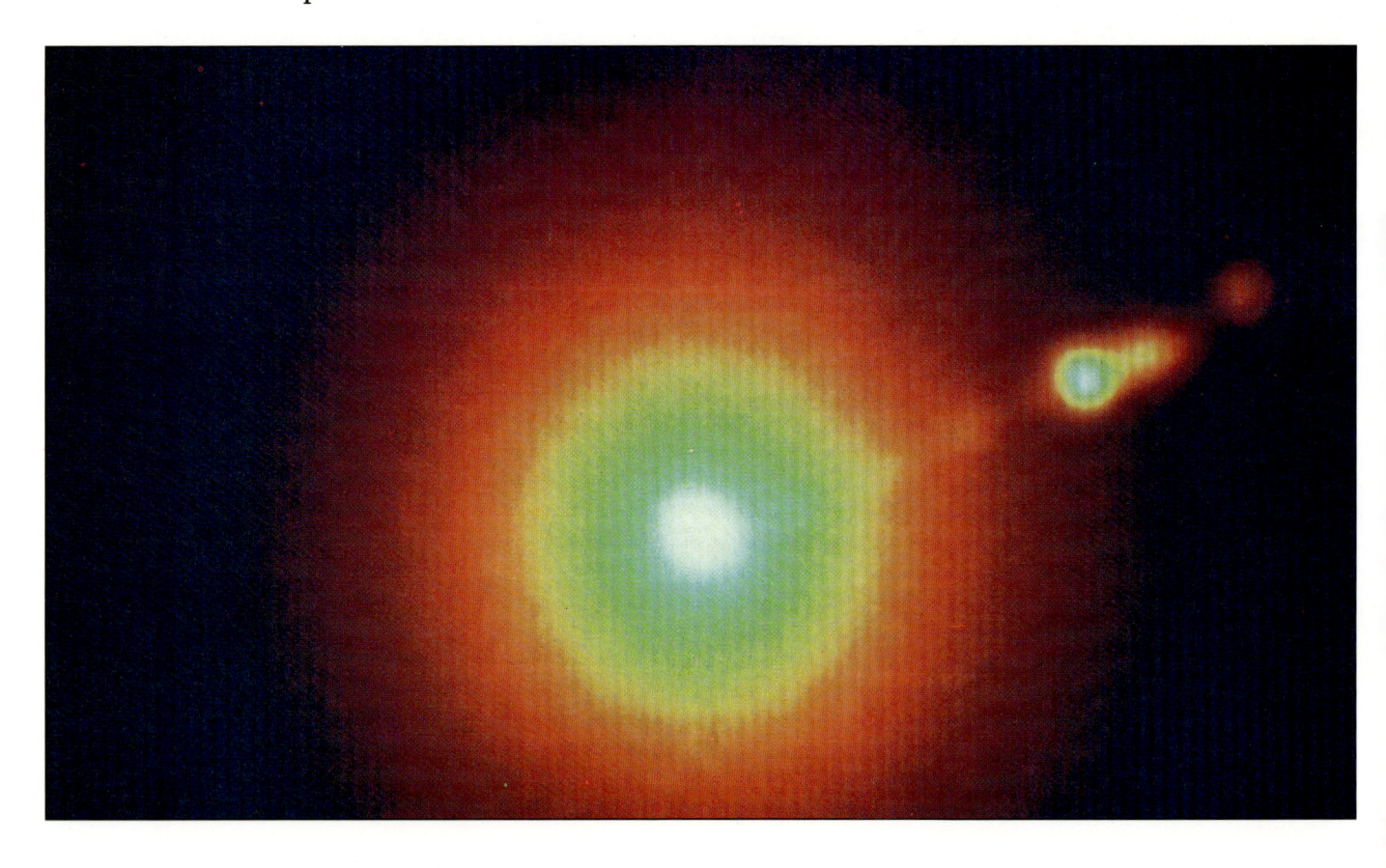

▲ ▶ The galaxy Centaurus A is crossed by a broad dust lane. In most of the elliptical central regions, there are mainly old yellow stars, with younger blue ones around the edges of the lane. The photograph (right) is a radio map of the same galaxy. The image reproduced above would fit within the pink circle in the middle.

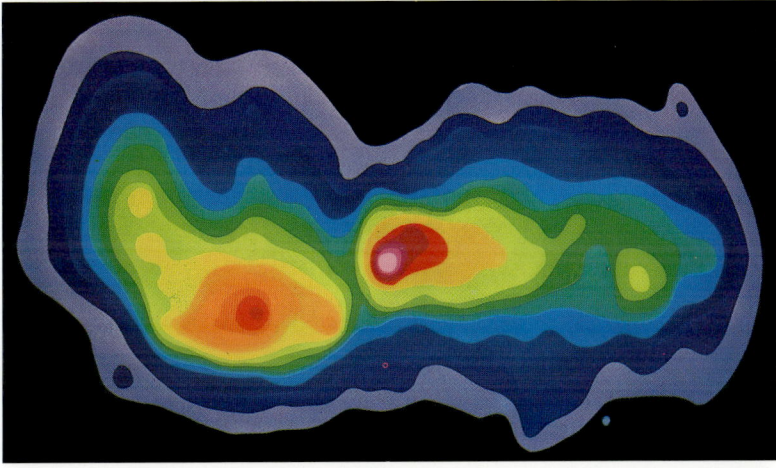

◀ M87, one of the nearest active galaxies, is notable for the long "jet" that extends from it. The jet is radiation given off by electrons moving rapidly in a magnetic field.

Quasars

The most intriguing of all the active galaxy-type objects in the Universe are quasars. The first two to be discovered were not given names, but coded as 3C 48 (in 1960) and 3C 273 (in 1962). They were strong radio sources whose positions in the sky matched those of two faint blue stars. The spectra of the stars, however, were quite unlike any seen before.

Astronomers now know the reason for this. The lines in the spectrum of these star-like objects are shifted to the red by an enormous amount. In other words the sources must be very far away. 3C 273 proves to be more than 2,000 million light-years away. No star can be seen from this distance. Therefore it cannot be an ordinary star, even though it looks like one. And for it to be visible from such a distance, it must be hundreds of times brighter than an ordinary galaxy. Astronomers have since discovered more than 1,500 other quasars, or quasi-stellar radio sources.

Quasars do not shine steadily like ordinary galaxies. They vary in brightness over periods of days or years. For this reason they cannot possibly be as big as an ordinary galaxy. For if a quasar changes in brightness in a year, say, it cannot be more than one light-year across. And if it changes in brightness in a day, it cannot be more than a light-day across. From its variation in brightness, 3C 273 works out to be less than one-hundredth of a light-year across, which makes it only one ten-millionth the size of a typical galaxy!

However, it seems that quasars are not separate bodies. They appear to be eruptions at the centre of massive galaxies. The rest of the galaxies are too faint to be visible at the distances involved.

The power house
What kind of energy source could make quasars the size of the Solar System put out the power of hundreds of galaxies? There seems only one possibility – a massive black hole.

A black hole is created when aging stars collapse. It is a region of space with super-high gravity, which swallows matter like a cosmic vacuum cleaner. It is surrounded by a rapidly rotating disc of hot gas. Matter attracted by the hole's enormous gravity acquires great amounts of energy. This is released as radiation when the matter ploughs into the disc.

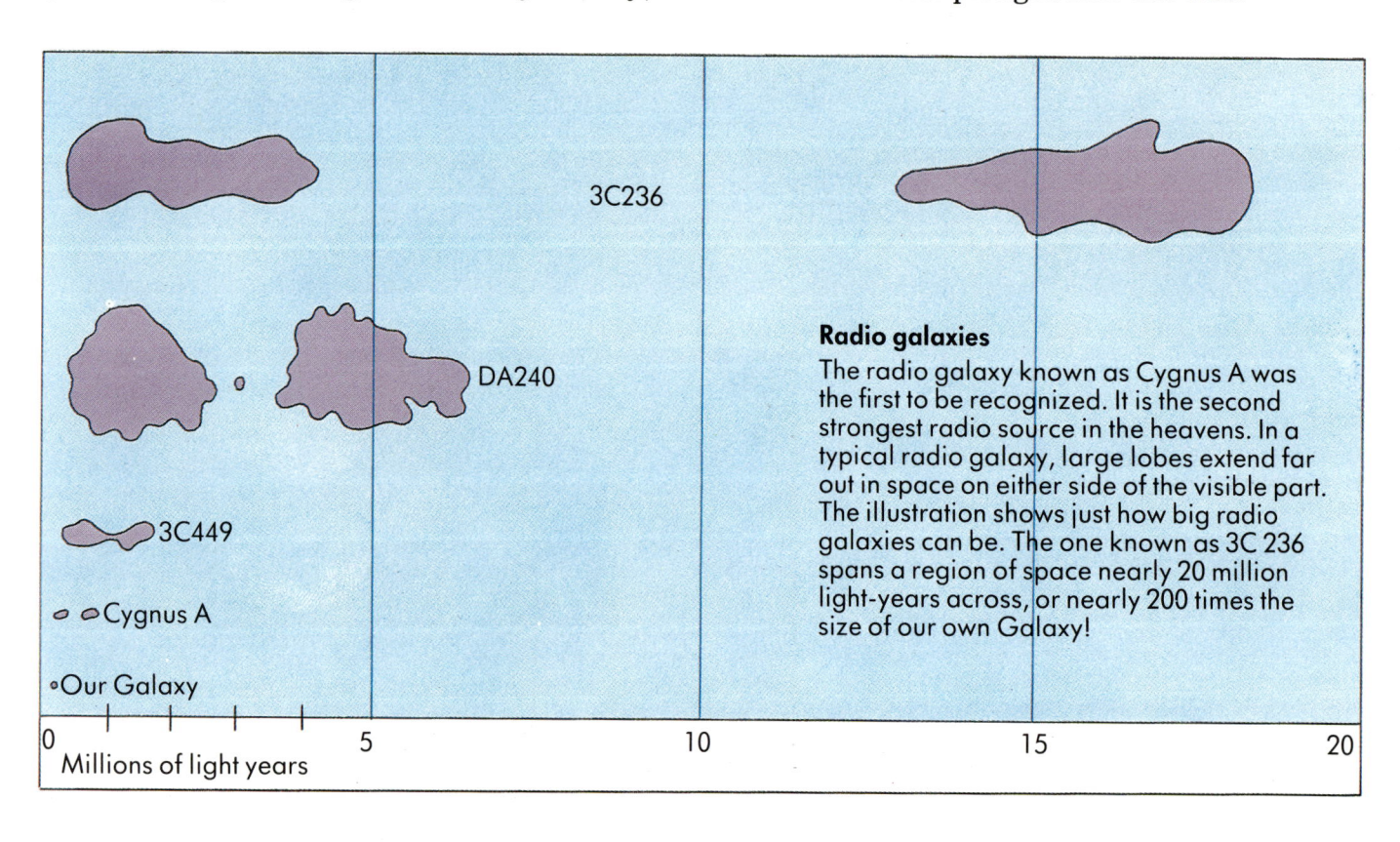

Radio galaxies
The radio galaxy known as Cygnus A was the first to be recognized. It is the second strongest radio source in the heavens. In a typical radio galaxy, large lobes extend far out in space on either side of the visible part. The illustration shows just how big radio galaxies can be. The one known as 3C 236 spans a region of space nearly 20 million light-years across, or nearly 200 times the size of our own Galaxy!

The Double quasar 0957 + 561

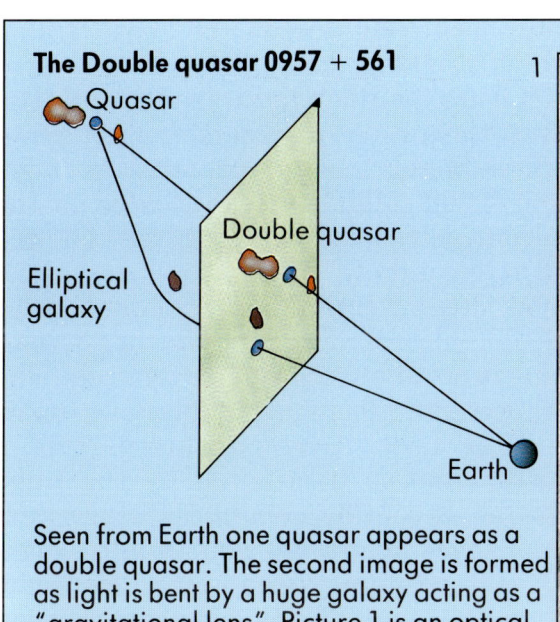

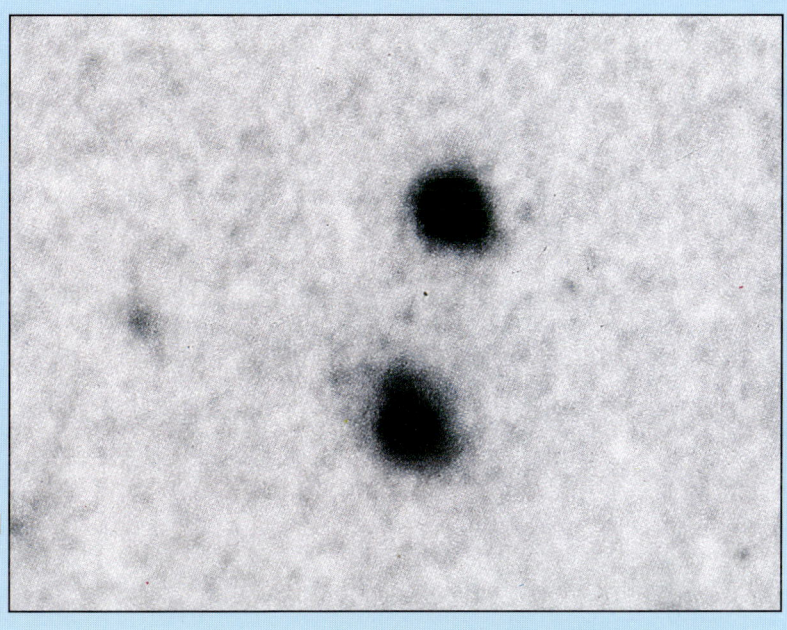

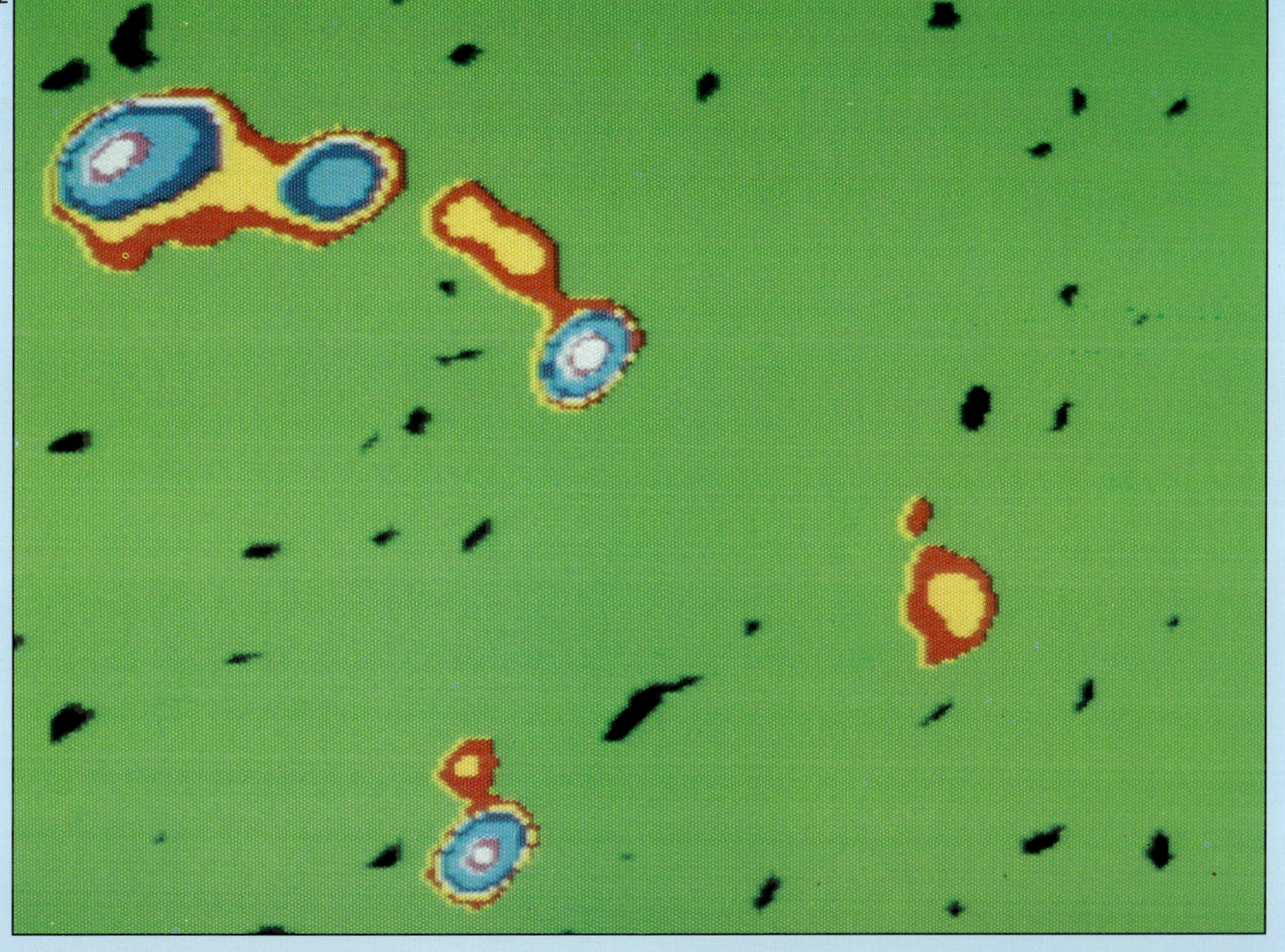

Seen from Earth one quasar appears as a double quasar. The second image is formed as light is bent by a huge galaxy acting as a "gravitational lens". Picture 1 is an optical photograph; 2 is a radio picture.

Clusters of galaxies

Just as stars cluster together in space in galaxies, so the galaxies themselves tend to cluster together. Our own Galaxy, the Milky Way, has two close companion galaxies. They are the Large and Small Magellanic Clouds. All three galaxies form part of a cluster of about 30 galaxies known as the Local Group.

This Group contains three large spirals: the Milky Way, M33 and the Andromeda galaxy. M33 lies about 2.4 million light-years away, about 200,000 light-years farther than Andromeda. The Magellanic Clouds are two of four small irregular galaxies in the group. Most of the galaxies, however, are ellipticals and they are smaller still.

Two of the small elliptical galaxies are surrounded by clusters of millions of stars, known as globular clusters. The small ellipticals are found in the southern constellations of Sculptor and Fornax.

The only large elliptical in the Local Group, Maffei I, is probably as massive as our own Galaxy. It lies 3.3 million light-years away. The Local Group is quite a small cluster of galaxies. The nearest major cluster contains between 1,000 and 2,000 galaxies. It is located in the constellation Virgo and is centred on the powerful active galaxy M87.

But most clusters contain only between about 100 and 400 galaxies. The grouping of galaxies also seems to occur on an even larger scale. The clusters apparently form part of massive superclusters. Our Local Group forms part of a supercluster that is dominated by the huge Virgo cluster. It contains about 100 clusters in a region of space about 250 million light-years across. A cluster of galaxies in Hercules, some 600 million light-years away, is part of an even larger supercluster. It is so large that it spans a third of the sky.

▼ Part of the Virgo cluster of galaxies, which lie about 50 million light-years away. In some clusters the galaxies are linked by clouds of hydrogen gas. The radio telescope image (below) shows that gas (blue) joins the galaxy M81 (top) to another small galaxy.

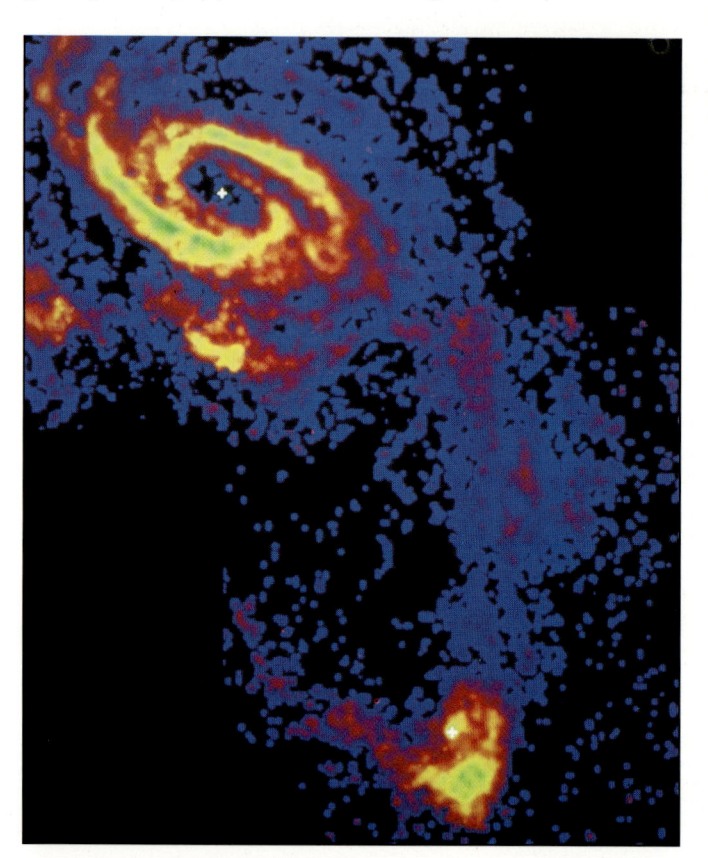

▲ Abell 1060 is a typical cluster of galaxies.

▶ Our Galaxy is one of the biggest members of the Local Group (numbers), which in turn forms part of the local supercluster (letters).

Local Group
1 Draco system
2 Large Magellanic Cloud
3 Small Magellanic Cloud
4 Ursa Minor system
5 Leo I
6 Leo II
7 The Galaxy
8 NGC 682
9 IC 1613
10 IC 1643
11 NGC 185
12 NGC 147
13 M83
14 M31
15 M32

Local supercluster
A Virgo III cloud
B Virgo II cloud
C Crater cloud
D Virgo I cloud
E Leo cloud
F Canes Venatici cloud
G Canes Venatici spur

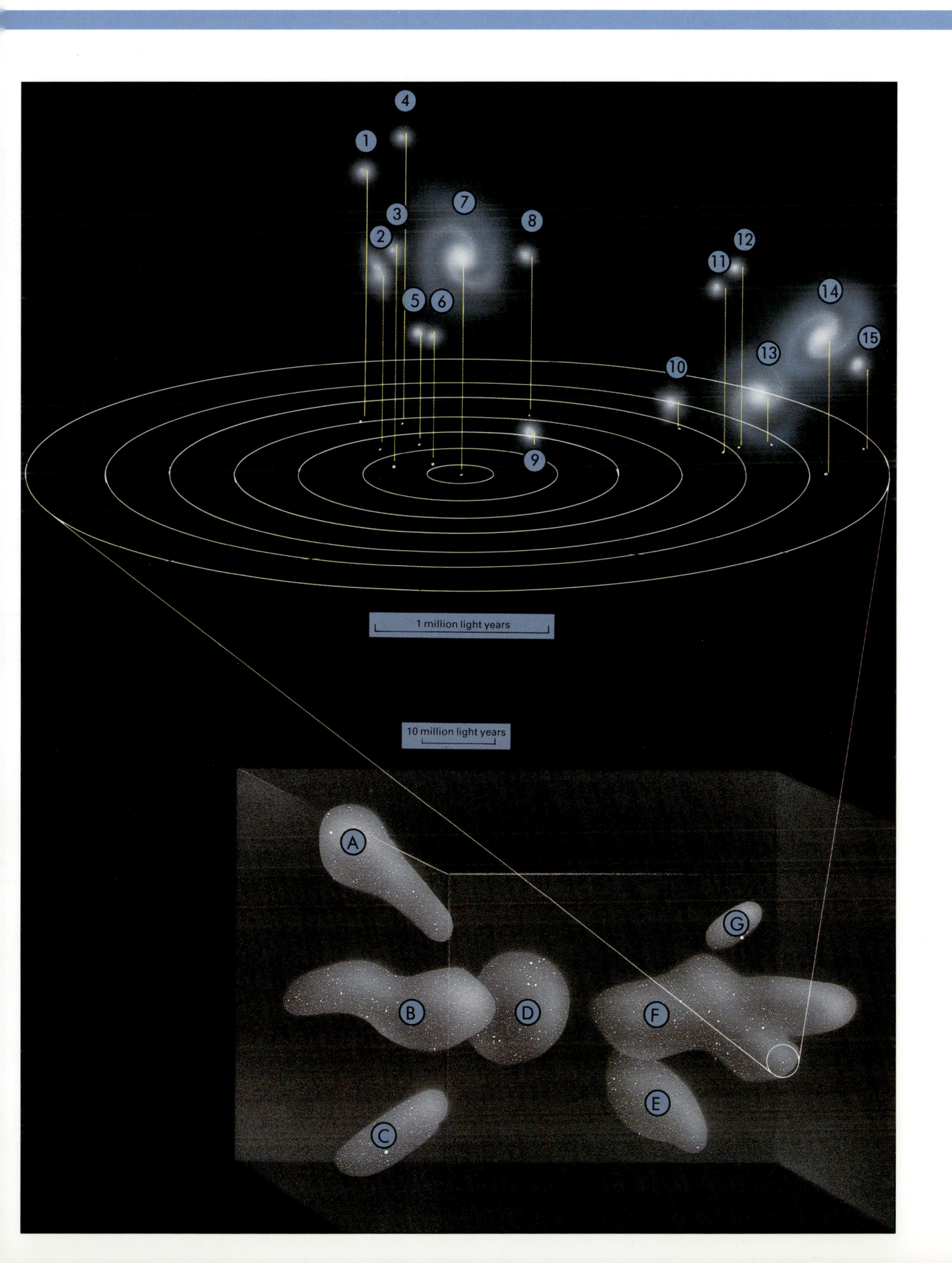

1 million light years

10 million light years

Big Bang, Big Crunch

Spot facts

- Astronomers reckon that the speed of a galaxy increases by about 55 km per second for each megaparsec (3,260,000 light-years) of distance. This value is the Hubble constant.

- From their red shifts, the most distant quasars appear to be travelling at over 270,000 km per second, or more than 90 per cent of the speed of light. This places them more than 13 billion light-years away.

- The first atomic nuclei began forming just three minutes after the Universe was born.

- When the Universe is 100 times older than it is now, all the stars will be dead and the galaxies will be fading.

- It requires only about 1 gram of matter in every 40 million million cubic kilometres of space to "close" the Universe – stop it expanding.

▶ The spiral galaxy NGC 4319, is pictured here in false colours. Like most galaxies, it is racing away from us at tremendous speed.

How did the Universe begin, and when? How has it evolved? How will it end, and when? These are the kinds of questions astronomers have been trying to answer for centuries. Cosmologists, the astronomers who study such things, now think they have many of the answers. They see in the expansion of the Universe evidence that it began in a gigantic explosion. And they have worked out how it has evolved from the first millisecond of its existence to the present day. They are not so sure how the Universe will end. But the signs are that it will go on expanding for ever until the stars and galaxies fade away.

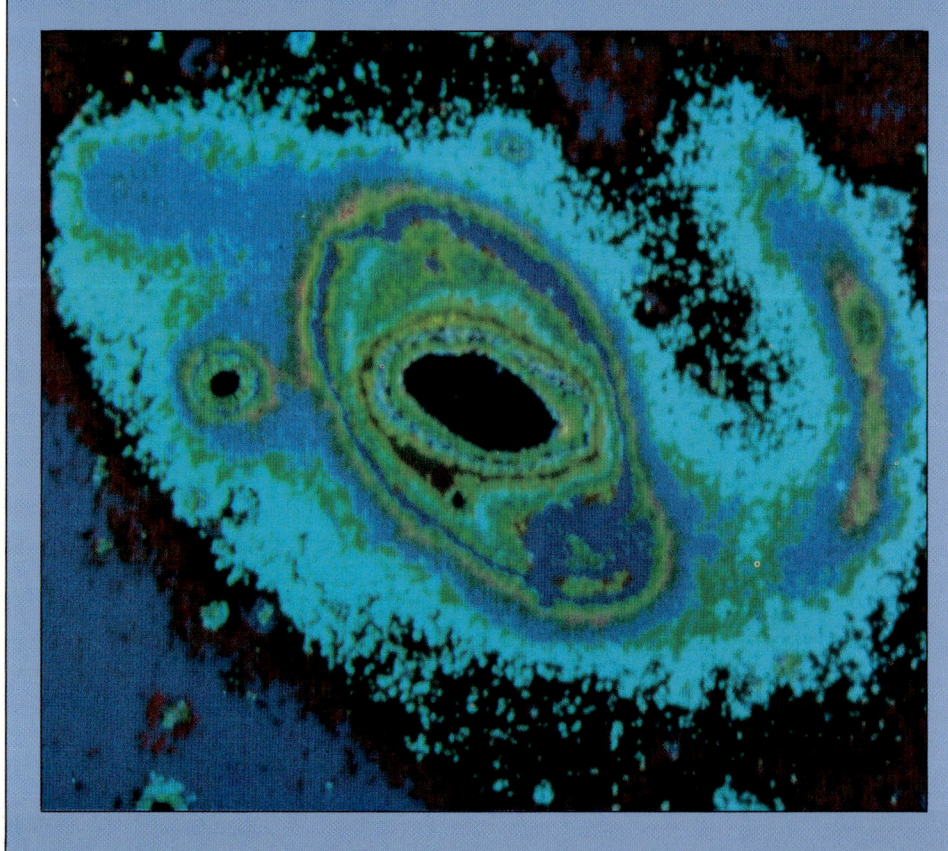

An expanding Universe

In 1914 the American astronomer Vesto Slipher began studying the spectra of galaxies. He found that all of them, with the exception of the Local Group, had red shifts. The red shifts indicated that all the galaxies were moving away from us. At the time no one was sure if the galaxies were part of the Milky Way or not. In the 1920s, however, fellow American Edwin Hubble showed without doubt that the galaxies did lie outside it. So, with all the galaxies rushing away, it appeared that the whole Universe was expanding. All evidence since has confirmed that this is happening.

So it follows that in the past the Universe must have been smaller. Working backwards, there must have been a time when all the matter in the Universe was packed together in one place. This is the reasoning behind the most widely accepted theory about how the Universe began. It is thought that the Universe came into being as a result of an explosion, or Big Bang, which set in motion the expansion that we observe. Astronomers have worked out that the Big Bang must have occurred about 15,000 million years ago.

During his study of the spectra of galaxies, Hubble also discovered another interesting thing. The farther galaxies are away, the greater is their red shift and the faster they are moving. And he worked out a relationship between speed and distance, which became known as the Hubble constant.

Measuring the amount of red shift and applying the Hubble constant provides the only method of finding out the distance to the farthest galaxies and the remote quasars.

▲ Almost all the galaxies appear to be moving away from us. So does this mean that our Galaxy is the centre of the Universe? The answer is no. What is happening is that every galaxy is moving away from every other galaxy as the Universe expands.

To see how this kind of thing could happen, think of the surface of a balloon as the Universe and dots on the surface as galaxies. When you blow up the balloon, the "Universe" expands and the "galaxies" move farther away from each other.

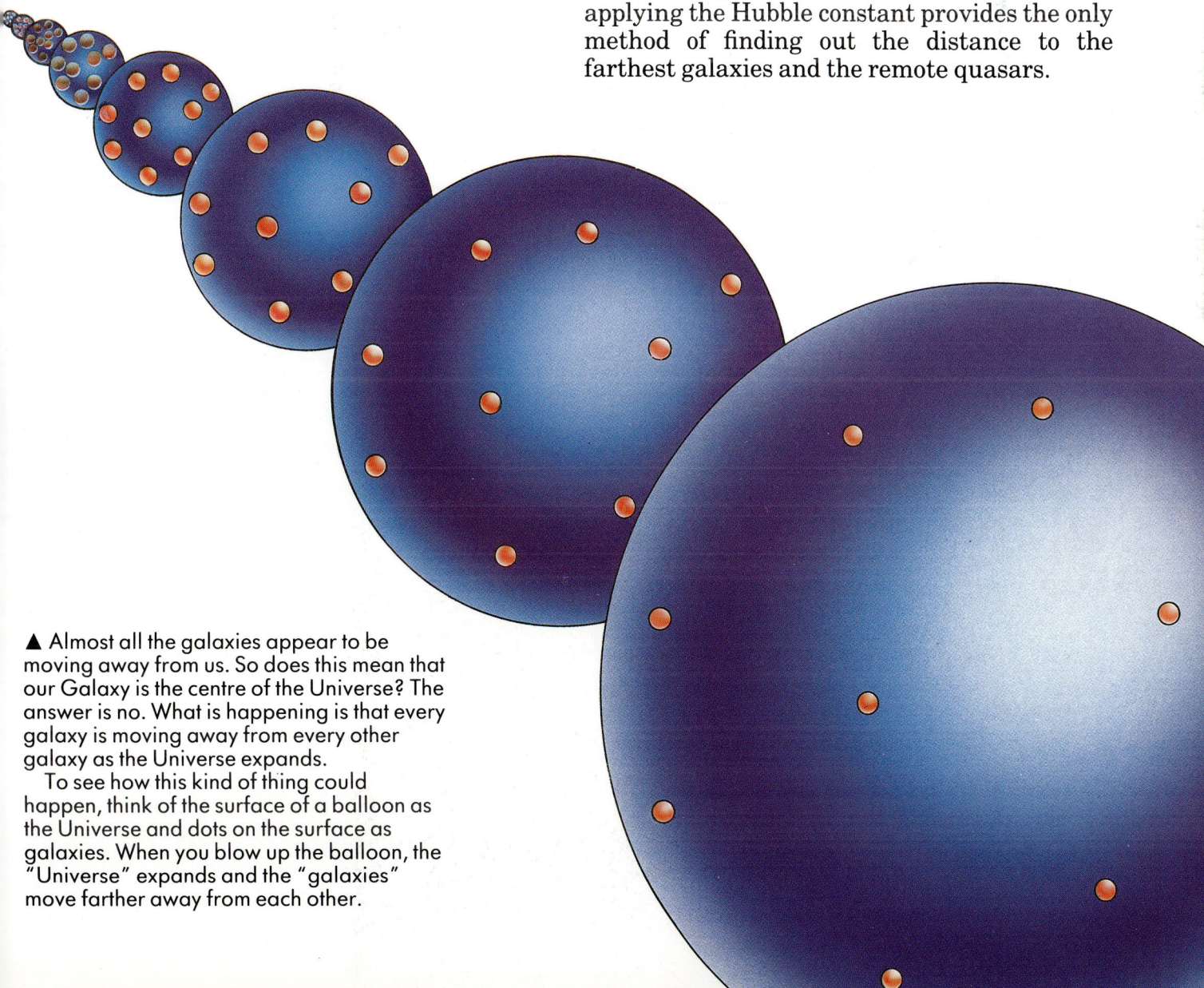

After the Big Bang

We cannot think of the Big Bang as being like an ordinary explosion, which scatters material into the surrounding space. There was nothing before the Big Bang – no matter, no energy, no space and no time. The Universe did not exist. Matter, energy, space and time came into being with the Big Bang. The Universe was born then, some 15,000 million years ago.

Astonishingly, astronomers think they know how the Universe has evolved from the very beginning. A millionth of a second after the Big Bang, the temperature of the Universe was over 10 million million degrees. It was filled with energy in the form of photons – little "packets", or particles of radiation.

The birth of matter

Under suitable conditions, high-energy photons can turn into particles of matter. And that is what happened in the early stages of the Universe. Most of the particles turned back into radiation. But some remained to form the atomic particles that make up the Universe as we know it today.

All of the subatomic particles – protons, neutrons and electrons – had been formed by the time the Universe was 10 seconds old. From the instant it was created, the Universe began expanding and cooling. After about three minutes, its temperature had dropped to about 1,000 million degrees. Then protons and

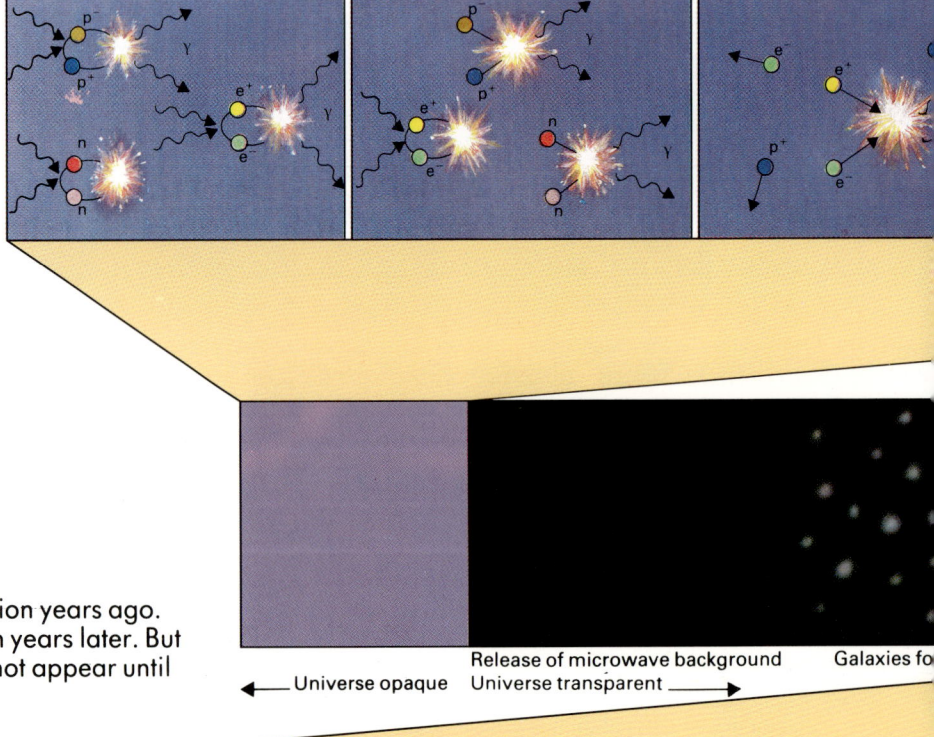

▶ When the Universe was created in the Big Bang, it was fantastically hot and filled with photons (particles of radiation). Heavy particles such as neutrons (n) and protons (p) formed when photons collided (A). But they were destroyed almost immediately, changing back into radiation (B). Likewise, most electrons (e) were destroyed (C). But a few protons, neutrons and electrons survived (D). It was now just 10 seconds after the Big Bang. After about three minutes the protons and neutrons began coming together to form the nuclei of atoms. First came two forms of heavy hydrogen, called deuterium (E) and tritium (F). Finally came helium (G). Hydrogen and helium are the most plentiful elements in the Universe.

▼ The Earth was born about 4,600 million years ago. Primitive life began about 1,500 million years later. But early members of the human race did not appear until about 3 million years ago.

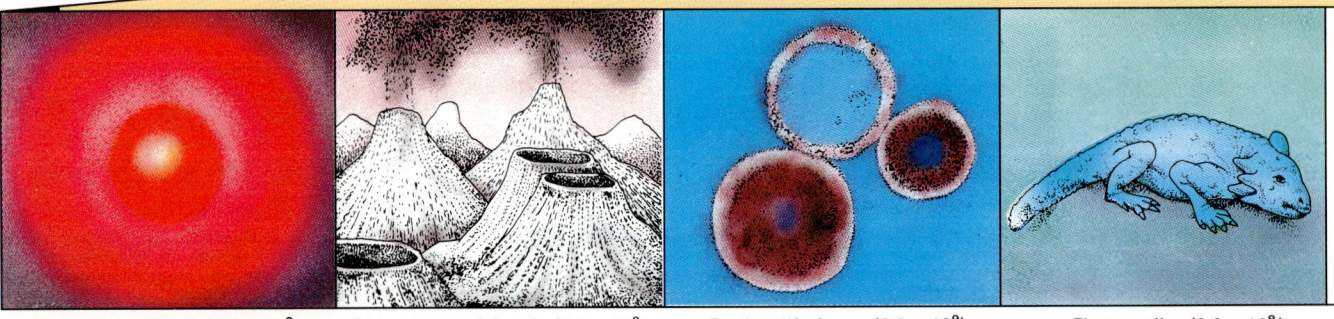

← Universe opaque   Release of microwave background   Universe transparent →   Galaxies fo

Formation of Earth (4·6 × 10⁹)   Oldest terrestrial rocks (3·6 × 10⁹)   Earliest life-forms (3·0 × 10⁹)   First reptiles (3·0 × 10⁸)

neutrons began combining to form the central cores, or nuclei, of atoms, such as helium. The Universe was then made up of radiation and matter, in the form of protons, helium nuclei and electrons.

After several hundred thousand years the temperature had fallen to about 3,000°C. The protons were now able to capture and hold on to electrons. The protons became atoms of hydrogen, and the helium nuclei became helium atoms. With fewer particles about and the Universe greatly expanded, radiation could travel over vast distances without being either absorbed or deflected. And the Universe had become transparent.

Background radiation

Radiation has continued travelling since that time. It has been spreading out through larger and larger volumes of space as the Universe has continued to expand. Astronomers have worked out that such radiation would now give space an overall temperature of about −270°C, or 3K – three degrees above absolute zero.

In 1965 two American scientists discovered that radiation at this temperature did indeed fill space. They were Arno Penzias and Robert Wilson. Their discovery of what is often called fireball radiation provided convincing evidence that the Big Bang theory is correct. The discovery also won them a Nobel Prize.

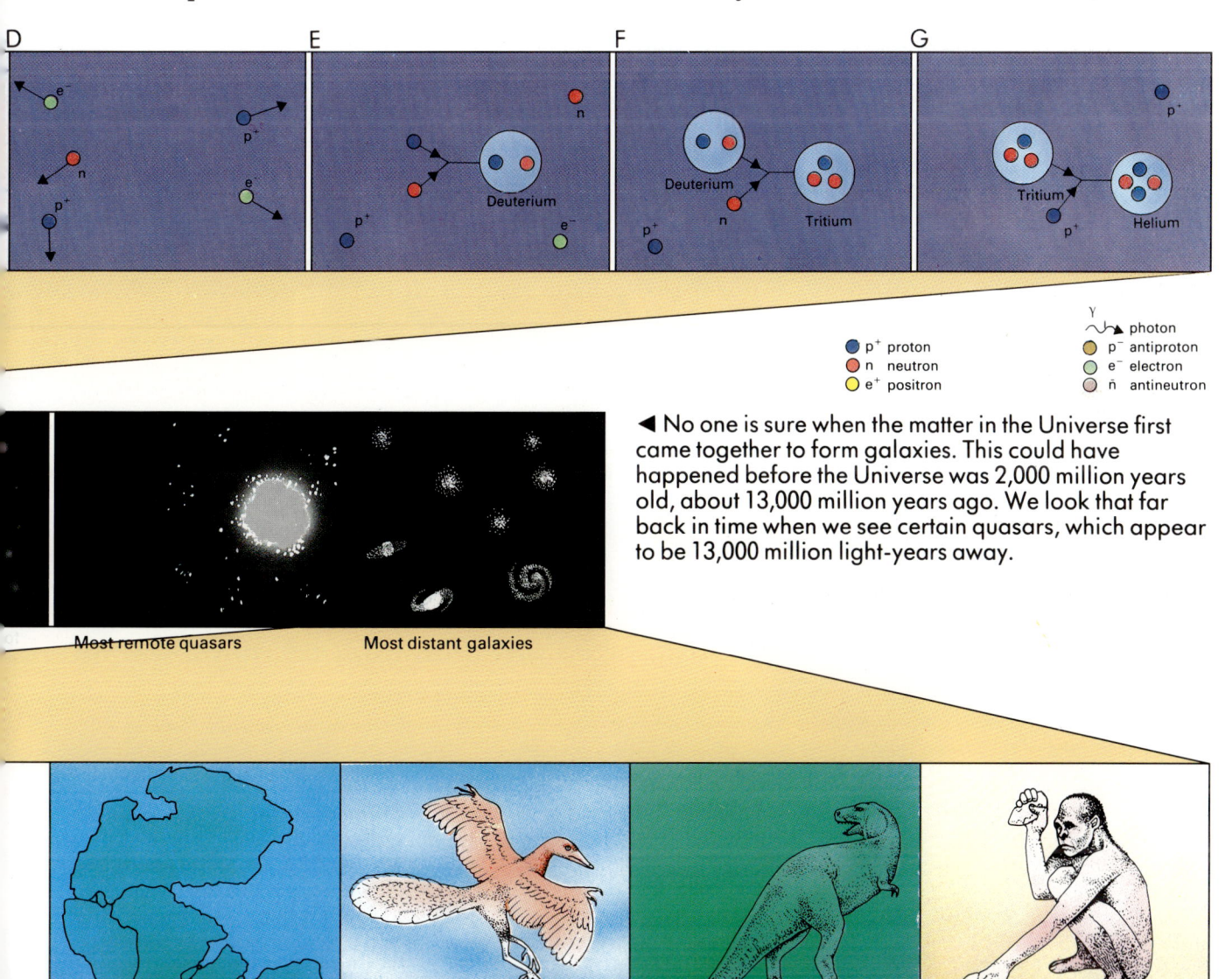

- p^+ proton
- n neutron
- e^+ positron
- γ photon
- p^- antiproton
- e^- electron
- $\bar{n}$ antineutron

◀ No one is sure when the matter in the Universe first came together to form galaxies. This could have happened before the Universe was 2,000 million years old, about 13,000 million years ago. We look that far back in time when we see certain quasars, which appear to be 13,000 million light-years away.

Most remote quasars Most distant galaxies

Separation of continents (2.0×10^8)   First bird (2.0×10^8)   Death of dinosaurs (6.5×10^7)   First humans (1×10^6 years ago)

Open or closed?

Most astronomers agree about how the Universe began in a Big Bang and how it has evolved since then. They are not so sure what will happen in the future. Certainly for many billions of years the Universe will continue to expand. But will it expand for ever?

Most astronomers think it will. They say we have an open Universe. The only thing that could prevent the galaxies flying apart for ever would be gravity. For gravity to be powerful enough to do this, the Universe must have a certain mass. But there appears to be nowhere near enough mass in the stars and galaxies to halt the expansion.

Dark matter
However, astronomers know that stars and galaxies are not the only matter in the Universe. There is also dark matter which we are largely unable to detect. There is dark matter in the dust clouds in space and in dead burned-out stars. There is also matter hidden in the abyssal depths of black holes.

Other possible sources of dark matter are the atomic particles called neutrinos. Recent experiments have indicated that they might have a slight mass. If they have, they would greatly increase the density of the Universe because there are so many of them.

Crunch, bang, crunch...
If there is sufficient dark matter to halt the expansion of the Universe, then we have a closed Universe. Eventually the Universe will collapse in on itself and end in a Big Crunch. But things may not end here. Another Big Bang may be triggered off that will set the Universe expanding once again. In turn will come another Big Crunch, yet another Big Bang, and so on. We will have an oscillating Universe.

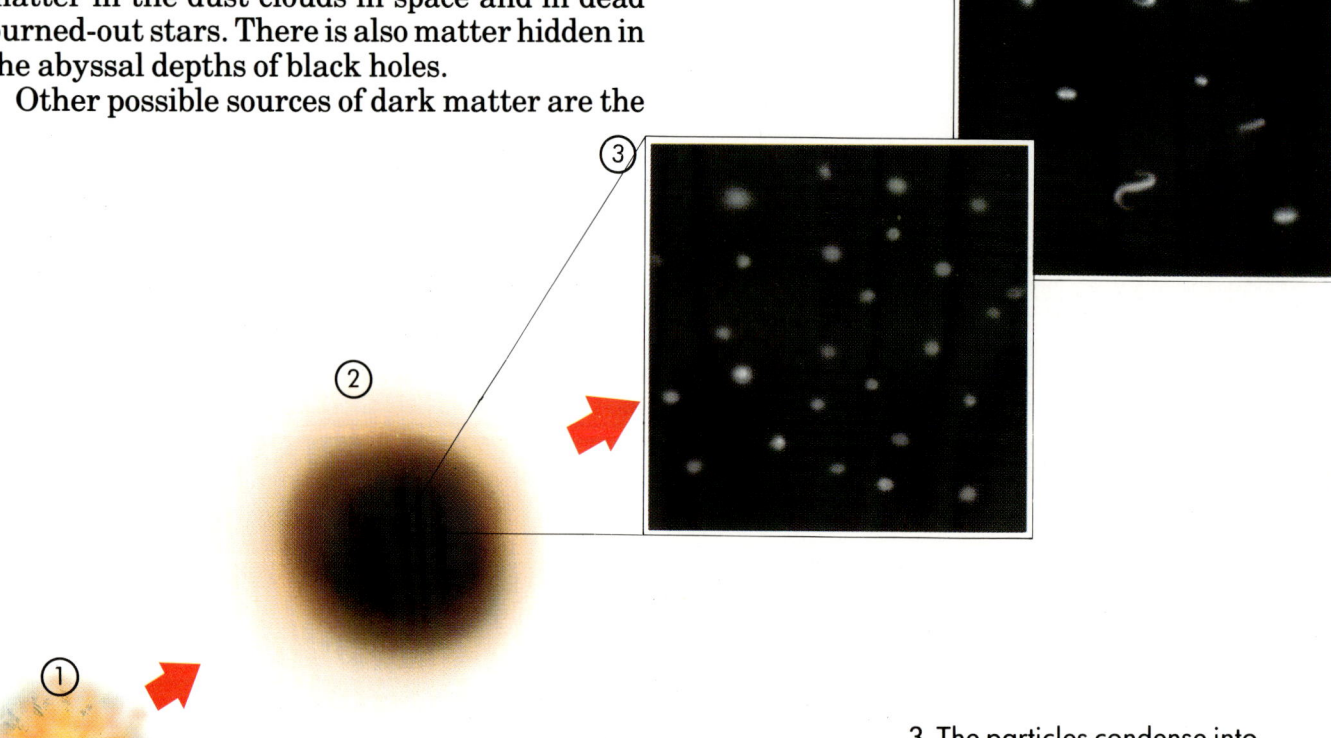

The pictures on this page show stages in the development of the Universe from the time of the Big Bang. Most astronomers are agreed on these stages (1-4).

1 The Universe is created, and time and space begin.
2 A dense opaque cloud of particles and radiation fills the cooling, expanding Universe.
3 The particles condense into galaxies, and stars begin to shine.
4 After 15,000 million years, the Universe reaches the stage it is today. It is filled with spiral, elliptical and irregular galaxies rushing headlong through space, bringing about further expansion. What happens next will depend on whether the Universe is open or closed.

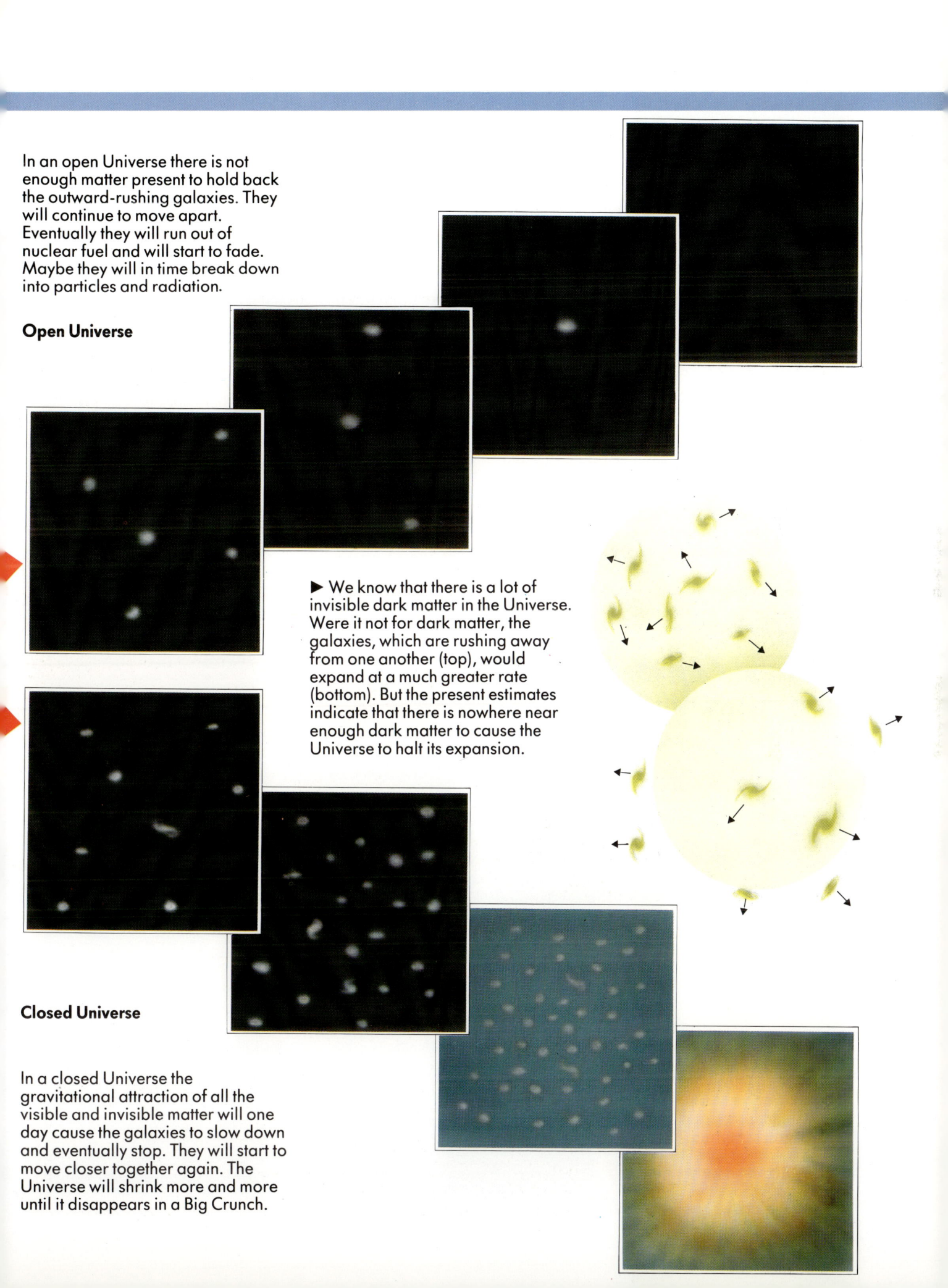

In an open Universe there is not enough matter present to hold back the outward-rushing galaxies. They will continue to move apart. Eventually they will run out of nuclear fuel and will start to fade. Maybe they will in time break down into particles and radiation.

Open Universe

▶ We know that there is a lot of invisible dark matter in the Universe. Were it not for dark matter, the galaxies, which are rushing away from one another (top), would expand at a much greater rate (bottom). But the present estimates indicate that there is nowhere near enough dark matter to cause the Universe to halt its expansion.

Closed Universe

In a closed Universe the gravitational attraction of all the visible and invisible matter will one day cause the galaxies to slow down and eventually stop. They will start to move closer together again. The Universe will shrink more and more until it disappears in a Big Crunch.

Into space

Spot facts

- *The world's most powerful rocket, Russia's Energia, can put into orbit satellites weighing up to 100 tonnes.*

- *Explorer 1, the first US satellite, circled the Earth 58,376 times during its 12 years in orbit (1958-1970).*

- *The US satellite Lageos, launched in 1976, will not fall back to Earth for 10 million years, when the Earth's surface will look quite different from how it does today.*

- *Some spy satellites, such as the US Big Bird, take photographs that are able to show clearly the letters and figures on car number plates.*

- *After a journey of 7,000 million km, the space probe Voyager 2 was guided in August 1989 to within 5,000 km of Neptune, then the most distant planet in the Solar System.*

▶ Carrying a communications satellite, an Atlas-Centaur rocket blasts off at night from Cape Canaveral in Florida. In a few hours the satellite will be in orbit nearly 36,000 km above the Earth's Equator.

High above the Earth in the airless world of space, hundreds of artificial moons, or satellites, circle silently in orbit. Some satellites pick up telephone messages and TV programmes from one country and beam them down to others. Some take photographs of the swirling clouds of fierce hurricanes and transmit them back to meteorologists. Others survey the Earth's surface with electronic "eyes" that can reveal otherwise invisible features.

A few spacecraft have escaped from the Earth and are aiming for a distant rendezvous with other planets. Others are heading out of our Solar System and beginning an aeons-long voyage to the stars.

Beating gravity

Every piece of matter in the Universe has an attraction for every other piece of matter. This attraction arises from gravitational forces. The English scientist Isaac Newton worked out the basic principles of gravity about 300 years ago. He realized that gravity holds the Universe together. It holds the Earth and the other planets in their paths, or orbits, around the Sun; the Sun and the other stars in orbit around our Galaxy; and so on.

On Earth gravity keeps our feet firmly on the ground and the gases in the atmosphere. It makes anything we throw up in the air soon come back down again to the ground. The Earth's gravity is very powerful, so how can we beat it and launch objects into space?

Newton worked out how gravity could be beaten by speed. However, to beat gravity an object must be launched from the Earth at the colossal speed of 28,000 km/h. At this speed it will be able to circle around the Earth. Because there is no air in space, there is nothing to slow the object down, and it will continue circling at the same speed, in orbit. It will become an artificial satellite of the Earth.

▶ A cross-section of the Earth's atmosphere. The air is thickest at the bottom. It thins out with increasing height until it merges into space. But faint traces of air remain even at 200 km high.

▼ Isaac Newton drew this diagram to show how to beat gravity. If you throw a ball faster and faster, it will travel farther and farther before falling back to Earth. At a very high speed indeed, the ball will "fall around the Earth" and enter orbit.

Into orbit

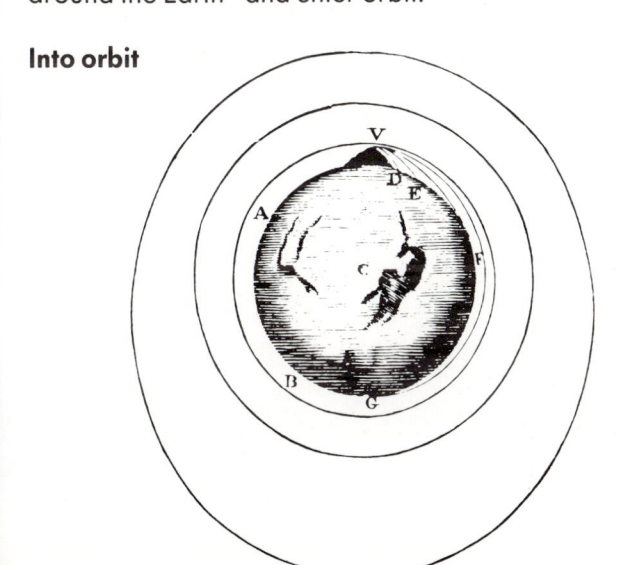

Earth's atmosphere

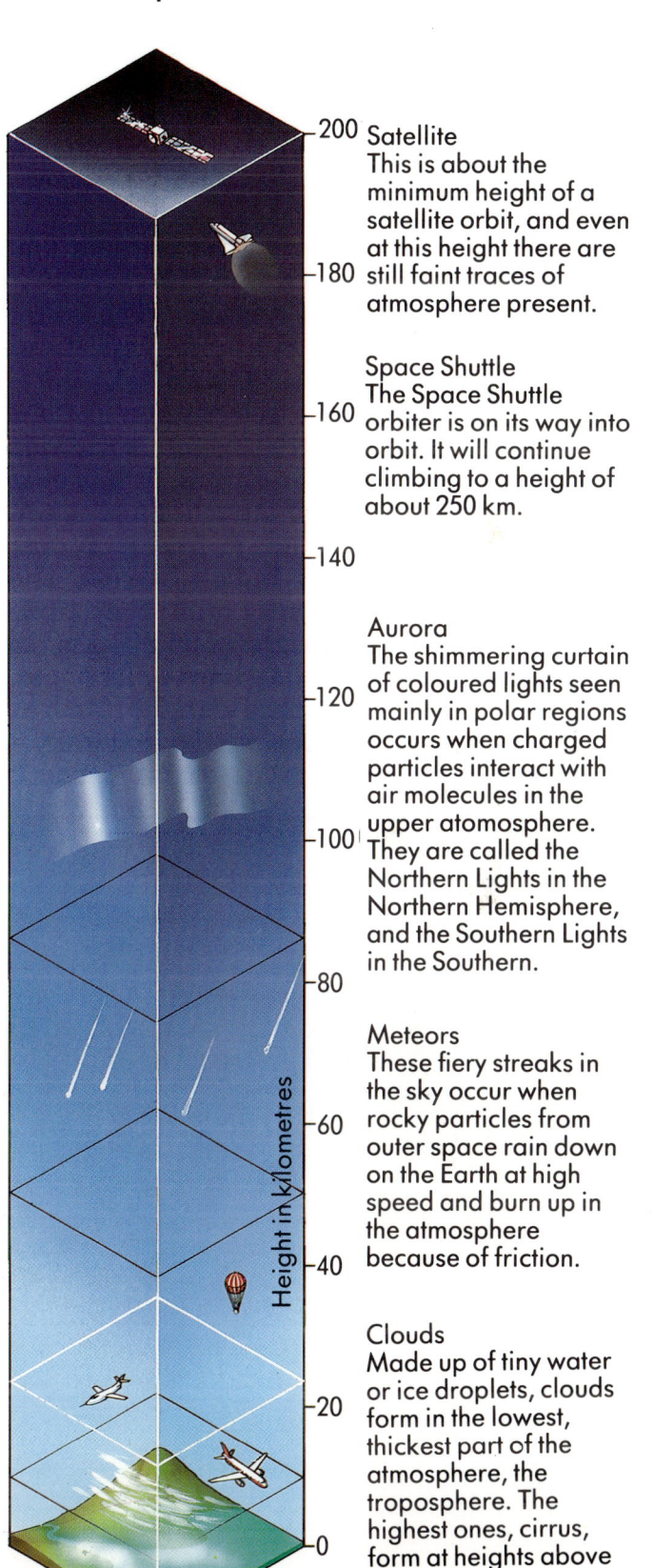

Satellite
This is about the minimum height of a satellite orbit, and even at this height there are still faint traces of atmosphere present.

Space Shuttle
The Space Shuttle orbiter is on its way into orbit. It will continue climbing to a height of about 250 km.

Aurora
The shimmering curtain of coloured lights seen mainly in polar regions occurs when charged particles interact with air molecules in the upper atomosphere. They are called the Northern Lights in the Northern Hemisphere, and the Southern Lights in the Southern.

Meteors
These fiery streaks in the sky occur when rocky particles from outer space rain down on the Earth at high speed and burn up in the atmosphere because of friction.

Clouds
Made up of tiny water or ice droplets, clouds form in the lowest, thickest part of the atmosphere, the troposphere. The highest ones, cirrus, form at heights above 6000 m.

199

Comsats and weather satellites

Perhaps the most useful kinds of satellites are those that relay, or pass on, communications between countries. These communications satellites, or comsats, handle all kinds of electronic communications in the form of microwaves, or very short radio waves, including telephone calls, radio broadcasts, TV programmes, telex and facsimile (fax) messages.

Signals are beamed up to the comsats and received back from them by huge dish aerials at transmit/receive ground stations. These are linked into each country's communications system by cable or microwave radio links.

Intelsat (International Telecommunications Satellite Organization) is the biggest worldwide satellite communications network. It has over 110 member nations, and launches and maintains powerful comsats, such as *Intelsat 6*, over the Atlantic, Pacific and Indian Oceans. The comsats are in geostationary orbit: they circle over the Equator at a height of 35,900 km. In this orbit they circle the Earth every 24 hours. In other words they keep pace with the Earth as it turns, and therefore appear fixed in the sky.

Russia maintains a large comsat network

The first Sputnik

Russia thrust the world into the Space Age on 4 October 1957, when it launched *Sputnik 1* with a modified Sapwood intercontinental ballistic missile (ICBM). It was an aluminium sphere measuring 58 cm across and weighing 83 kg. It sent back simple radio signals from its four long aerials. Its orbit took it as low as 220 km above the Earth, and the whiff of atmosphere there gradually caused it to slow down. It fell back to Earth after 92 days.

◀ A satellite dish in a remote village in India. It receives signals from a communications satellite in geostationary orbit, 35,900 km high. The satellite beams down regular television programmes for entertainment and also specialist programmes for education and instruction in, for example, farming and family health.

▼ A GOES weather satellite and an image taken by *GOES 4* in geostationary orbit over the eastern Atlantic Ocean. It shows a hurricane (David), spiralling over the Caribbean islands of Hispaniola and Puerto Rico. The central core of the hurricane measures more than 400 km across.

Geostationary orbit

Most satellites that need to reach geostationary orbit 35,900 km high are first launched into low orbit. Then their on-board motor fires to send them into a transfer orbit. When they are 35,900 km high, the motor fires again to direct them into a circular geostationary orbit.

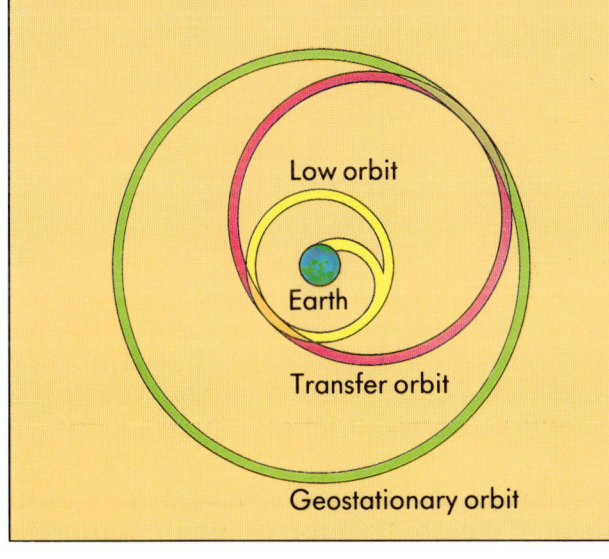

known as Orbita, which uses Molniya satellites. They do not circle in geostationary orbits, but in orbits that are eccentric, or highly elliptical. These orbits take them as high as 40,000 km over Russia but as low as 600 km on the other side of the Earth. In this way they are "in sight" of Russian ground stations for most of the time.

Weather forecasting has been revolutionized by the use of satellites. They are able to scan the whole Earth and the atmosphere continuously, day and night. They can show how weather systems are developing anywhere in the world, in places where there are no ground weather stations. They take cloud pictures, measure water and air temperature, and relay weather data.

Some weather satellites circle in geostationary orbit, where they view nearly a whole hemisphere. The US GOES and European Meteosat are examples. Other satellites are launched into a polar orbit, over the North and South Poles. They can scan the whole Earth every 12 hours as it spins beneath them. The US NOAA series of satellites are in polar orbit.

Earth-survey satellites

Mapmakers, town planners, mineral prospectors, farmers and foresters are among the many groups of people who have benefited from another kind of satellite. This is the Earth-survey, or Earth-resources satellite.

At the beginning of the Space Age, ordinary photographs of the Earth's surface from orbit revealed much useful information. Earth-survey satellites are able to gather even more by scanning the surface in light of different wavelengths, such as infrared.

The best-known series of such satellites has been the US Landsats, of which five were launched, the first in 1972 and the last (*Landsat 5*) in 1984. *Landsat 5* orbits at an altitude of about 700 km. Using an oscillating-mirror system, it scans the Earth's surface in 185-km square blocks in green and red visible light and at four infrared wavelengths.

The French SPOT and the European Space Agency's *ERS-1* are also Earth-survey satellites. *ERS-1* uses radar for scanning the surface and also monitors weather and climate.

Information comes back from the satellites in the form of electronic data. Computers process the data and display it as images. They can manipulate the data in various ways and display it in false-colour pictures. The colours are chosen so as to pick out certain features of the landscape. This is possible because every kind of feature has a different "spectral signature". This means that it reflects different wavelengths in its own particular way.

Landsat imaging

SPOT
The French Earth-resources satellite SPOT (satellite probatoire pour l'observation de la Terre) has better resolution than Landsat. It can spot details as small as 10 m – about the size of a bus. Like Landsat, it scans at different visible and infrared wavelengths.

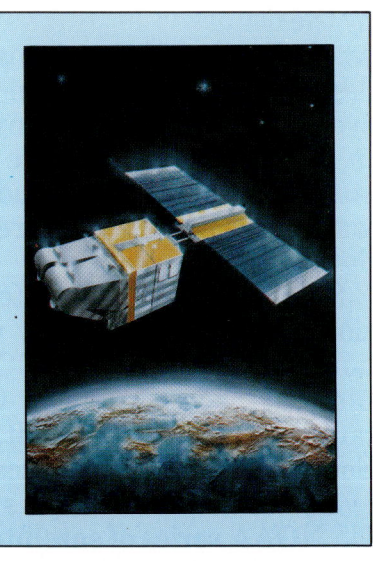

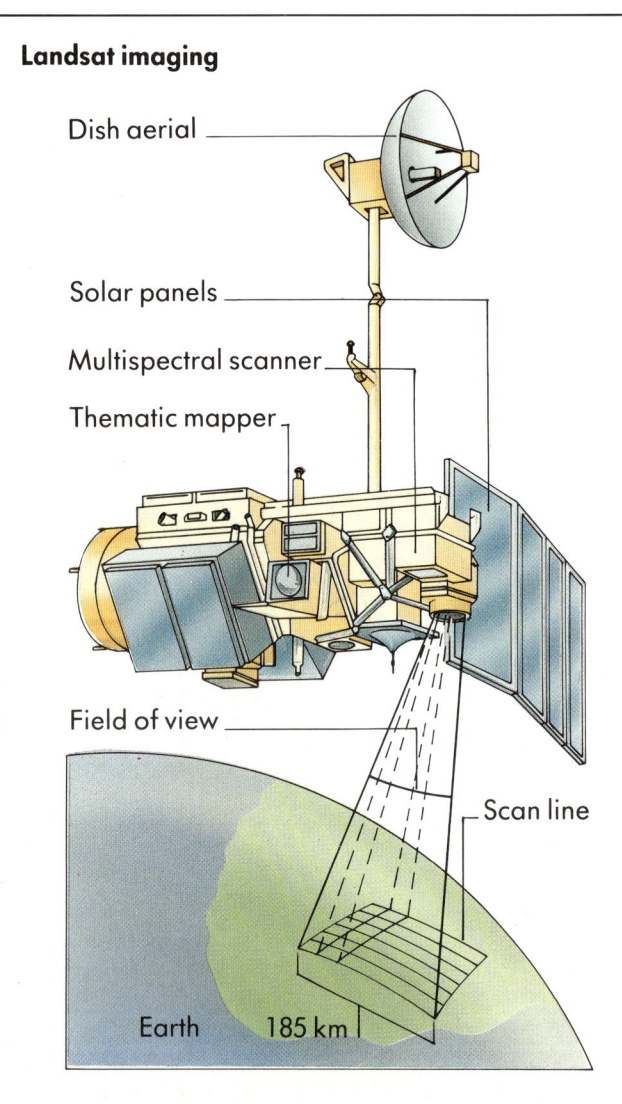

The *Landsat 5* satellite (top) pictures the Earth using two scanning systems, the Thematic Mapper and the Multispectral Scanner. Both scan the Earth in 185-km swathes at several wavelengths. The received data can be displayed in false colours on screen (above) or on film (opposite) in a number of different ways.

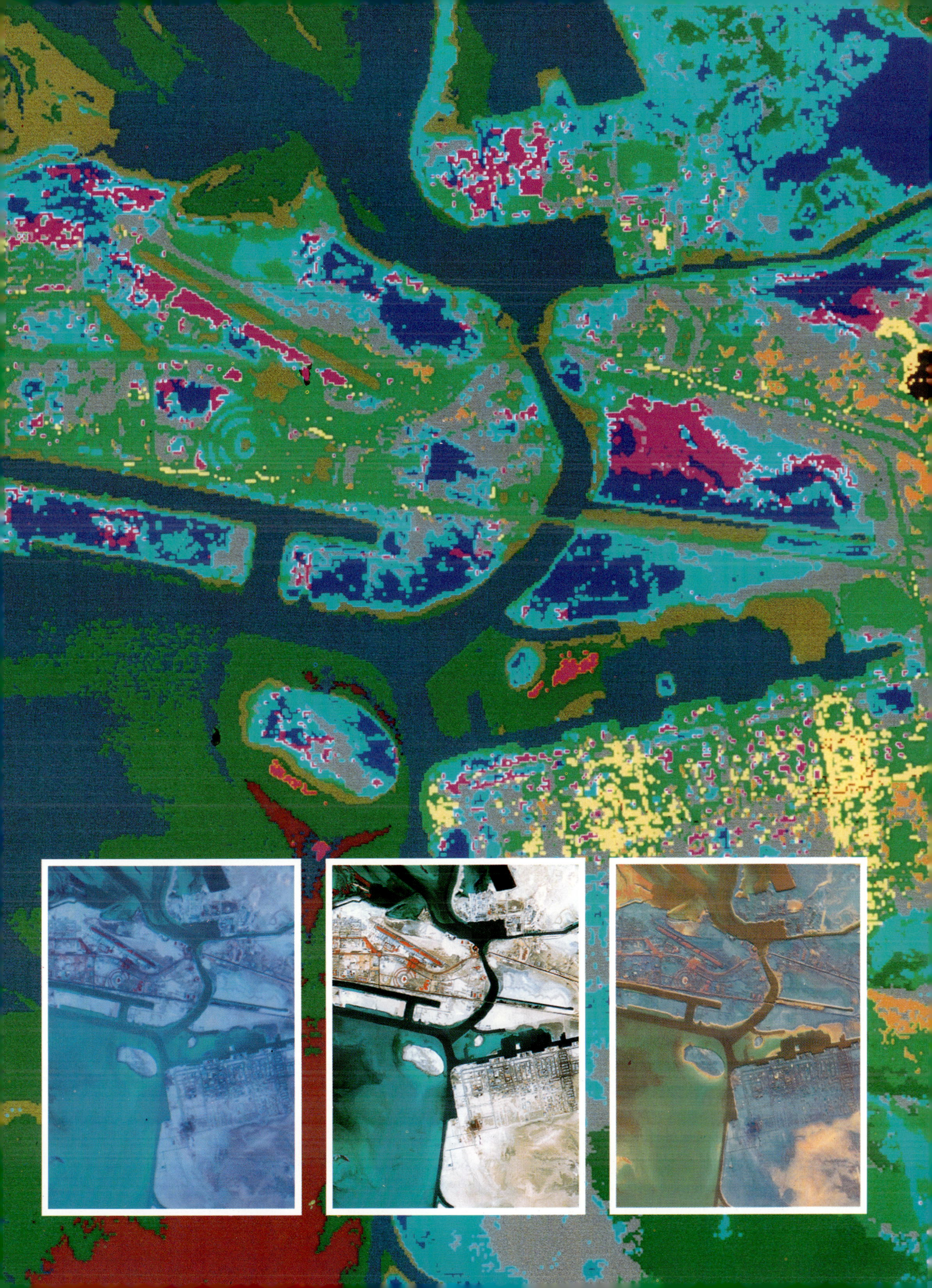

Probes to other worlds

If you launch a spacecraft with a speed of 28,000 km/h, it will go into orbit as a satellite. But it is still tied to Earth by gravity. You have to launch a spacecraft with a much higher speed if you want it to escape from the Earth's gravity completely. This speed, called escape velocity, is no less than 40,000 km/h. This is nearly 20 times the speed of the supersonic airliner Concorde.

A spacecraft that escapes from the Earth is called a space probe. The first probes were sent to the Moon. Russia first achieved success in 1959, when its probe *Luna 1* crashlanded there. Later, Russian and US lunar probes went into orbit around the Moon and landed on it. Exploration of the planets from space began in 1962, when the US probe *Mariner 2* flew close (35,000 km) to Venus. Since then probes have visited all the planets in our Solar System except Pluto. They have also flown to meet a regular visitor to Earth's skies, Halley's comet.

Targeting the planets

There are all kinds of problems involved in sending probes to the planets. One is distance. Even our nearest planetary neighbour, Venus, never comes closer to us than 42 million km. And the most distant planet at the moment, Neptune, lies more than 4,000 million km away. Using the rockets we have at present, it takes a probe several months to reach even Venus and our next nearest neighbour, Mars. It takes years to reach the more distant planets: Jupiter, Saturn, Uranus, and so on.

Another problem is aiming the probe. It must be aimed at a point in space in the target planet's orbit so that it will arrive at the same time as the planet. If the probe leaves the Earth in slightly the wrong direction or at slightly the wrong speed, then it could miss its target by hundreds of thousands of kilometres.

Keeping in touch

Maintaining communications with a probe over many millions of kilometres is also a major problem. The probe must be tracked precisely so

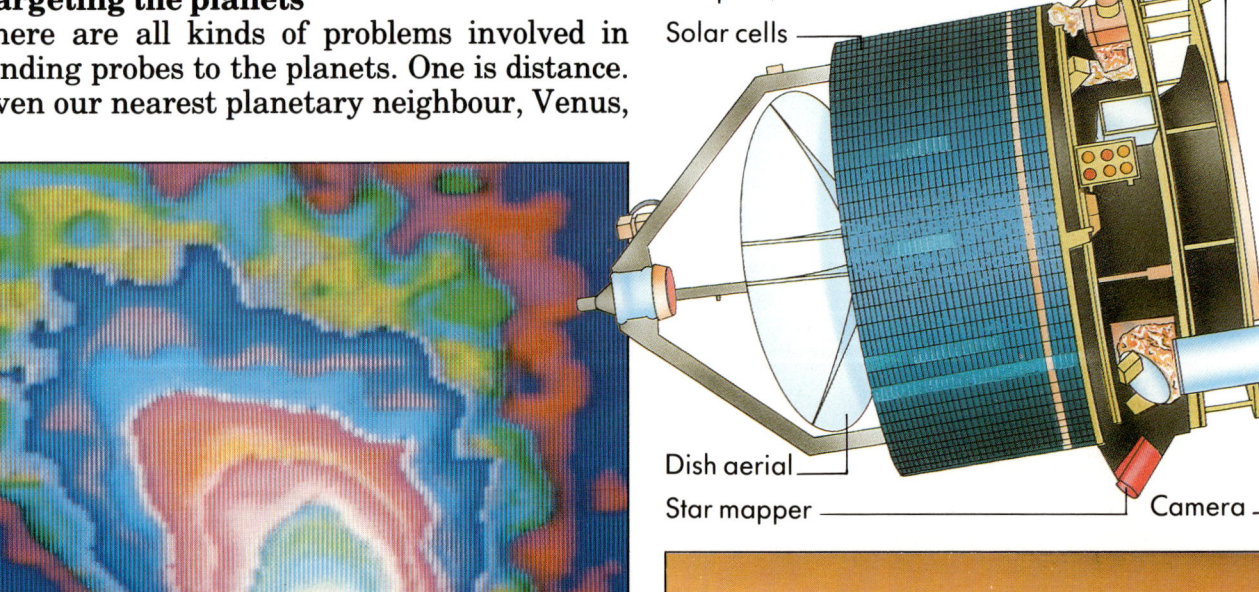

Giotto
Bumper shield
Solar cells
Dish aerial
Star mapper
Camera

▲ A false-colour image of Halley's comet, produced from data returned by the European space probe *Giotto* in March 1986. At the time the probe was less than 1,000 km away. It took a beating from the rocky debris around the comet, but managed to survive.

that radio signals can be beamed in the right direction in space. To send and receive signals, large dish aerials are used. NASA, the National Aeronautics and Space Administration of the United States, communicates with US space probes through its Deep Space Network. It comprises tracking stations at Goldstone in California, at Madrid in Spain and at Canberra in Australia. They use dish aerials up to 70 m across.

Because the probes are so far away, there is a time-lag between the sending and receiving of signals between the tracking station and the probes. When *Voyager 2* sent radio signals back from Neptune in 1989, they took over four hours to reach the Earth. The transmitter on the probe has a power output about the same as the bulb in a refrigerator. Yet NASA scientists were able to convert the signals into remarkably clear pictures.

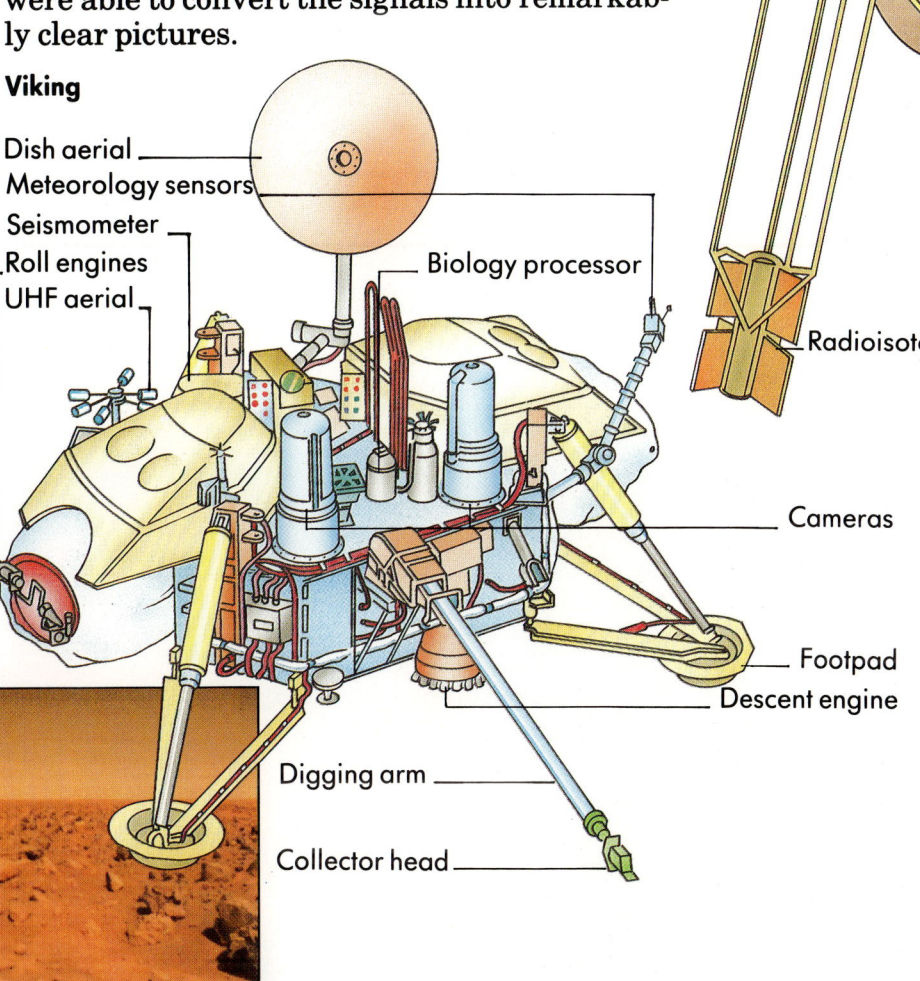

Viking

▲ *Pioneer 10* and its twin, *Pioneer 11*, were the first probes to journey through the asteroid belt to the giant planet Jupiter, in 1973 and 1974, respectively. In addition, *Pioneer 11* continued on to Saturn, which it reached in 1979. Because they travelled so far from the Sun, the Pioneers carried nuclear batteries (RTGs) to power their instruments.

◄ Two Viking probes landed on the Red Planet Mars in 1976, and took close-up pictures of its surface (photo). The pictures revealed that the surface is rust-red in colour, and even the sky has a reddish tinge. The landing probes carried automatic soil samplers, which tested the soil for signs of life, but in vain. They reported temperatures up to −30°C and winds gusting up to 120 km/h.

205

Voyager 2

The US spacecraft *Voyager 2* has been the most successful probe ever launched. It began its journey of discovery in 1977, reaching Jupiter in 1979, Saturn in 1981, Uranus in 1986 and Neptune in 1989. And what sights it saw: raging storms on Jupiter, sulphur volcanoes on Jupiter's moon Io; tiny "shepherd" moons of Saturn that keep its rings in place; crazy landscapes on Uranus's moon Miranda; and geysers of liquid gas on Neptune's moon Triton.

Voyager 2's "grand tour" of the outer planets was made possible by two things. The planets were aligned in space in a favourable way. This happens only once every 175 years. Also, *Voyager 2* used gravity-assist to direct and accelerate it from planet to planet.

The gravity-assist technique was first used with *Mariner 10* in 1973 to enable the probe to visit both Venus and Mercury. On a gravity-assist mission, a probe is targeted close to a planet. The planet's gravity then makes the probe speed up and curve around the planet before being slung in another direction.

Voyager 2 is now heading out of our Solar System towards interstellar space. It set out from Earth two weeks before its sister craft *Voyager 1*. *Voyager 1* visited only Jupiter and Saturn before heading for the stars.

It is just possible, in aeons to come, that the Voyagers might be found by intelligent beings from another planet in another solar system. In case they are, they both carry record discs called Sounds of Earth. On these discs are recorded greetings from Earth people in 60 languages, sounds from nature and the human world and, in code, a selection of photographs. Helpfully, instructions on how to play the discs are given in pictorial form on the record covers.

▶ The path *Voyager 2* took through the Solar System on a 12-year, four-planet mission of discovery that cannot be repeated for more than a century and a half. *Voyager 2* set out from planet Earth on 20 August 1977. It arrived at its last port of call, Neptune, on 25 August 1989. Its 7-billion km journey had been so well planned that it was able to swoop to within 5,000 km of Neptune's cloud tops. Now it is heading towards a region called the heliopause, which marks the boundary between the Solar System and interstellar space. In 300,000 years time it should pass within a few light-years of Sirius, the brightest star in the sky.

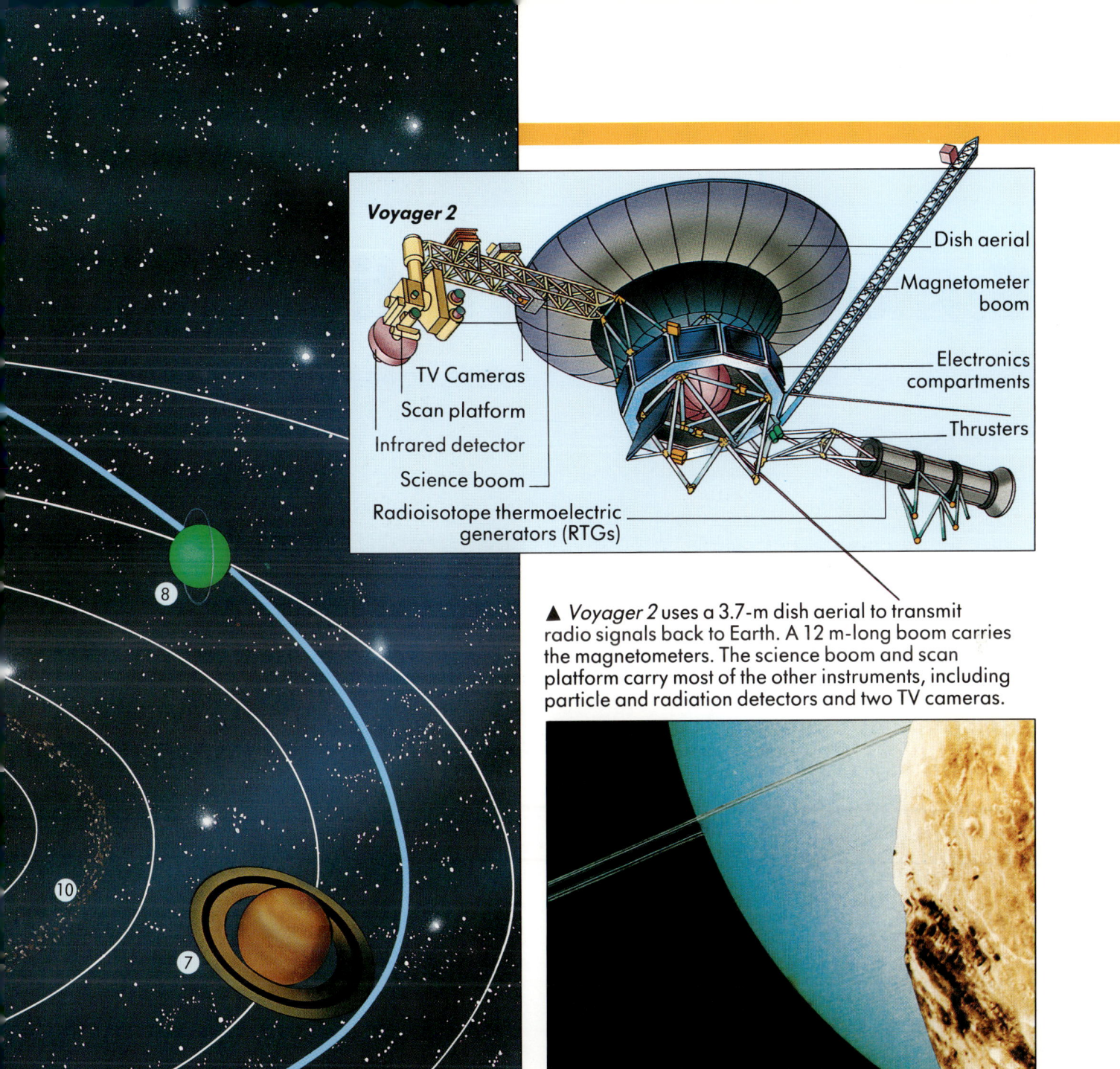

Voyager 2

- Dish aerial
- Magnetometer boom
- Electronics compartments
- Thrusters
- TV Cameras
- Scan platform
- Infrared detector
- Science boom
- Radioisotope thermoelectric generators (RTGs)

▲ *Voyager 2* uses a 3.7-m dish aerial to transmit radio signals back to Earth. A 12 m-long boom carries the magnetometers. The science boom and scan platform carry most of the other instruments, including particle and radiation detectors and two TV cameras.

1 Sun
2 Mercury
3 Venus
4 Earth
5 Mars
6 Jupiter
7 Saturn
8 Uranus
9 Neptune
10 Asteroids

Voyager 2 images

Voyager 2 sent back tens of thousands of images during its remarkable mission. It spied the multicoloured disc of giant Jupiter (far left), Saturn's glorious rings and several of its moons (left). At Uranus, *Voyager 2* spotted a system of faint rings (above), seen here looking across the rugged surface of the moon Miranda. Neptune (top left) was revealed as a blue cloud-flecked planet with dark spots, which are probably huge storms.

Space transportation

Spot facts

- Neil Armstrong was the first human being to step down on to the Moon on 20 July 1969. "That's one small step for a man," he said, "one giant leap for mankind."

- The two eight-track crawler transporters built to carry the Saturn V/Apollo stacks to the launch pad are now used to carry the Space Shuttle stack. They are still the world's largest land vehicles, measuring 40 m long, nearly 35 m wide and 8 m high.

- During critical phases of a Shuttle mission, such as launching, the on-board computer system has to perform over 300,000 operations every second.

▶ Astronaut Eugene Cernan takes the lunar roving vehicle for a test drive during the *Apollo 17* Moon-landing mission in December 1972. The collapsible "Moon buggy" was powered by electric motors and had a top speed of 16 km/h.

Engineers in Russia and the United States began designing suitable vehicles for transporting human beings into space almost as soon as the Space Age began in 1957. But the first manned spacecraft, such as Vostok, Mercury and Gemini, were cramped, uncomfortable capsules. The next-generation Apollo and Soyuz spacecraft had only marginally more room. But the Apollo craft nevertheless supported crews of three astronauts on daring flights to the Moon and back, not once but nine times. Not until 1981 did the modern era of space travel begin, at least for the Americans, using the Space Shuttle, the world's first reusable spaceship. The Russians launched a shuttle craft in 1988.

Early days

A Russian Air Force major, Yuri Gagarin, made the first human flight in space on 12 April 1961. He orbited the Earth once in a *Vostok 1* capsule, landing by parachute after ejecting. Cosmonaut Gagarin was aloft for 108 minutes.

The Americans, racing to catch up, managed to launch Alan Shepard in a Mercury capsule named *Freedom 7* on a 15-minute suborbital flight into space on 5 May 1961. Not until 20 February 1962 did John Glenn become the first American in orbit, circling the Earth three times in the Mercury capsule *Friendship 7*. They both splashed down at sea.

The Russians launched the first multiple crew into space in a modified Vostok craft called *Voshkod 1* in October 1964. The second-generation US spacecraft was the Gemini. It was named after the constellation of the zodiac, whose English name is the Twins. It was an apt name because the Gemini craft carried a crew of two. Twelve highly-successful Gemini flights took place in 1965 and 1966, during which the astronauts practised spacewalking, manoeuvring in orbit, and other techniques that would be needed on the Apollo missions to the Moon.

▲ Virgil Grissom in the Mercury capsule *Liberty Bell 7* blasts off the pad at Cape Canaveral, Florida, on 21 July 1961 to begin a 15-minute suborbital flight.

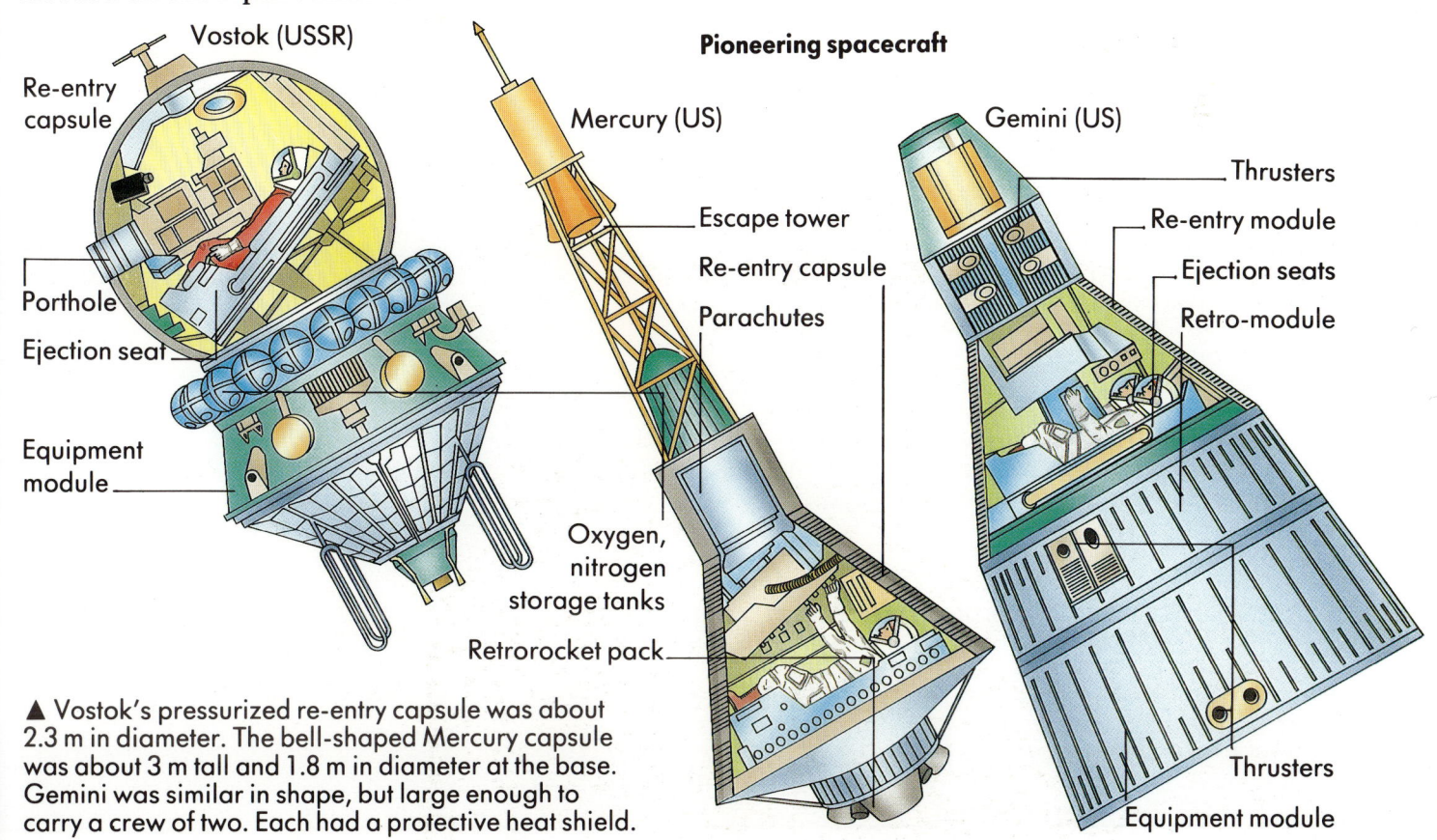

Pioneering spacecraft

▲ Vostok's pressurized re-entry capsule was about 2.3 m in diameter. The bell-shaped Mercury capsule was about 3 m tall and 1.8 m in diameter at the base. Gemini was similar in shape, but large enough to carry a crew of two. Each had a protective heat shield.

Apollo

On 25 May 1961 the US President John F. Kennedy urged the American people to undertake the greatest adventure in the history of humankind. "I believe that this nation should commit itself", he said, "to achieving the goal, before this decade is out, of landing a man on the Moon and returning him safely to Earth."

This plea gave birth to the Apollo Moon-landing project. In the event, the Americans achieved not just one, but two landings before the decade was out. The first landing, on 20 July 1969, saw *Apollo 11* astronauts Neil Armstrong and Edwin Aldrin walking on the Moon's Sea of Tranquillity. They were followed over the next three-and-a-half years by five more crews, from *Apollo 12, 14, 15, 16* and *17*. *Apollo 13* aimed for a Moon landing but was nearly blasted apart *en route*. The crew just managed to make it safely back to Earth. The *Apollo 17* astronauts left the Moon on 14 December 1972. No humans will return there this century.

To launch human beings to the Moon and bring them back safely was an enormous undertaking. It required a great technological effort and also the creation of some gigantic

▲ A Saturn V rocket thunders away from the launch pad on 21 December 1968, carrying three astronauts in *Apollo 8* to the first human encounter of the Moon.

To the Moon and back

The technique Apollo used to reach the Moon was called lunar orbit rendezvous. The Apollo spacecraft lifted off atop a Saturn V rocket (1,2,3). It was then accelerated out of Earth orbit (4) and configured for the outward journey (5,6,7). Retrofire (8) took it into lunar orbit, where the lunar module (LM) separated (9) and dropped down to land (10). After the landing mission, the top part of the LM took off (11) and rendezvoused with the mother ship (12). A burn of the main engine (13) boosted the craft out of lunar orbit for the return to Earth. Before re-entry the service module was jettisoned (14). The command module, travelling at nearly 11 km a second and with heat shield blazing (15), plunged through the atmosphere. The air slowed it down, then parachutes opened to lower it to a gentle splashdown (16).

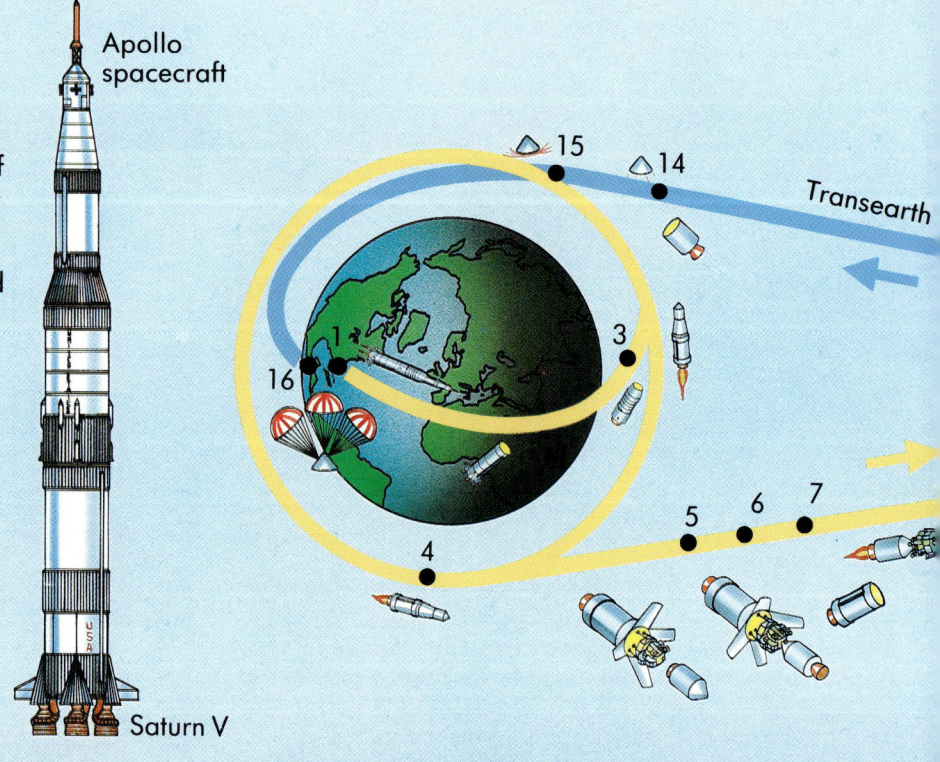

Apollo spacecraft

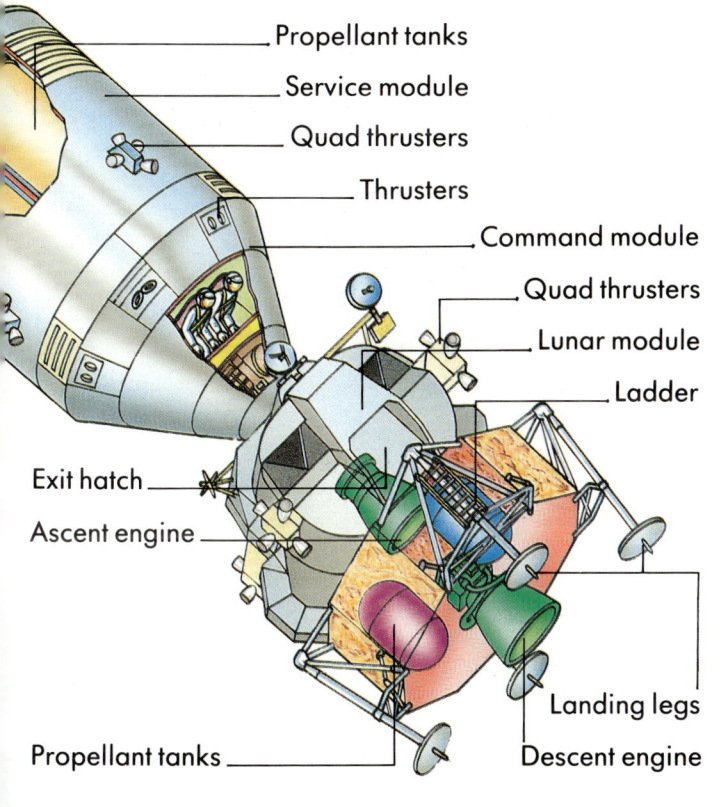

structures and equipment. To launch the 45-tonne Apollo spacecraft to the Moon required a mammoth rocket, the Saturn V, which stood 111 m high and weighed 3,000 tonnes.

To assemble such a giant required a massive building, the Vehicle Assembly Building, at the launch site, the Kennedy Space Center in Florida. This building, now used to assemble the Space Shuttle, measures 160 m high, 158 m wide and 218 m long.

A three-module design was adopted to fit in with the technique chosen for the Moon landing, called lunar orbit rendezvous. The main part of the craft was the pressurized command module (CM), which housed the crew of three. This was the only part to return to Earth. For most of the mission it was attached to the service module (SM), the combined unit being termed the CSM. The third unit was the lunar module (LM), the spacecraft used to ferry two of the crew to and from the Moon's surface.

◀ The three modules of the Apollo spacecraft, linked up for the journey to the Moon. The spacecraft measured about 17 m long overall and up to about 4 m across. It had an Earth weight of about 45 tonnes.

▼ On the *Apollo 15* mission James Irwin is pictured with the lunar module and lunar roving vehicle. Behind him are the Apennine Mountains.

◀ Sites of the six Apollo landings. *Apollo 11* landed on the Sea of Tranquillity; *Apollo 12* on the Ocean of Storms; *Apollo 14* at Fra Mauro; *Apollo 15* and *17* on the edge of the Sea of Serenity; and *Apollo 16* in the lunar highlands.

211

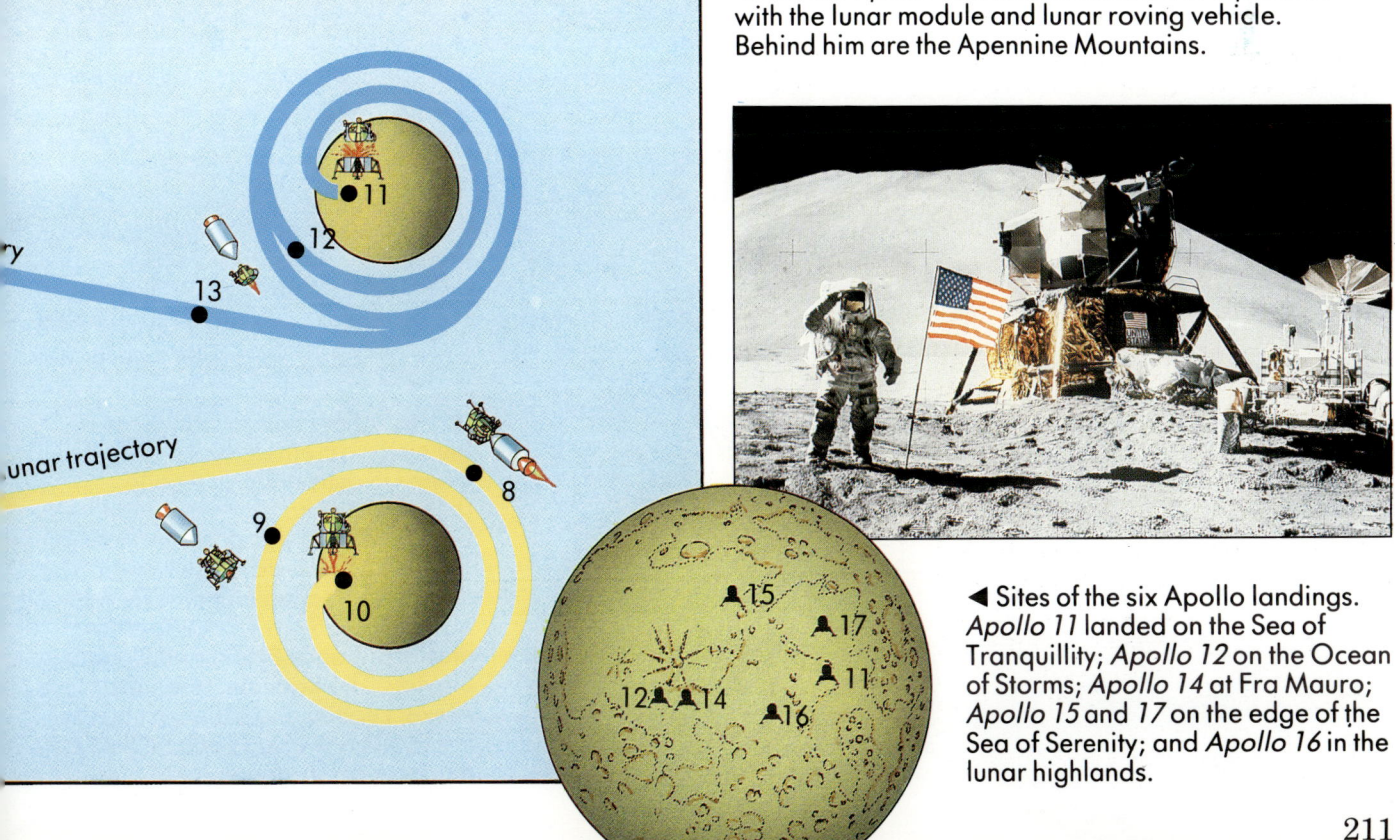

Soyuz

Russian space scientists developed a successor to their Vostok and Voshkod spacecraft at much the same time as the Americans developed Apollo. The Russian craft, called Soyuz, made its maiden flight in April 1967. But it ended in tragedy when its pilot, cosmonaut Vladimir Komarov, was killed during landing. He was the first known in-flight casualty of the Space Age. Even today Russian cosmonauts make their journeys into space in Soyuz spacecraft. Russia has a shuttle craft, called *Buran*, but it is not yet used as a ferry.

Soyuz, like Apollo, is made up of three modules and measures nearly 8 m long and up to 2.7 m across. At the front is the orbital module, in which the crew can work in orbit. It has a hatch by which it can link up, or dock, with Russia's space stations. The crew usually occupy the centre section, the command and re-entry module. At the rear is the instrument module, which contains equipment, fuel and rocket motors. The front and rear modules separate before re-entry. The re-entry module is braked first by the atmosphere, then by parachutes and finally, just before landing, by

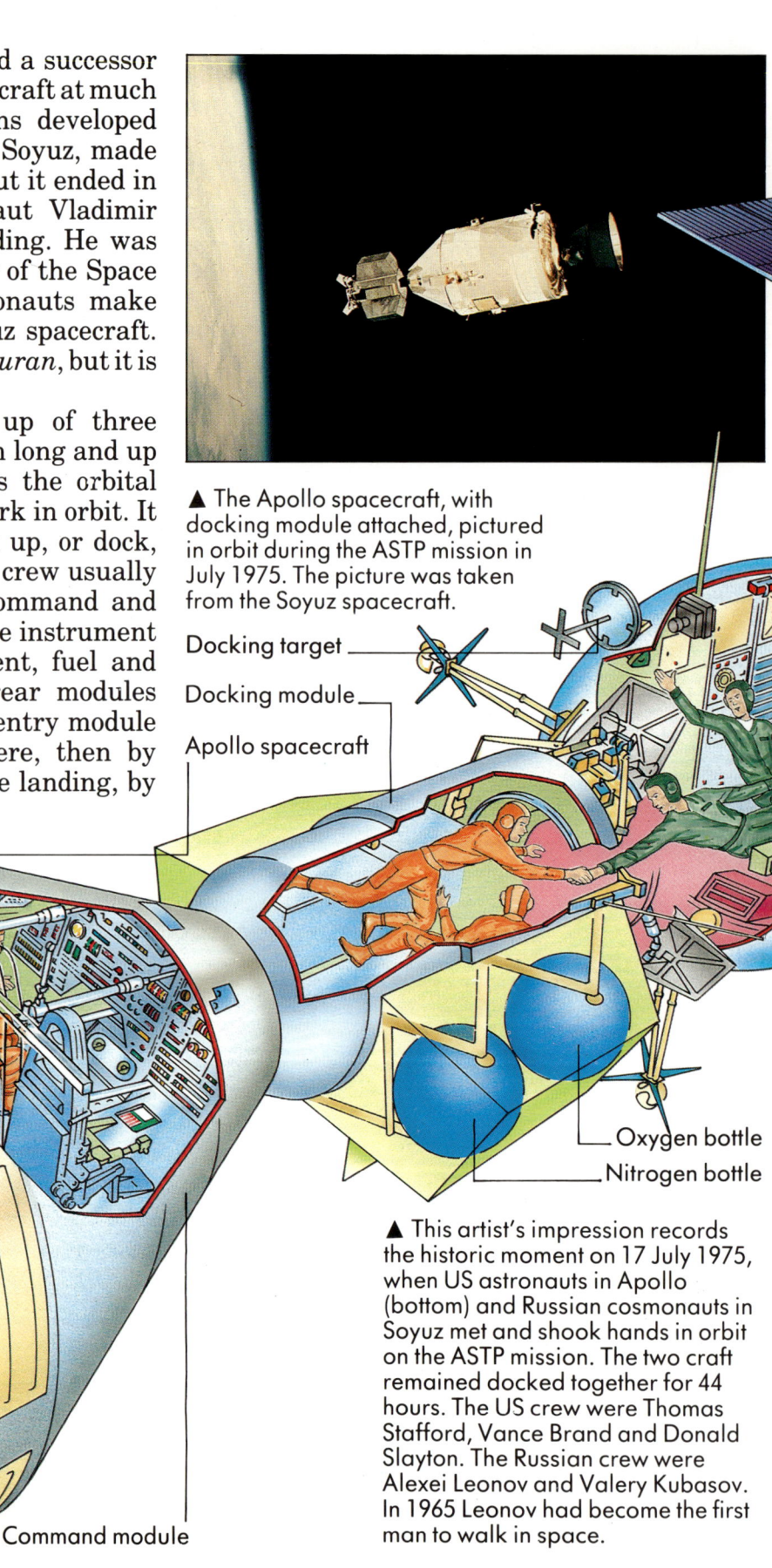

▲ The Apollo spacecraft, with docking module attached, pictured in orbit during the ASTP mission in July 1975. The picture was taken from the Soyuz spacecraft.

Docking target
Docking module
Apollo spacecraft

Oxygen bottle
Nitrogen bottle

▲ This artist's impression records the historic moment on 17 July 1975, when US astronauts in Apollo (bottom) and Russian cosmonauts in Soyuz met and shook hands in orbit on the ASTP mission. The two craft remained docked together for 44 hours. The US crew were Thomas Stafford, Vance Brand and Donald Slayton. The Russian crew were Alexei Leonov and Valery Kubasov. In 1965 Leonov had become the first man to walk in space.

Command module

Handshake in orbit

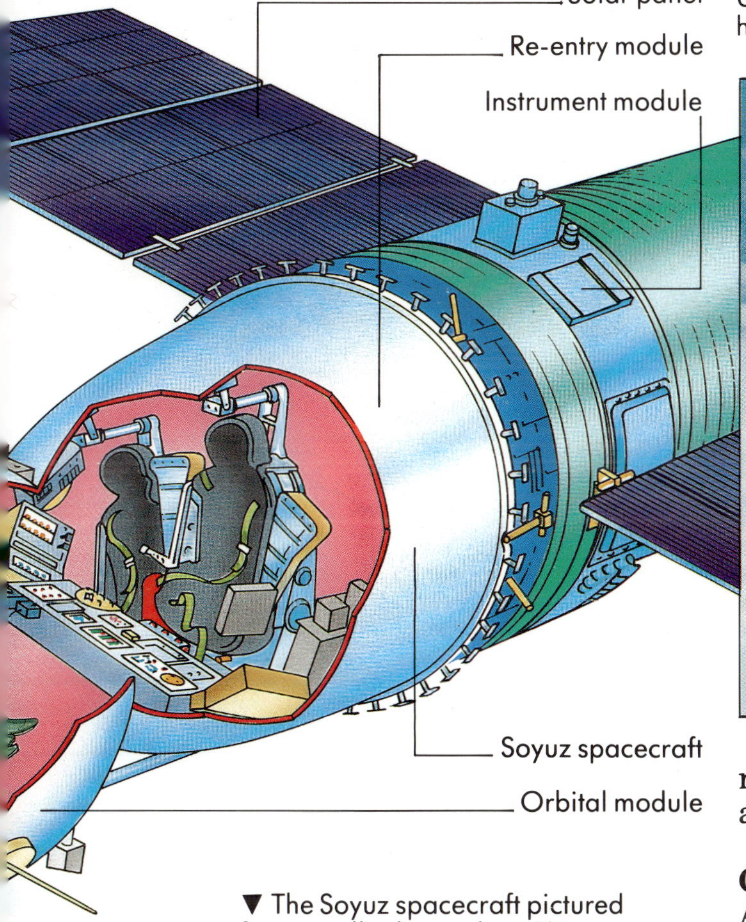

▼ A Soyuz spacecraft and launch vehicle on the launch pad at the Baikonur Cosmodrome in central Asia. The gantry structure is being retracted prior to a launch. On the pad the vehicle stands 45 m tall. Unlike US rockets, Russian launch vehicles are put together horizontally and then tilted upright on the pad.

- Solar panel
- Re-entry module
- Instrument module
- Soyuz spacecraft
- Orbital module

▼ The Soyuz spacecraft pictured from Apollo during the ASTP mission. At the front of the craft (left), on the spherical orbital module, is the docking mechanism.

retrorockets. The crew stay inside the module all the way down.

Cooperation in the cosmos

At the beginning of the Space Age Russia and the United States were battling with each other for supremacy in space. This became known as the Space Race. In 1975, however, old rivalries were set aside, at least temporarily, when the two space powers mounted a joint mission, the Apollo-Soyuz Test Project (ASTP).

To allow the two craft to dock together, a special docking module was designed. It had two docking ports: one fitted the Apollo and the other the Soyuz docking systems. Apollo flew into orbit with the docking module attached on 15 July 1975 from the Kennedy Space Center in Florida. Soyuz was launched a few hours later from its usual launch site, the Baikonur Cosmodrome in Central Asia. On 17 July they met up and docked with each other. They remained docked for nearly two days, while the astronauts dined, relaxed and carried out a number of experiments together.

The Shuttle system

All of the rocket launch vehicles used in the first two decades of space flight were expendable. This means that they could be used only once. This represented a very wasteful kind of technology. It made more sense, surely, to design a vehicle that could be used again and again. This thinking led to the birth of the US Space Shuttle, which made its maiden flight into orbit in 1981.

Three main elements make up the Space Shuttle transportation system. The main one is the winged orbiter, which carries the crew and the payload (cargo). It rides into space on top of a huge external tank, which carries fuel for its engines. Strapped to the sides of the tank are twin solid rocket boosters.

These elements are put together in the cavernous Vehicle Assembly Building at Complex 39 at the Kennedy Space Center in Florida, the main shuttle launch site. They are mounted

▼ On 12 April 1981 orbiter *Columbia* soars from the launch pad on the first Space Shuttle mission. On board are test pilots John Young and Robert Crippen. They touch down 54 hours later at the Edwards Air Force Base in California after a flawless flight.

▼ The Space Shuttle orbiter is launched like a rocket, acts like a spacecraft in space, and lands on a runway like an ordinary aircraft. It is blasted from the pad by its three main engines and the twin solid rocket boosters (SRBs), which together develop a thrust of more than 3,300,000 kg. On its way into orbit, about 250 km high, it sheds its boosters and external tank in turn. To return to Earth, the orbiter fires retrorockets and drops from orbit. Friction with the air slows it down and makes its heat shield blaze. It makes an unpowered runway landing, like a glider, and is later flown back by jumbo jet to its launch site.

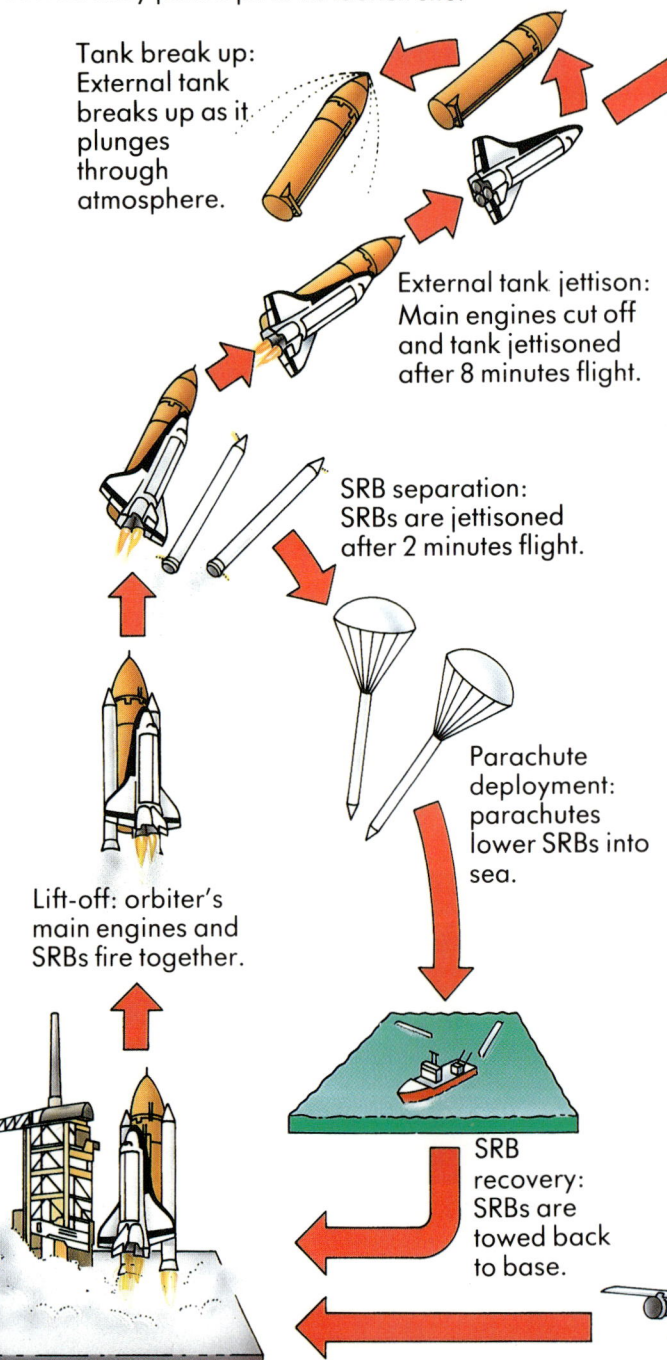

Tank break up: External tank breaks up as it plunges through atmosphere.

External tank jettison: Main engines cut off and tank jettisoned after 8 minutes flight.

SRB separation: SRBs are jettisoned after 2 minutes flight.

Parachute deployment: parachutes lower SRBs into sea.

Lift-off: orbiter's main engines and SRBs fire together.

SRB recovery: SRBs are towed back to base.

From lift-off to landing

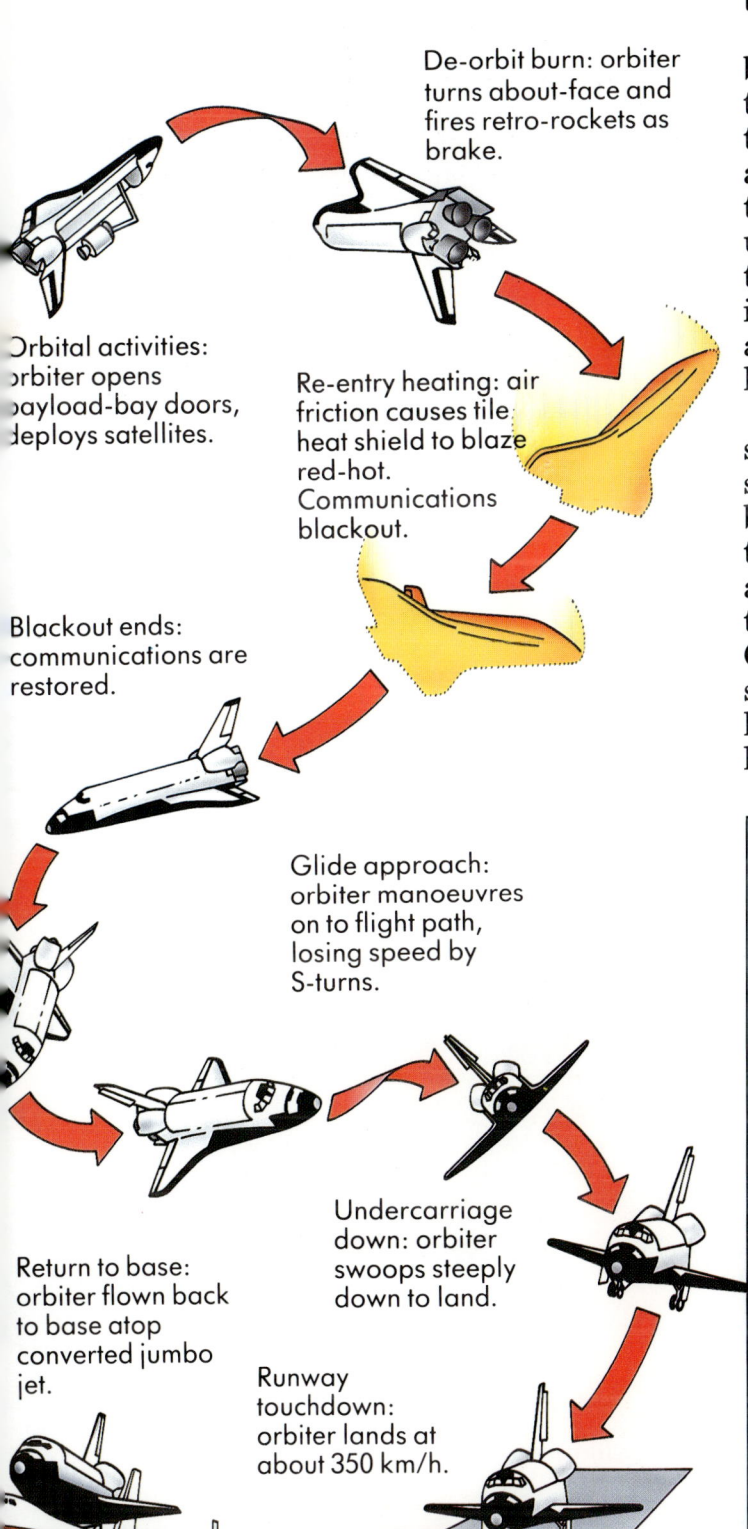

De-orbit burn: orbiter turns about-face and fires retro-rockets as brake.

Orbital activities: orbiter opens payload-bay doors, deploys satellites.

Re-entry heating: air friction causes tile heat shield to blaze red-hot. Communications blackout.

Blackout ends: communications are restored.

Glide approach: orbiter manoeuvres on to flight path, losing speed by S-turns.

Undercarriage down: orbiter swoops steeply down to land.

Return to base: orbiter flown back to base atop converted jumbo jet.

Runway touchdown: orbiter lands at about 350 km/h.

vertically on a mobile launch platform, which is carried out to the launch pad by a huge crawler transporter.

The solid rocket boosters fire for two minutes before falling away. They parachute down to the ocean, where they are recovered. They are then towed back to the Space Center to be used again. The main engines meanwhile continue to thrust the Shuttle ever faster, ever higher, until, after eight minutes of flight, the external tank runs out of fuel. This is then discarded, and is not recovered. Two small engines then fire to accelerate the orbiter to orbital velocity (28,000 km/h) and place it in orbit.

Later, at the end of the Shuttle mission, these same engines fire again as retrorockets. They slow down the orbiter so that gravity can pull it back to Earth. It re-enters the atmosphere travelling at about 25 times the speed of sound, and air friction rapidly slows it down. It glides in to land, usually at Edwards Air Force Base in California, at a speed of about 350 km/h. A specially-converted Boeing 747 carrier jet is on hand to transport it back to the Kennedy launch site, to be prepared for its next mission.

Russia's shuttle

On 15 November 1988 a Russian space shuttle lifted off from Baikonur Cosmodrome in central Asia. The winged orbiter, named *Buran*, was making its maiden, unmanned flight. *Buran* looks much like the US Shuttle orbiter. It rides into space on the world's most powerful launch vehicle, Energia.

Shuttle hardware

The orbiter is the key part of the Space Shuttle system. It is ingeniously designed to be part rocket, part spacecraft and part aircraft, and it performs all these functions well. A fleet of four orbiters was originally planned: *Columbia*, *Challenger*, *Discovery* and *Atlantis*.

On making its second flight into orbit in November 1981, *Columbia* became the first launch vehicle ever to return to space. *Challenger* met a tragic end on its tenth flight in January 1986, when it exploded 73 seconds after lift-off, killing its crew of seven. Shuttle operations were suspended until September 1988 to allow modifications of the hardware and of management procedures to take place. A replacement orbiter, *Endeavour*, was commissioned for operation in 1993.

The crew of up to seven ride in the forward fuselage of the orbiter, pressurized with air at normal pressure. Two astronauts fly the craft from a cockpit at the front of the upper deck. The cockpit looks much like that of a modern airliner, but has more switches, instruments and controls. It also incorporates three video display units connecting with the orbiter's powerful computer system.

The orbiter carries its payload (cargo) in the huge payload bay, which measures 18 m long and 4.5 m across. Because it is so large, the bay can accommodate two or more satellites at the

Space Shuttle construction

- Remote manipulator system
- Getaway Specials
- Payload bay
- Satellite pods
- Rudder
- OMS engine
- OMS/RCS propellants
- Main engines
- Body flap
- RCS thrusters
- Separation motors
- SRB (solid rocket booster)
- Delta wing
- Elevons

▶ The delta-winged Space Shuttle orbiter is about the size of a medium-sized airliner, with a length of 37 m and a wingspan of nearly 24 m. On the launch pad it weighs about 90 tonnes. The biggest part of the Shuttle stack is the external tank, which measures 47 m long. It holds some 2 million litres of liquid hydrogen and liquid oxygen, the propellants for the orbiter's three main engines. Strapped to the tank are two boosters, which burn solid propellants, a mixture of powdered aluminium (fuel) and ammonium perchlorate (oxidizer). The solid rocket boosters (SRBs) are 45.5 m long and nearly 4 m in diameter. They are made up of thick steel segments, locked together by pins. The joints are sealed by sets of rings.

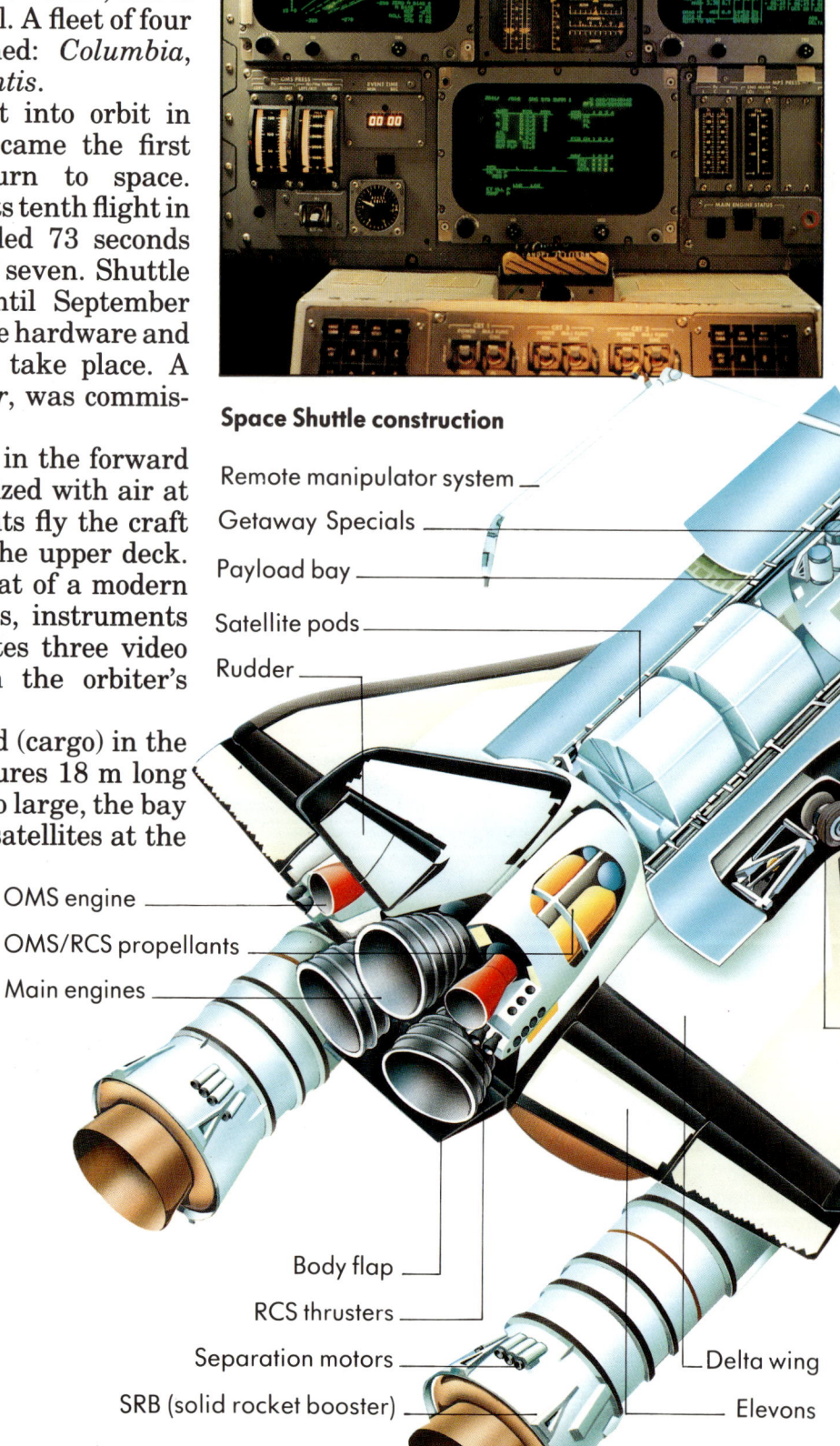

◀ A close-up of the instrument console in the cockpit of the Space Shuttle orbiter, showing the three cathode-ray tube displays. They are tied into the powerful computer system that operates the orbiter. The pilot and commander can call up all kinds of data on to the screens via computer keyboards.

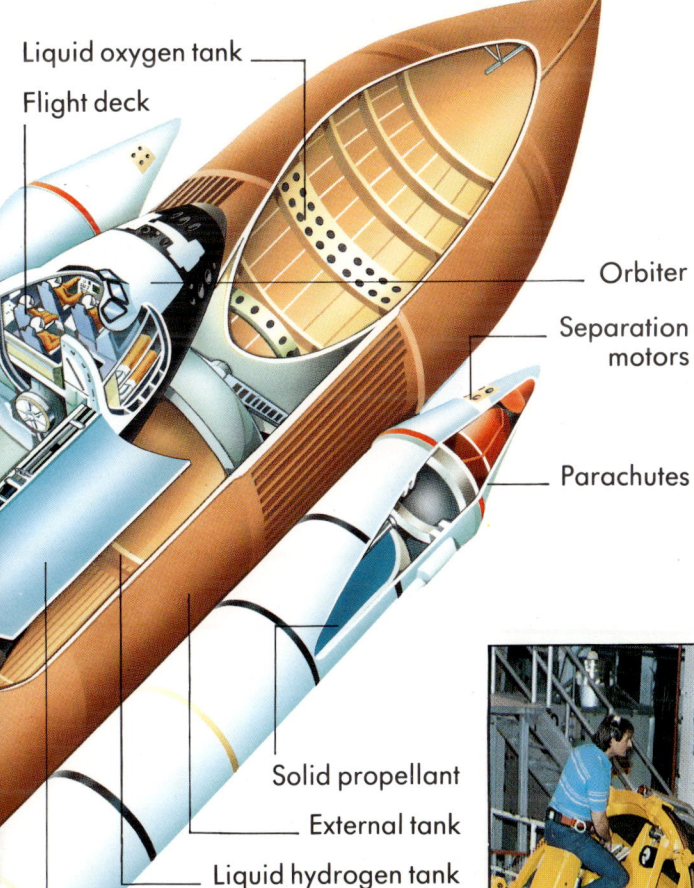

Liquid oxygen tank
Flight deck
Orbiter
Separation motors
Parachutes
Solid propellant
External tank
Liquid hydrogen tank
Radiator panel
Undercarriage
Carbon insulation

▶ A replacement main engine being manoeuvred into position in the tail pod of orbiter *Discovery* in preparation for Space Shuttle mission STS-26. The mission, which eventually took place in September 1988, was the first since *Challenger* exploded in the Florida skies just after lift-off on 28 January 1986. That tragedy, in which seven astronauts died, forced a complete rethink of Shuttle design and operational procedures to ensure greater safety in the future.

same time. On some missions it carries a single large payload, such as the European-built Spacelab, a fully equipped scientific laboratory. Around the major payloads, the bay also has room for so-called Getaway Specials. These are experiments by small research teams that can "hitch a ride" into space at low cost.

The tail pod houses the three main engines and the two engines of the orbital manoeuvring system (OMS). The OMS engines fire to inject the orbiter into orbit and to brake the craft prior to re-entry. The pod also carries sets of thrusters for the reaction control system (RCS), by which the pilots can change the position, or attitude, of the craft in space.

To prevent the aluminium airframe of the orbiter overheating when re-entering the atmosphere, it is covered with insulation. Much of the orbiter is covered with a layer of ceramic tiles made of silica. Over 30,000 tiles are required, each one individually tailored for a particular location. On the nose and wing edges, where temperature can soar to over 1,500°C, a carbon refractory material is used.

217

Shuttle operations

On most Shuttle missions the main objective is to launch satellites. Launch operations are carried out from a console at the rear of the orbiter flight deck. From there astronauts can look through windows into the payload bay.

Satellites may be launched in a number of ways. They can be sprung out of a pod or rolled from the bay rather like a frisbee. They can also be literally placed in orbit by the Shuttle's "crane", the robot arm of the remote manipulator system. The Hubble Space Telescope, for example, was launched in this manner in 1990. The robot arm, which is 15 m long, extends from inside the payload bay. It has flexible joints and a snare device at the "hand" end to grip the satellites.

The Shuttle usually goes into orbit at a height of about 250 km. This is much too low for many of the satellites it launches. Most communications satellites, for example, need to orbit at 35,900 km. So these satellites have a booster rocket attached, which fires to lift them to high orbit. Shuttle-launched probes destined to explore the planets likewise carry a powerful booster (called the Inertial Upper Stage) to accelerate them to escape velocity.

The robot arm is also put to good use in retrieving satellites from orbit. This technique was first used in 1984 to capture a satellite called *Solar Max*, which had malfunctioned only a few months after launch four years before. After the satellite had been captured, spacesuited astronauts repaired it. It was then relaunched by the arm and operated successfully for six years before falling from orbit.

Shuttle astronauts also carry out a certain amount of experimental work in orbit. However, most scientific work on the Shuttle takes place during missions in which the scientific laboratory *Spacelab* is being carried.

▲ George Nelson carrying out an experiment into crystal growth on *Discovery* during the STS-26 Shuttle mission in September 1988.

◀ On an earlier mission, 41-C in April 1984, Nelson inspects the *Solar Max* satellite, which has just been captured. He is hitching a ride on *Challenger*'s robot arm.

▶ Also on mission 41-C, the arm is used to place in orbit the *Long Duration Exposure Facility* (*LDEF*), carrying a host of experiments.

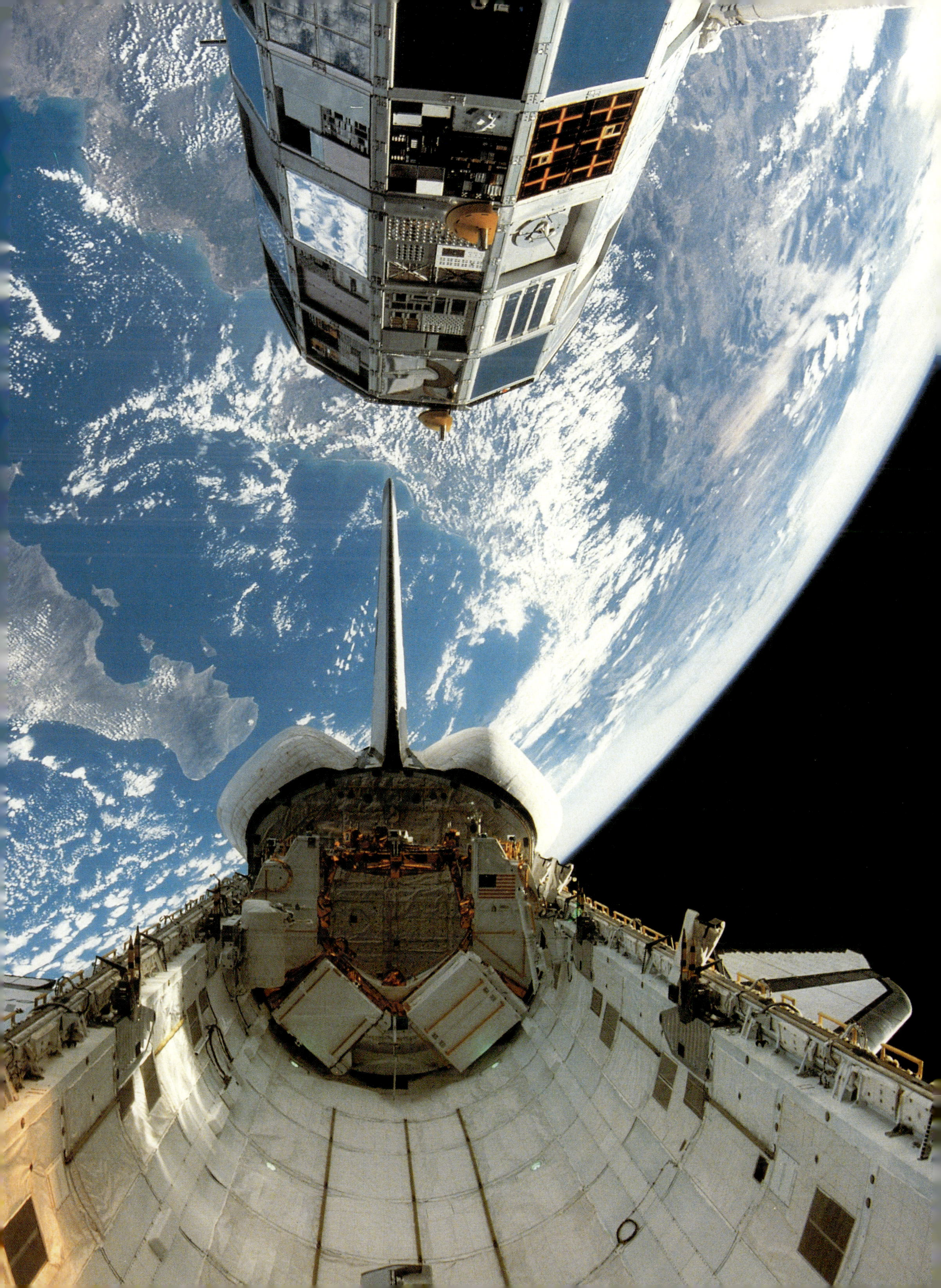

Humans in space

Spot facts

- The world's first space traveller was not human. She was a dog called Laika, meaning "Barker". She flew aboard the second artificial satellite, Sputnik 2, which the Russians launched in November 1957.

- The Space Shuttle lavatory is flushed by air, not water, and is fitted with foot restraints and a seat belt.

- Astronauts sometimes experience flashes of light, even with their eyes closed. They are caused by cosmic rays from outer space striking the retina of the eyes.

- On 21 December 1988 cosmonauts Vladimir Titov and Musakhi Manarov returned to Earth after a record 365 days 22 hours in space in the Mir space station.

▶ The days when astronauts have to wear cumbersome spacesuits in orbit are long gone. For most of the time they live in "shirt-sleeve", air-conditioned comfort. Occasionally they sport some really way-out gear, as the crew did here on the STS-26 Shuttle mission in 1988!

Human beings cannot by themselves survive the alien world of space. In space there is no air for breathing or protection; it is full of dangerous radiation and bits of rock travelling at high speed; the temperature is either scorching hot (in sunlight) or deathly cold (in shade).

But by designing suitable transport and living accommodation, human beings have shown that they can live quite happily in space. They find that their bodies can tolerate the strange state of weightlessness for at least a year. And, protected in pressurized spacesuits, they can venture outside their spacecraft to "walk" and work in space: carrying out experiments and mending satellites.

Surviving the hazards

At the beginning of the Space Age no one had the remotest idea whether flesh-and-blood human beings would be able to survive the hazards of space flight. First they had to withstand the high g-forces during launch – the forces on their bodies caused by the fierce acceleration of the launch rockets. This would make their bodies up to eight times heavier than normal.

When, however, they entered orbit, the pull on their bodies would cease abruptly. They would become weightless. What effect would this have on their blood, their heart and on the other body organs? Would these organs fail?

To help them find out, space scientists subjected astronauts to high g-forces in giant centrifuges. They sent chimpanzees and dogs into space, first for brief suborbital trips and then into orbit and back. The results were encouraging. Human beings could survive high g-forces for a short time; animals could survive short periods of weightlessness.

But no one really knew what would happen to the first humans to brave the space frontier until 12 April 1961. On that day the Russian Yuri Gagarin soared into space, circled the Earth once, and returned safely to a hero's welcome. In 108 minutes this first cosmonaut travelled a distance of 40,000 km. He appeared unharmed by the g-forces and one hour of weightlessness. This gave Russian space scientists the confidence to launch a second cosmonaut. In August Gherman Titov, aged only 25, remained in space for more than a day without coming to any harm. The breakthrough had been made. Space no longer appeared to be such a barrier. Humankind had begun its journey to the stars.

▲ The Russian cosmonauts who pioneered space travel in 1961. Yuri Gagarin (right) made a one-orbit flight on 12 April; Gherman Titov made 17 orbits on 6 August.

▶ John Glenn enters the Mercury capsule *Friendship 7* on 20 February 1962. Within hours he will be speeding around the Earth in orbit at 28,000 km/h. He was the first US astronaut to confront and overcome the hazards of space flight.

Living in orbit

In orbit objects appear to have no weight. They do not fall if you let them go. The Earth's gravity seems to have disappeared. But it is still there. The spacecraft (and everything it contains) is actually falling towards the Earth under gravity. But it is travelling so fast (28,000 km/h) that the Earth beneath curves away at the same rate as it is falling. In other words it stays at the same height – in orbit. This state is properly called free fall, but is popularly termed weightlessness.

Weightlessness dominates everything you do in orbit – moving, eating, drinking, sleeping and going to the lavatory. For example, you cannot walk in orbit, because there is nothing to hold your feet down. You cannot pour liquid from a bottle – it just stays where it is. But you can suck it through a straw, because that depends on air pressure. To sleep, you have to zip yourself into a sleeping bag and fix it to something, otherwise you will just float away. Space lavatories are fitted with an air-flushing system to draw wastes away from your body once they have been excreted.

The body itself is affected by the weightless state in a number of ways, some of them serious. The study of these effects and their treatment is known as space medicine. For the first few days in orbit you will probably feel sick because the balance organs in your ears cannot make sense of the new sensations. Without gravity to pump against, your heart will begin to lose muscle tissue; so will your legs. Unless you take regular exercise, the muscles will waste away, making you feel weak when you return to Earth and gravity once again. Regular exercise is essential on long space missions.

Even more serious is a progressive loss of calcium from the body, which reduces the mass and strength of the bones. However, a careful diet and a strict exercise regime helps to combat these effects, allowing astronauts to remain in space for a year or more without suffering permanent body damage.

▼ Mealtimes can be fun on the Space Shuttle. Mike Lounge chases a spherical globule of raspberry drink during a dinner break on Shuttle mission STS-26. Astronauts Fred Hauck and Dave Hilmers look on.

▶ European Space Agency astronaut Wubbo Ockels fitted out for an experiment on a "space sled" during a Spacelab mission. He will later be accelerated on the sled-like device along a track and stopped suddenly. At the same time his eyes will be subjected to different sensations and his reactions will be monitored. This experiment is designed to investigate space sickness, or space adaptation syndrome, which affects the majority of astronauts for the first few days in space.

▼ (below left) Sleeping aboard the Shuttle orbiter. The sleeping quarters are on the mid-deck, and comprise a number of bunks, to which the astronauts attach their sleeping bags. When they are asleep, their arms tend to float upwards in the weightless conditions. If the bunks are full, astronauts fix their sleeping bags to the walls or anything suitable. Because the orbiter is quite noisy, they usually wear ear plugs.

▼ (below right) Guion Bluford gets in some exercise on a treadmill during an early Shuttle flight. On Shuttle missions taking exercise is not really necessary because they seldom last longer than a week. It is on long-stay missions in space stations that it becomes vital to take regular exercise to prevent the body muscles wasting away.

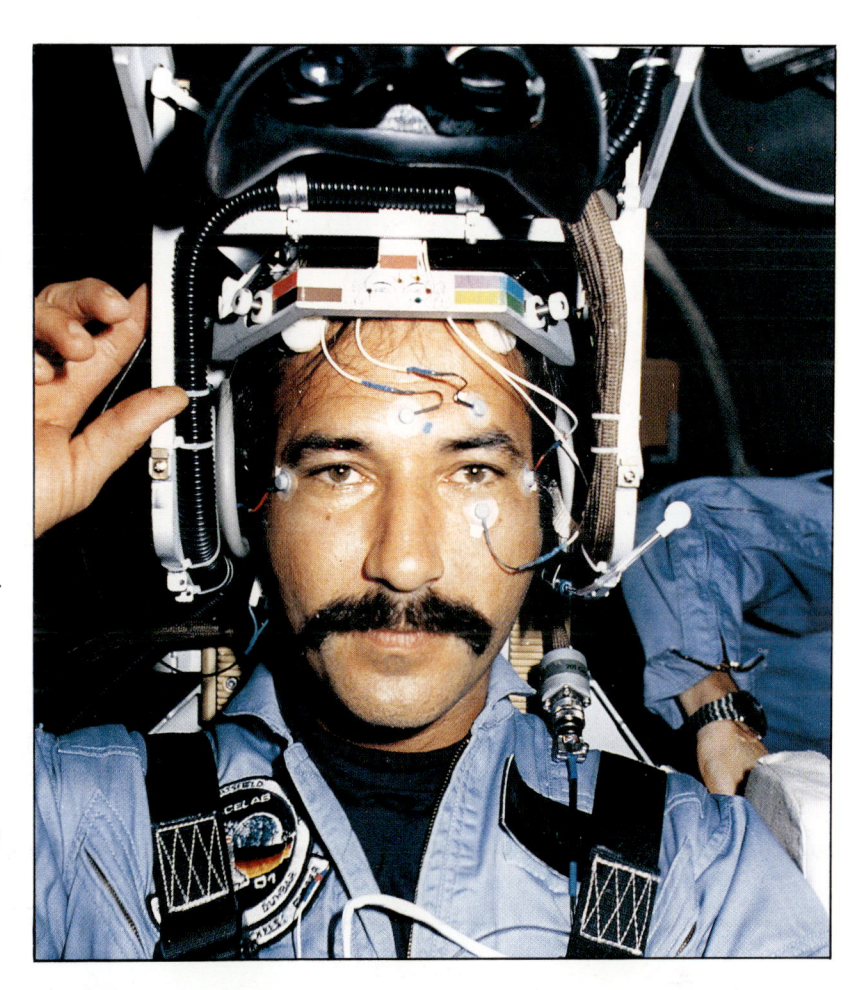

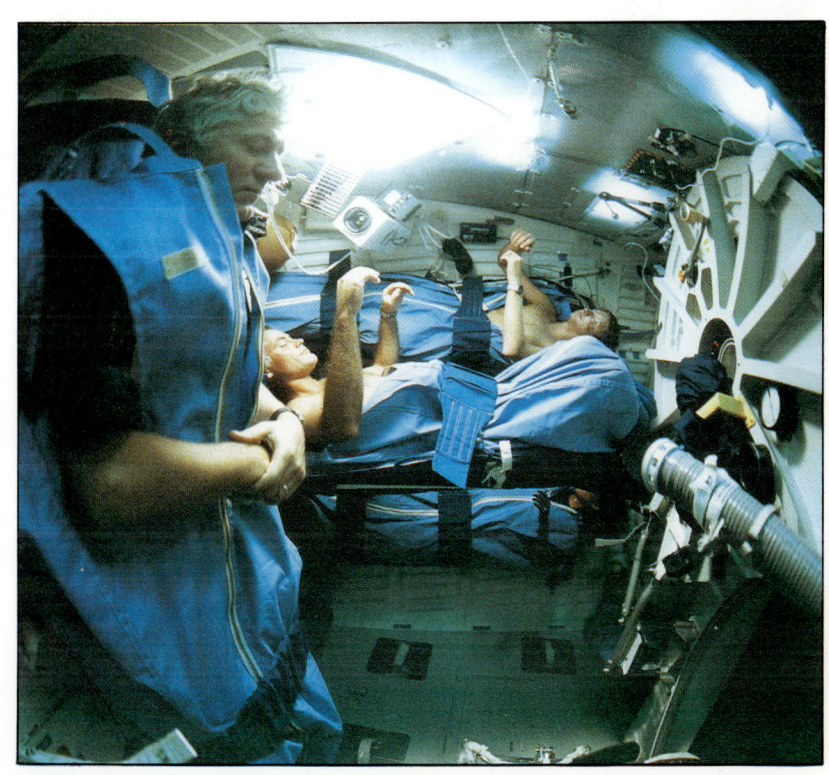

223

Spacesuits

On any manned spacecraft the most important system by far is the life-support system. This provides the means of keeping the astronauts alive and protecting them from space hazards.

The astronauts live in a pressurized cabin, whose metal walls act as a barrier to dangerous radiation and to micro meteoroids, the tiny swift particles that stream through space. It is pressurized to atmospheric pressure. They breathe an 80/20, nitrogen/oxygen mixture, much like the air on Earth. The air is circulated through a highly efficient air-conditioning unit, which absorbs odours and the carbon dioxide the astronauts breathe out. The air is kept at a comfortable temperature and humidity.

When the astronauts leave their spacecraft to go spacewalking, they wear a spacesuit that affords them the same level of protection as their pressurized cabin. The early spacesuits, worn by the US Mercury astronauts, for example, were simply modified versions of the pressure suits worn by high-flying jet pilots. When astronauts began spacewalking in the mid-1960s, specialist spacesuits were developed to offer extra protection from the direct exposure to space. They were umbilical suits, which drew oxygen from the life-support system of the spacecraft through a tube.

For the Apollo Moonlanding missions, the astronauts wore a spacesuit that was self-contained so that they were free to roam. The life-support equipment was in a backpack.

The Shuttle suit, properly called the extra vehicular mobility unit (EMU), evolved from it. It is made in two parts – trousers and top; the top part has a rigid aluminium frame and a built-in life-support backpack.

In the early days the whole spacecraft often had to be depressurized before the astronaut could open a hatch and float into space. Modern craft, however, have an airlock, a chamber inside the crew cabin which can be independently depressurized.

Astronauts enter the airlock and breathe pure oxygen for some time before they suit up. This is to clear their blood of nitrogen. Otherwise the nitrogen would bubble out when they wore their suits, which operate at reduced pressure. This would give them dangerous cramp attacks known as "the bends". When the astronauts are suited up, they depressurize the airlock, open the exit hatch and float out into space. They become human satellites.

◀ George Nelson gets kitted up for the STS-26 Shuttle mission in September 1988. He is wearing a newly-designed pressurized flight suit. All Shuttle astronauts now wear these suits on their journey into orbit. They came into use following the *Challenger* disaster to give astronauts added protection in the event of a cabin depressurization during lift-off.

Apollo EVA suit

The spacesuit the Apollo astronauts wore on their daring EVAs on the Moon was multi-layered. Over the astronaut's water-cooled "long johns" was a comfort layer, a pressure "bladder" and a restraint layer. On top was a 17-layer outer suit to provide protection against meteoroid particles and sunlight. A backpack carried oxygen, power and communications equipment.

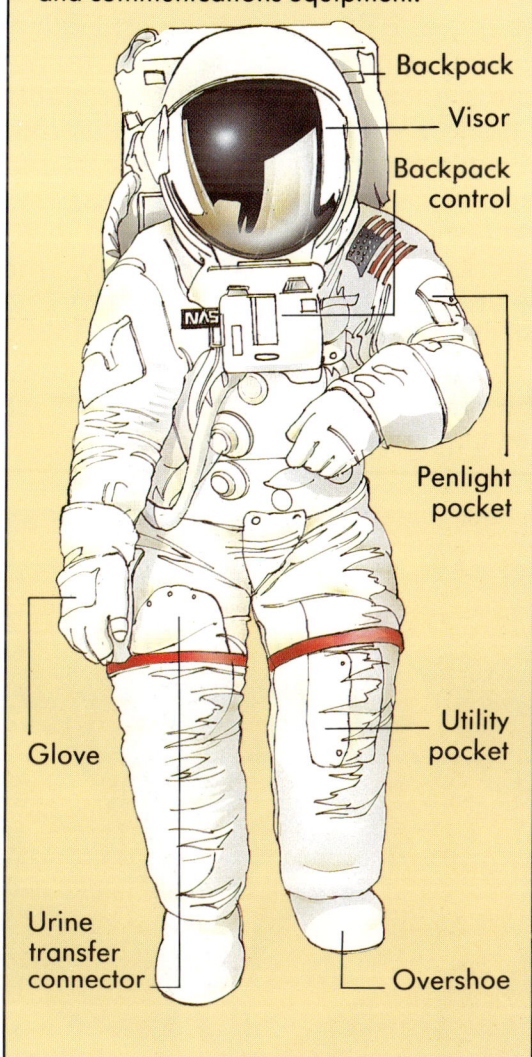

- Backpack
- Visor
- Backpack control
- Penlight pocket
- Glove
- Utility pocket
- Urine transfer connector
- Overshoe

▶ An astronaut puts on the two-part Shuttle spacesuit, or EMU. Next to her skin she wears water-cooled "long johns". She steps into the trousers first, then dons the upper torso, which is fitted with a life-support backpack. Torso and trousers join together by means of an airtight seal at the waist.

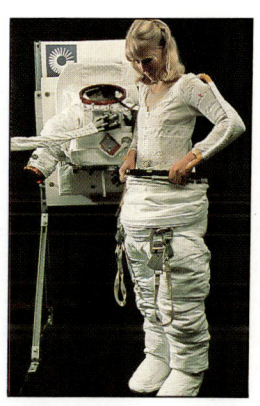

225

Spacewalking

Astronauts began leaving the comparative safety of their spacecraft and floating in space in 1965. This extravehicular activity (EVA) has been popularly termed spacewalking. EVAs are always risky because the astronauts have only a few thin layers of fabric and plastic between them and the lethal space environment. A small rip in their spacesuit would bring an agonizing death in seconds.

Surface EVAs took place on the Moon during the Apollo missions of the late 1960s and early 1970s. The astronauts wore self-contained spacesuits and roamed far and wide across the lava plains and rugged highlands of the Moon. On the last three missions they had wheeled transport in the shape of the lunar roving vehicle, or Moon buggy.

Useful EVA in orbit did not begin until the *Skylab* space station mission of 1973. *Skylab* was damaged during launch, losing a solar panel and some vital insulation. In orbit, exposed to the Sun, it began to overheat. But the first crew ferried up to the station carried out two daring EVAs and managed to erect a sunshade over the damaged area. The mission was saved and became spectacularly successful.

Over the years since then, the long-stay residents in Russia's space stations have carried out many in-orbit EVAs to effect essential repairs to their craft. Some of the EVAs have been vital. In July 1990, for example, cosmonauts Anatoli Solovyov and Alexsandr Balandin made two long EVAs on *Mir* to check and repair their Soyuz ferry craft and close a faulty airlock.

Since the Space Shuttle was introduced, many scheduled and a few unscheduled EVAs have taken place. EVAs have been scheduled, for example, to support experiments taking place in the payload bay. Shuttle astronauts have also made EVAs to repair satellites that have been captured from orbit and secured in the payload bay. This happened with *Solar Max* in 1984 and *Leasat 3* a year later.

While working in the payload bay, the astronauts are usually tethered to a safety line or "ride" on the orbiter's robot arm. Often they are actively involved in satellite capture, "flying" the jet-propelled backpack known as the manned manoeuvring unit (MMU).

The first spacewalk

On 18 March 1965 the Russian cosmonaut Alexei Leonov opened the airlock of his spacecraft, *Voshkod 2*, and "walked" out into space. No one had done this before. His spacewalk lasted nearly 10 minutes.

▲ On Shuttle mission 51-A in November 1984 astronauts Dale Garner and Joseph Allen helped capture two communications satellites.

▶ Bruce McCandless test-flies the jet-propelled manned manoeuvring unit (MMU) in February 1984 during the 41-B Shuttle mission.

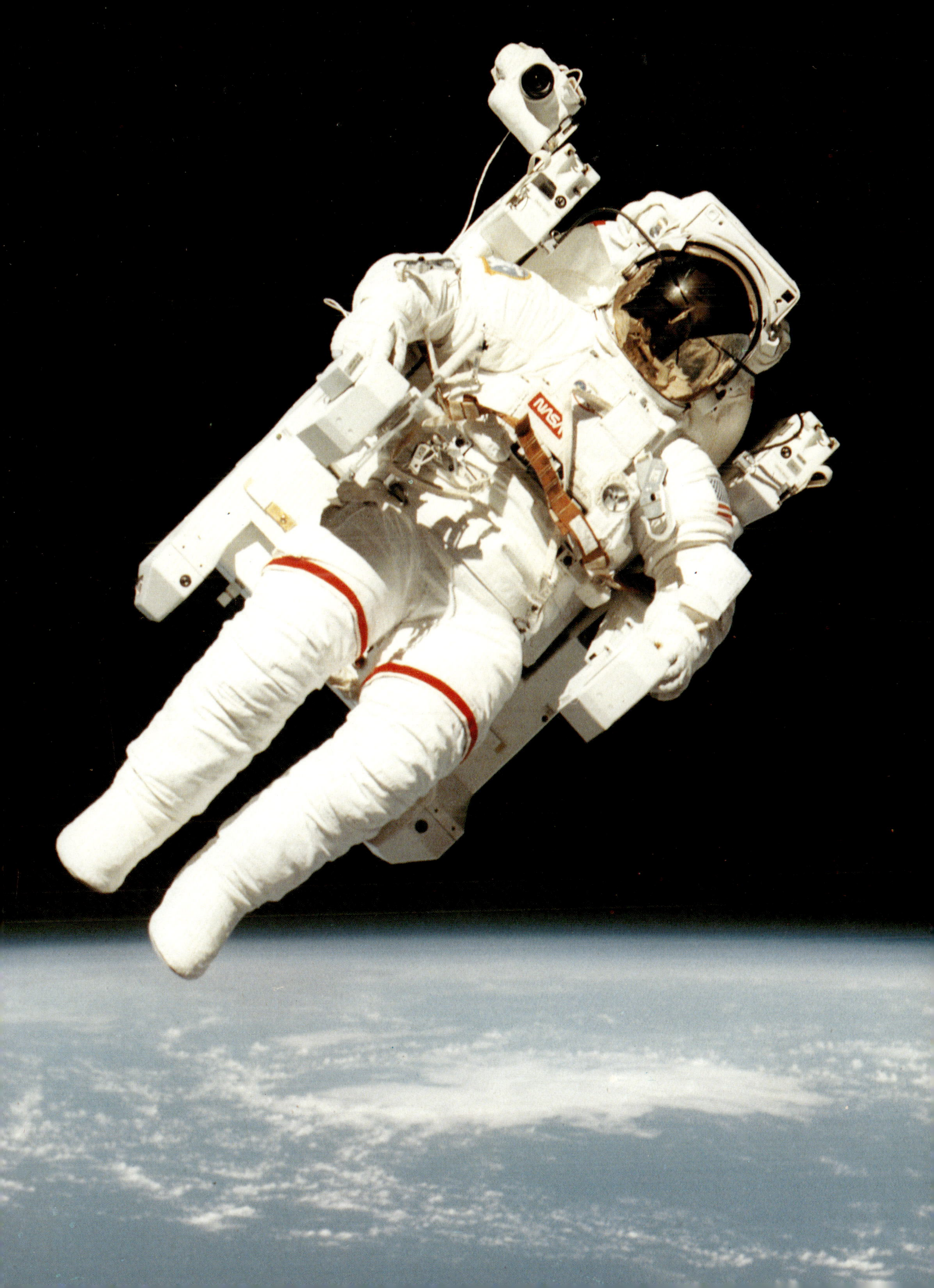
NASA

Space stations

Spot facts

- The Skylab astronauts took with them two spiders, Anita and Arabella, to see if they could spin webs in zero-gravity conditions. In fact they spun quite good webs.

- Skylab, launched in May 1973, fell out of orbit in July 1979, after circling the Earth 34,981 times.

- On the first Spacelab mission, in November 1983, Ulf Merbold from West Germany became the first foreign national to fly on a US spacecraft.

- In December 1987 cosmonaut Yuri Romanenko returned to Earth after a record 326 days in space in the Mir space station. He was so fit that the day after landing he managed to jog for 100 m.

▶ Astronaut Jack Lousma, enjoying a shower during the second manned mission to Skylab. It was the first spacecraft to feature such a luxury. The problem was that, when an astronaut took a shower, fine droplets of water splashed everywhere around the cabin and had to be vacuumed up by a colleague.

In the early days of space flight astronauts did not spend long in space. Even the Apollo missions to the Moon took less than two weeks. The early spacecraft were too small and were not equipped for lengthy trips. There was also little room for astronauts to carry out scientific studies or experiments.

Later, bigger craft, such as the Russian Salyuts and the US *Skylab*, provided more spacious living and working accommodation for long periods. *Skylab* and the later Salyuts proved spectacularly successful. They were the first space stations. Russia's latest station, *Mir*, is being built up, module by module, into an increasingly large complex. The European-built space laboratory Spacelab is also carrying out valuable work in orbit.

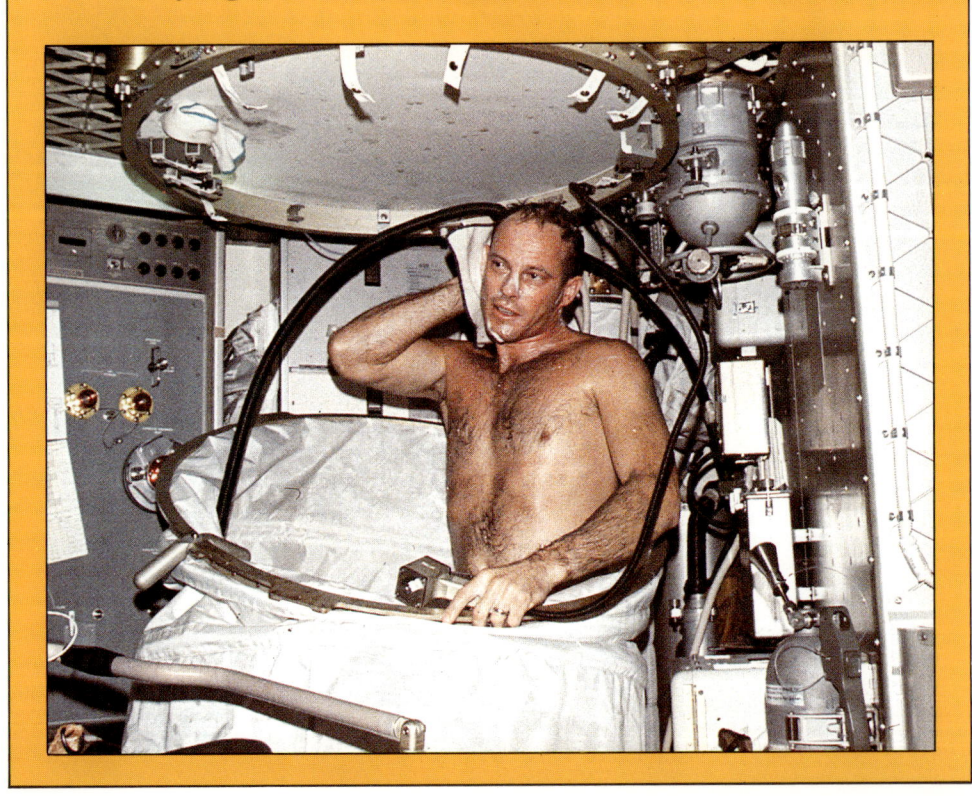

Salyut

The first space station, Russia's *Salyut 1*, was launched into orbit on 19 April 1971, and was first inhabited six weeks later by the three-man crew of *Soyuz 11*. They spent nearly 24 days in *Salyut 1*, smashing all space duration records. Tragically they were killed when returning to Earth when their cabin accidentally depressurized at high altitude.

It was a bad start for the Salyut space-station programme, which did not meet with real success until *Salyut 6* was launched in 1977. It was built of cylinders of different sizes, the biggest some 4 m across. It measured nearly 15 m long. It differed from earlier Salyuts in having two docking ports, one at each end.

In December 1977 and January 1978 two Soyuz craft flew up and docked at these ports, making the first triple link-up in space history. It showed the way ahead. A few days later, a remote-controlled cargo ship called Progress docked automatically with Salyut at a port vacated by one of the Soyuz. It brought up fresh supplies of fuel, food and mail, much welcomed.

By using automatic Progress ferries, Russia solved the problem of supporting the cosmonauts on long-stay missions. These missions grew longer and longer – up to 140 days in 1978 and 185 days in 1980. In 1982 *Salyut 7* took over, and the space duration record continued to tumble. In 1984 came a 236-day mission – nearly eight months. And the cosmonauts still did not suffer any long-term ill-effects from the prolonged weightlessness. This was good news for the future of human space travel.

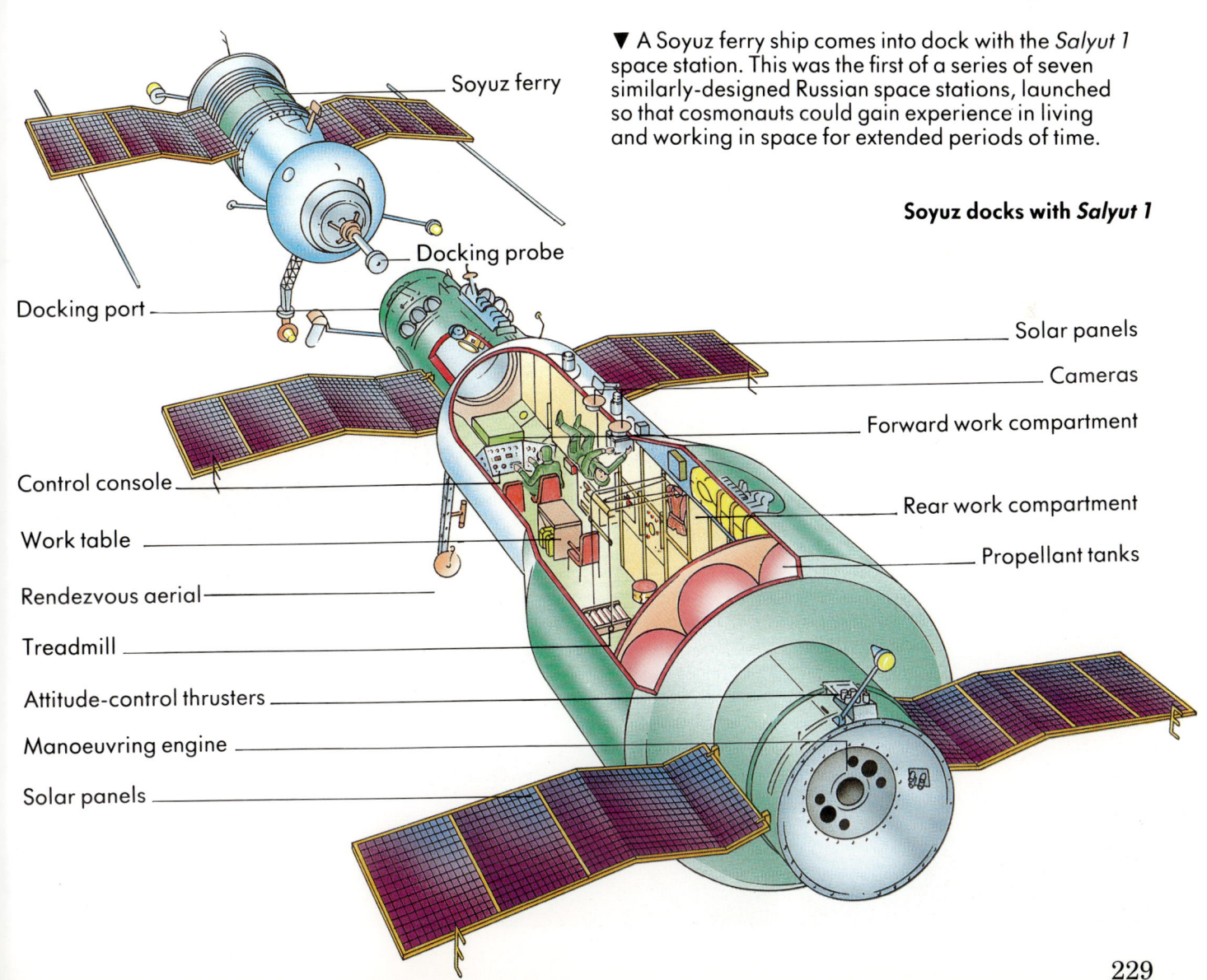

▼ A Soyuz ferry ship comes into dock with the *Salyut 1* space station. This was the first of a series of seven similarly-designed Russian space stations, launched so that cosmonauts could gain experience in living and working in space for extended periods of time.

Soyuz docks with Salyut 1

Skylab

The first US space-station project, called *Skylab*, was very much a makeshift affair. It used bits of Apollo hardware, left over when the number of Moon landings was reduced. The main unit was the third stage of a Saturn V rocket, to which other units were attached.

The completed "sky laboratory" was lifted into orbit on 14 May 1973 by a Saturn V rocket. Three teams of three astronauts visited the space station over the next nine months for periods of 28, 59 and 84 days respectively, travelling in Apollo spacecraft. They smashed all space duration records and proved that human beings could make their homes in space for long periods.

The third rocket stage formed the unit called the orbital workshop (OWS). The main living and working accommodation for the crew occupied the empty liquid hydrogen tank of the rocket. The smaller empty liquid oxygen tank provided storage space for waste. The upper part of the OWS led, through the airlock module, to the multiple docking adapter (MDA). This was a unit equipped with ports at which the Apollo spacecraft could dock.

▼ *Skylab* as it finally appeared in orbit, photographed by the departing third crew in February 1974 after a record-breaking 84 days in space. The picture shows the makeshift sunshields that the astronauts erected over the damaged orbital workshop.

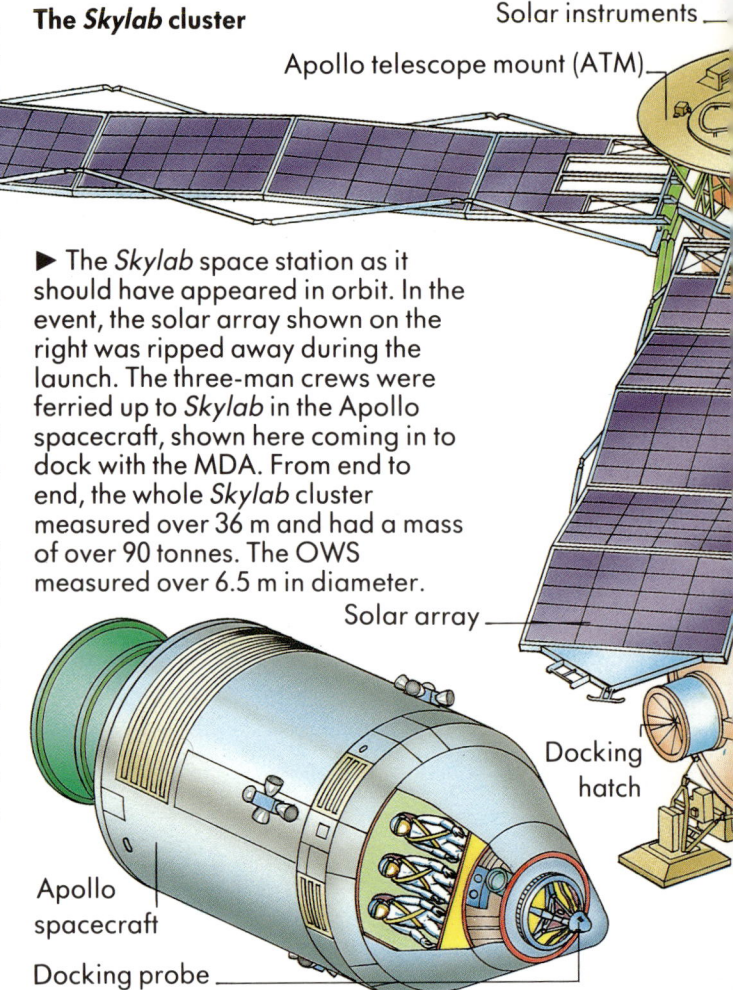

The *Skylab* cluster

▶ The *Skylab* space station as it should have appeared in orbit. In the event, the solar array shown on the right was ripped away during the launch. The three-man crews were ferried up to *Skylab* in the Apollo spacecraft, shown here coming in to dock with the MDA. From end to end, the whole *Skylab* cluster measured over 36 m and had a mass of over 90 tonnes. The OWS measured over 6.5 m in diameter.

▼ The cavernous forward compartment of the orbital workshop provided plenty of room for the astronauts to carry out gymnastic feats. Here, Gerald Carr, one of the final crew, is performing for the camera.

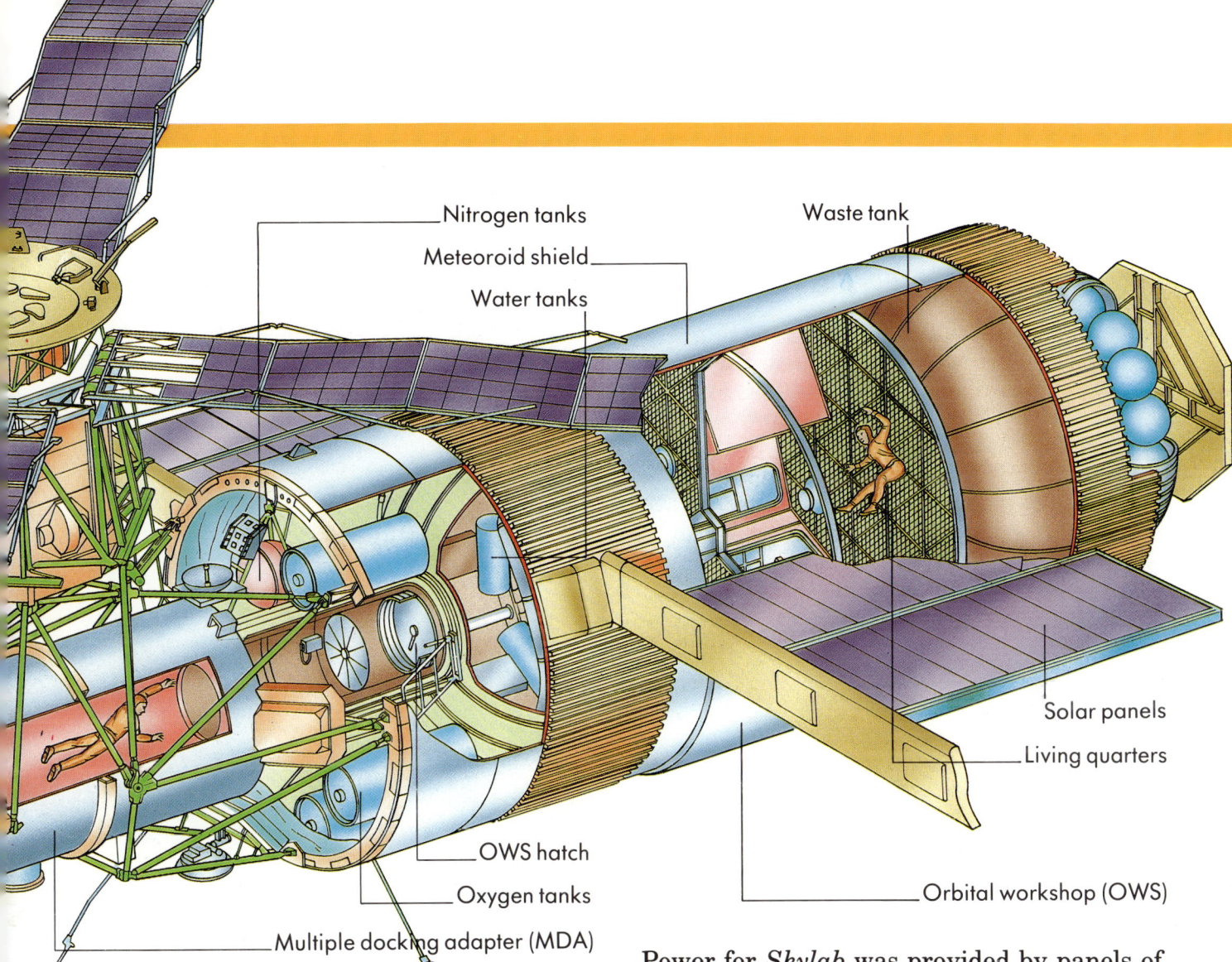

Diagram labels: Nitrogen tanks, Meteoroid shield, Water tanks, Waste tank, Solar panels, Living quarters, Orbital workshop (OWS), OWS hatch, Oxygen tanks, Multiple docking adapter (MDA)

▼ On the first *Skylab* visit, Dr Joseph Kerwin inspects Charles Conrad's mouth. Medical inspections were carried out regularly on *Skylab* to monitor the effects of prolonged weightlessness.

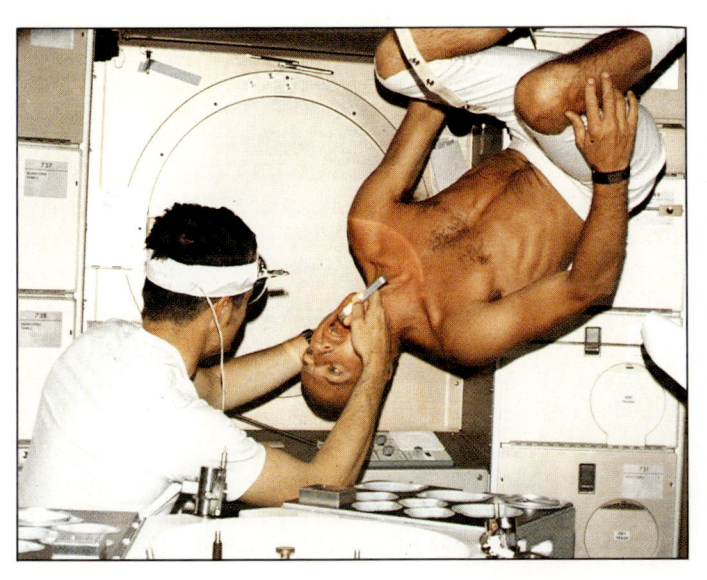

Power for *Skylab* was provided by panels of solar cells on the OWS and on a sail-like structure mounted on the MDA. This structure, which also housed a package of instruments for studying the Sun, was called the Apollo telescope mount. There should have been two solar panels on the OWS, but one was ripped off during launch. A section of insulation was also ripped off the OWS and had to be repaired by the first team of astronauts.

The astronauts had a heavy work load, carrying out all manner of observations and experiments. Their observation of the Sun produced the most spectacular results. They also carried out Earth-survey observations at different wavelengths, demonstrating the great potential of such remote sensing. In engineering, they experimented with melting and crystallizing materials to produce new compounds. All the while they used themselves as guinea pigs for space medicine experiments, to monitor how their bodies reacted to long periods of weightlessness. To help combat muscle wastage, they exercised on a bicycle ergometer.

Spacelab

European space science took a great leap forward in November 1983, when the Space Shuttle carried into orbit Spacelab, the first specialist space laboratory. Spacelab was designed and built by the European Space Agency. It fits into the payload bay of the Shuttle orbiter and remains there while in space. Like the Shuttle itself, Spacelab is designed to be reusable.

There are two main units in Spacelab. One is a pressurized laboratory module, and the other is an unpressurized pallet, or platform, for carrying instruments that need to be exposed to the space environment. The standard configuration is the so-called long module and pallets, shown in the picture at far right. The long module is a two-segment cylinder about 7 m long and 4 m across. It is fitted out with laboratory equipment and instruments in standard-sized racks along the sides. It has a powerful computer system to analyse results on the spot. But many results are relayed to scientific centres back on Earth.

Some Spacelab investigators are professional astronauts from NASA, called mission specialists, who have a strong scientific back-

◀ Spacelab scientists carry out a variety of biological studies in orbit. On the *Spacelab 3* mission, in April 1985, there were two dozen rats and a pair of squirrel monkeys as well as a human crew of seven. The scientists studied the effects on these animals of weightlessness. Here mission specialist William Thornton is observing one of the squirrel monkeys. But who is really upside-down: man or monkey?

ground. Others are non-astronaut scientists from both the United States and Europe, who have particular expertise in the subjects being studied. They are known as payload specialists.

The investigators carry out experiments and observations in many branches of science and engineering. They photograph the Earth and make telescopic observations of the heavens. They study living things, from flies to monkeys, to see how they react to weightlessness. They also conduct medical experiments on themselves, taking daily blood samples, for example. They also carry out tests to gain a greater insight into such problems as space adaptation syndrome, or space sickness.

▼ Spacelab in orbit some 270 km above the Earth. The laboratory is shown in its common configuration of long module and pallets. Spacelab scientists only work in the laboratory; they eat and sleep in the mid-deck living quarters of the orbiter, to which Spacelab is linked by a pressurized tunnel.

▲ European astronaut Ulf Merbold, pictured on the "ceiling" of Spacelab on its first flight in November 1983. Dr Merbold was one of two payload specialists.

▼ A busy scene on board Spacelab during the *Spacelab D1* mission of October 1985, dedicated to West Germany. The crew of eight included Guion Bluford (US, left) and Richard Furrer (West Germany).

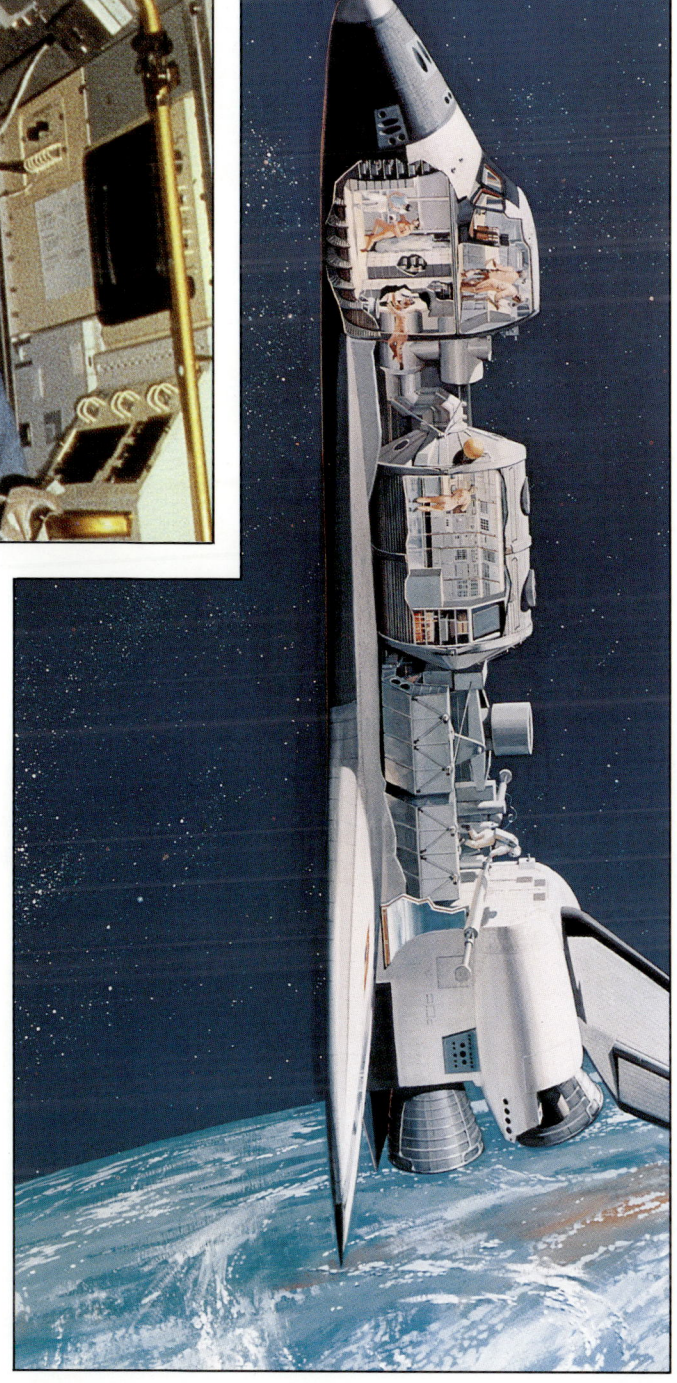

Mir

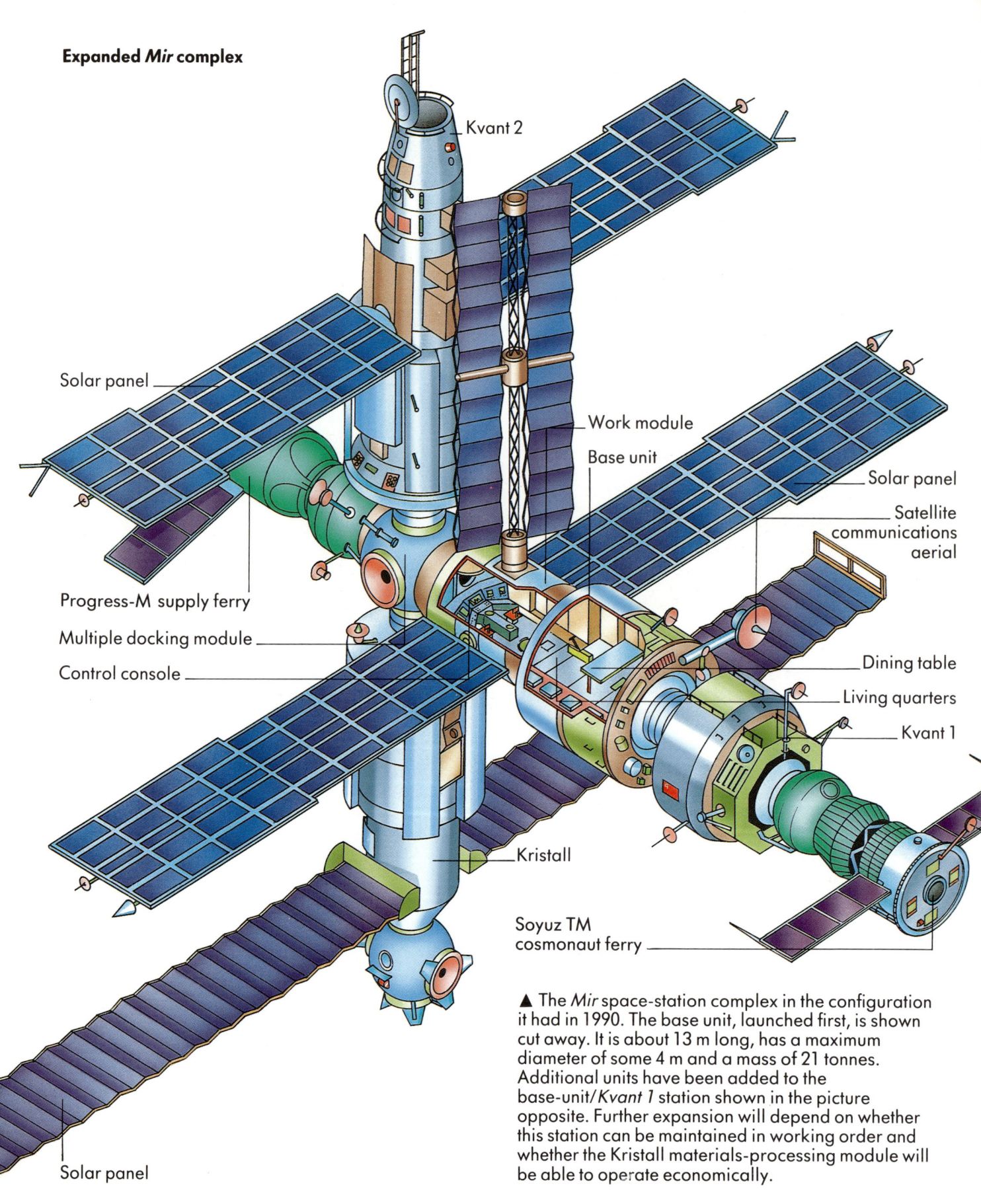

Expanded *Mir* complex

Labels: Kvant 2, Solar panel, Work module, Base unit, Solar panel, Satellite communications aerial, Progress-M supply ferry, Multiple docking module, Control console, Dining table, Living quarters, Kvant 1, Kristall, Soyuz TM cosmonaut ferry, Solar panel

▲ The *Mir* space-station complex in the configuration it had in 1990. The base unit, launched first, is shown cut away. It is about 13 m long, has a maximum diameter of some 4 m and a mass of 21 tonnes. Additional units have been added to the base-unit/*Kvant 1* station shown in the picture opposite. Further expansion will depend on whether this station can be maintained in working order and whether the Kristall materials-processing module will be able to operate economically.

Russia launched seven Salyut space stations between 1971 and 1982. In February 1986 it launched a new design called *Mir* (meaning "Peace"), which is still in orbit. The craft looks similar to the later Salyuts but is in fact quite different. For one thing the interior is much more spacious. It provides mainly living space for the crew, with separate cabins for each person. It is not cluttered, like the Salyuts, with experimental equipment.

Outside, the main new feature of *Mir* is a spherical docking module at the front end, with five docking ports. There is also a single docking port at the other end, making six in all. This design has allowed *Mir* to be expanded.

The first add-on unit docked automatically with the rear port in April 1987. It is called Kvant, after the physics term "quantum". In fact it is now called *Kvant 1*, because in December 1989 a large unit called the Re-equipment Module, or *Kvant 2*, docked at the other end. It was later repositioned to a sideport of the docking module.

Kvant 2 is, at 13.7 m, a metre longer than the base unit itself. It houses an experimental compartment, an airlock and a shower, the first on board *Mir*. It also carried up to *Mir* the first model of the Russian version of the American MMU, or manned manoeuvring unit. This jet-propelled backpack, named Icarus, runs on compressed air. It will be used for inspecting and repairing the *Mir* complex.

A third module, the Kristall materials-processing module, docked with the complex at the port opposite *Kvant 2*, in 1990. It is described as a mini-factory that is intended to produce flawless semiconductor crystals for use in electronics and ultra-pure drugs for use in medicine. The manufactured materials would probably be returned to Earth from time to time by unmanned Soyuz-type craft.

▲ The *Mir* space station in July 1987, as photographed by the crew of the departing *Soyuz TM-2*. By now the base unit has been expanded by the addition of the *Kvant 1* module, in place since April. Docked with *Kvant 1* in the picture is the newly arrived *Soyuz TM-3*.

Index

Page numbers in **bold** refer to pictures.

A
Abell 1060 **190**
aberration, chromatic 102
   spherical 102
absolute magnitude 168, 170, 172
absolute zero 17, 40
acceleration 78
activation, thin-layer 114
aerolites 151
Air Pump galaxy **182**
alchemy **30**, 111
Aldrin, Edwin 210
Allen, Joseph **226**
alpha particles 21
alternating current 62
aluminium **34**
aluminium fluoride 94
ammeter 92, 93
ammonia 138, 145
ampere 50
Ampère, André Marie 50, 92
amplitude 44
analysis 110–121
   chemical 112–113
   methods of 110
   qualitative 112
   quantitative 112
   spectroscopic 113, **113**
Andromeda galaxy 160, 182, **183**, 184, 190
angle of incidence 65
animals in space 220, 221, 232, **232**
annulus 124
anode 50, 51, 53
antinode **45**
Apennine Mountains **211**
Aphrodite Terra 131
Apollo 152, 155, 208, 210–211, **211**, **212**, 213, 224, 228
Apollo EVA suit 225, **225**
Apollo Moon landing project 210
   landing sites of **211**
   techniques of 210
Apollo-Soyuz Test Project (ASTP) 212, 213
apparent magnitude 168
Archer (*see* Sagittarius)
Arecibo 107
argon 52
Argyre Basin **132**, 134, **134**
Arizona Meteor Crater 146, **150**
Ariel 158, **159**
armalcolite 152
Armstrong, Neil 208, 210
Ascraeus Mons **132**
astatine 30
asteroids 126, 137, 146, 147
   orbits of 148, 149
   size of 148, 149
astrolabe **85**
astrology 164, 165
astronomical coordinates 163
astronomy 166
Atlantis 216
Atlas-Centaur rocket **198**
atmosphere, Earth's 199, **199**
atomic bomb **29**
atomic number 32
atoms 20–23, 26, 28
   bonds between 13
   electrical charge of **35**, 46, 47
   in crystals 34
   structure of 21, **21**, 22–23, **22**, **23**, **32**
   uranium **22**
   vibration of 13
atom-smashing 120–121
atom-smasher (*see* particle accelerator)
aurora 199
Aurora Borealis **124**

B
Baikonur Cosmodrome 213, 215
Bailly 152
balance, chemical **86**
   electronic **86**, 87
   equal-arm 86
   letter 87, **87**
   precision 86
   single-pan 87
   spring 87, **87**
   torsion 87
Balandin, Alexsandr 226
balloon, hot-air **16**
Barnard's star 166
battery 50, **50**

dry **51**
Bell, Alexander Graham 44
Bell, Jocelyn 100
bell, electric **61**
belts 137, 138, 140
"the bends" 224
Bessel, Friedrich 161
Betelgeuse **167**, 172
Big Bang 193, 194, 195, 196
Big Bird satellite 198
Big Crunch 196, 197
bimetallic thermometer **41**
binary star systems 168, 169
black holes 109, **170**, **178**, 180, 181, 188
blue shift 169
Bluford, Guion **223**, **233**
Boeing 747 215
Bohr, Niels 21, 22
boron 54
Boyle, Robert 17
Boyle's law 17
Brahe, Tycho 180
Brand, Vance 212
breccias 155
bromine 30
brown dwarf **178**, 179
Brownian motion 16
bubble chamber **20**, 24
bursters 109

C
Callisto 157, **157**
Caloris Basin 129
Cambridge Radio Astronomy Observatory 100
canali 128
capacitors **48**
Cape Canaveral 198, 209
car, air resistance of **17**
   solar 80
carbon 12
carbonaceous chondrites 151, **151**
Carr, Gerald **230**
capillarity 14
Cartwheel galaxy 186
casualties, astronaut 212, 216, 217, 229
Cassegrain reflector **103**
Cassini division 142, **142**
catalyst 39
cathode 50, 51, 53
cathode-ray oscilloscope 93, **93**
cathode-ray tube displays **216**
Cavendish Laboratory 94
celestial equator 163
celestial sphere 160, 162, 163
cell **50**
Celsius, Anders 41
Celsius scale 41
Centaurus A 186, **187**
centigrade 41
centrifugal force 77
centrifuges 221
centripetal force 77
Cepheids 172
Ceres 148
CERN 110, 120
Cernan, Eugene **208**
Challenger 216, 218, 224
charge-coupled devices 105
Charles's law 17
Charon **145**, 159
chemical analysis 112–113
chlorine 53
Christie, James 159
chromatography 112
   gas 113
chromosphere 123, **125**
circuit, electrical **50**
circuit-breaker **61**
clepsydra **85**
clocks 90–91
cloud chamber **24**
clouds 199
Coal Sack 174, **174**
Cock, Christopher 95
Cockcroft-Walton generator 121, **121**
coherent light 73, 118
colour 70–71
   mixing of 70, **70**
Columbia **214**, 216
comets 126, 146, 149, **149**
   Halley's **147**, 147–149
   orbits of 148–149
   size of 148–149
command module 211, 212
commutator 62, **62**, 63, **63**
compass 57

compounds 31
computer-assisted tomography (CAT) 115
computers in space 202, 208, 216
comsats 200
Concorde **33**
condensation 19
conduction, heat 43
Conrad, Charles **231**
constellations 164–165
   main 165
   northern hemisphere **164**
   of the zodiac **164**
convection, heat 43
convection current 43, **43**
convective zone **125**
Copernicus 108
Copernicus, Nicolaus 126
corona 123, **125**
coronagraph 123
cosmic year 183
cosmology 192
coulometer **112**
counterweight 87
covalent bonds 36, **36**, 37
Crab nebula 180, **180**
Crab pulsar 180
Crippen, Robert 214
Crux **169**
crystals 12, **12**
   ionic **35**, 36
   metal 34–35
   quartz **12**, 90, 91
   yttrium-aluminium-garnet 73
cubit 85, **85**
Curie, Marie **32**
current, electric 50–51
Curtis-Schmidt reflector **103**
Cygnus 161, 168
Cygnus A 188
Cygnus X–1 **181**

D
Dalton, John 21
Daniell cell **50**
dark halo 182
dark matter 196
decibels 44
Deep Space Network 205
Deimos 156, **156**
Delta Cephei 172
Democritos 21
Deneb 168
diamond 12
diffraction 45
diffusion 16, 19
digital thermometer **41**
digital watch 91, **91**
Dione 158
direct current 62
discharge tubes 53, **53**
Discovery 216, **217**, 218
distance 88–89
docking module 213
docking ports 213
Dog Star (*see* Sirius)
doping 54
Doppler effect **45**

E
Earth 126
   orbit of **127**
Earth-survey satellite 202, **202**
echo-sounders 116–117
eclipse **124**
ecliptic 163, **164**
Edison, Thomas A. 111
Edwards Air Force Base 214, 215
Einstein, Albert 123
elasticity 13
electric charges 47
electric currents 50, 52
electric field 47, **47**
electricity 46–55
electrolysis 53
electrolyte 53
electromagnet **56**, 60
   in motors 63
   uses of 60
electromagnetic spectrum 74–75, **107**
electromagnetic waves 74–75, 106, 107
electromagnetism 60–61
electrons 20, 46, 47, 124
   and the Big Bang 194
   and crystals 34
   early ideas about 21

236

and elements 30, 32
and generators 62, 63
and lasers 73
and molecules 36
and motors 62, 63
movement of 50, 51
orbitals of 22, **22**, 23, **23**
and radio galaxies 186
in semiconductors 54
electron-volts 121
electroplating **53**
elements, chemical 30–33
Enceladus **158**
Endeavour 216
Energia rocket 198, **215**
energy 80–81
epsilon 144
equations, chemical 38
equinoxes 163
Eratosthenes **155**
Eros 148
ERS–1 satellite 202
escape velocity 199, 204
Eta Carinae 166
Europa 157, **157**
European Space Agency 232
evaporation 18, 19
evening star 130
Exosat 108
experiments in space 218, 226, 228, 231, 232, 233
Explorer 1 198
extravehicular activity (EVA) (*see* spacewalking)
extravehicular mobility unit (EMU) (*see* shuttle suit)

F
faculae **125**
Fahrenheit, Gabriel 41
Fahrenheit scale 41
feldspar 12
Fermilab 120
fibre optics **75**
filament **50**, 52
filaments, solar **125**
fission, nuclear 28, **28**
flares 124
Fleming's rules **62–63**
fluorescent tubes 52, **52**
focal length **67**
focal point 66
focus, Cassegrain 103
  Newtonian 103
foliot 90
force, electrical and magnetic 77
  gravity 77
  strong 77
  weak 77
Franklin, Benjamin 48
Fraunhofer, Josef von 167
Fraunhofer lines 167
Freedom 7 209
free fall (*see* weightlessness)
frequency 44, 45
friction 77, 79
Friendship 7 209, **221**
fundamental 45
Furrer, Richard **233**
fuse, electric 52
fusion, latent heat of 18
  nuclear 29, **29**

G
Gagarin, Yuri 209, 221, **221**
galaxies 167, 182–191
Galaxy (*see* Milky Way)
galaxy, Air Pump 182
  Andromeda 182, **183**, 184, 190
  Cartwheel 186
  Centaurus A 186, **187**
  Maffei I 190
  Whirlpool **106**
galena **31**
Galilean satellites 156
Galileo Galilei **83**, 90, 101, 156
  telescope of **101**
Galle, Johann Gottfried 145
gamma rays 74, 108, 115
Ganymede 137, 157, **157**
Garner, Dale **226**
gases 10–11, **11**, 16–17
  change of state in 18, 19
  compression of 17
  heating of 43
gauge, micrometer screw 88, **88**
  strain **86**, 87

Geiger counter **27**
Gemini 208, 209, **209**
generator 62–63
  AC **62**
  DC **62**
  principle of **62**
geostationary orbit 200, **201**
Getaway specials 217
g-forces 221
gibbous 153
Gilbert, William 56
Giotto 149, 204, **204**
glacier **15**
Glenn, John 209, **221**
globular clusters 169
globules 174
glycine 174
GOES satellite 201
gold **31**, 33
graphite 12
gravitational lens **82**
gravity 76, 82–83, 126, 153
  and space travel 199, 222
gravity-assist technique 206
Great Bear 164, 165
Great Nebula in Orion 173
Great Red Spot 138, **138**, **139**
Grissom, Virgil 209

H
Hale Observatory 104
half-life 26
Halley, Edmond 147
Halley's comet 146, **147**, 204, **204**
harmonics, piano **45**
Harvard Observatory 172
Hauck, Fred **222**
heat 40–41, 42–43
heat shield 209
heliopause 206
helium 30, 122, 123, 140, 145, 179, 194
Hellas 134, **134**
Henbury craters 150
Hercules cluster 190
Herschel, William 42, 144
hertz 44
Hertz, Heinrich 44, 75
Hertzsprung, Ejnar 170
Hertzsprung-Russell diagram 170, **170–171**
Hilmers, Dave **222**
Hipparcos 11
holograms 72, **118**
  making of **118–119**
Hooke, Robert 95
Horsehead nebula 174, **175**
Hubble, Edwin **184**, 193
Hubble constant 192, 193
Hubble Space Telescope 109, **109**, 218
Huygens, Christiaan 68
hydrogen **23**, 30, 136, 137, 140, 145
  bonds **36**
  and electrolysis 53
  isotopes of 26, **29**
  and the Universe 194
hydrogen sulphide 138

I
Iapetus 158
IBM 94
Icarus 235
ice 10, **10**, **15**, **19**
icebergs 10, 40, 41
images, real 66, 67
  virtual 66, 67, 96
Imperial measures 85
Inertial Upper Stage 218
Infrared Astronomy Satellite 100, 109
infrared rays 42, 75, 89, 202
instrument module 212
insulation, heat 43
integrated circuit 54
Intelsat 200
interference, light wave 68, **69**
International Ultraviolet Explorer 109
Io 152, **156**, 157, 206
ionic bonds 36, **37**
ions 35, 36
iron meteorites 150–151
Irwin, James **211**
Isaac Newton telescope 105
Ishtar Terra 131
isotopes 26, 28, 29, **29**

J
Jacobus Kapteyn telescope 105

Jannsen, Zacharias 95
Jansky, Karl **107**
jet 186, 187
Jewel Box (*see* Crux)
Jodrell Bank 107
Joule, James Prescott 80, **81**
joules 80
Jupiter 126, **127**, 136, 137–139, **139**, 206, **206**

K
kaon 24, **25**
Kelvin, Lord 41
Kelvin, scale 41
Kennedy, John F. 210
Kennedy Space Centre 211, 213, 214
Kepler, Johannes 180
Kerwin, Dr. Joseph **231**
kinetic theory 11
Kitt Peak Observatory **100**, **102**, **103**, 104, **104**
Komorav, Vladimir 212
Kristall materials-processing module 235
Kubasov, Valery 212
Kvant 235, **235**

L
laboratories 111, **111**
Labyrinth of the Night **133**
Lageos satellite 198
Laika 220
Landsats 202, **202**
Large Magellanic Cloud 160, **180**, 182, 184, 190
laser beams 89
lasers 72–73, 118–119
laser welding 18
latent heat 18, 19
lava 155
Leasat 3 satellite 226
Leavitt, Henrietta **172**
Leblanc, Nicholas 111
Leeuwenhoek, Anton van 95
length 88–89
lenses 66–67
  achromatic 95, 102
  concave 66, **67**
  converging 102
  convex 66, **67**
  Fresnel 67
Leonov, Alexei 212, **226**
Liberty Bell 209
life-support system 224
light 64–75
  coherent 73, 118
  colours of 70–71
  mixing 70
  polarized 96, 97
  reflection of 65, 71
  refraction of 65, 71
  speed of 64
  theories of 68
  ultraviolet 74, **74**
  visible 75
  waves of 68–69
  white 69
light bulb **50**, 52
lighthouse **67**
lightning 49
lightning conductors 48
light-years 160, 161, 168
linear accelerators 120
lines of force **58**, 74
liquid crystal display (LCD) 91, **92**
liquids 10–11, **10**, **11**, 14–15 **14**, **15**
  change of state in 11, 18, 19, **19**
  heating of 43
  viscous **14**, 15, **15**
Little Green Men 100
Local Group 190, **191**, 193
Long Duration Exposure Facility (LDEF) **219**
Lounge, Mike **222**
Lousma, Jack **228**
luminosity (*see* absolute magnitude)
Luna 1 204
Lunar Alps 155
lunar module 211, **211**
lunar orbit rendezvous **210–211**
Lunar roving vehicle **208**, **211**, 226

M
M13 169
M31 (*see* Andromeda galaxy)
M33 **185**, 190
M42 (*see* Great Nebula in Orion)
M81 **190**
M87 186, **186**, 190
magnesium **32**

237

magnetic domains 58
magnetic fields 58, 59, **59**
    and generators 62, 63
    and motors 62, 63
    and moving-coil meters 92
magnetic force **57**
magnetic levitation train **60**
magnetic poles 57
magnetic variation 57
magnetism 56–61
    Earth's **57**
    and electricity 59, 60
magnetometers 207
magnets, making of 58
magnifying glass **67**
Maiman, Theodore H. 72, 118
main sequence 170
Manarov, Musakhi 220
Manned Maneuvering Unit (MMU) 226, **227**
maria 129
Mariner 2 204
Mariner 10 129, 130, 134, 206
Mariner Valley **132**, 134
Mars 126, 128, 132–135, 204, **204**, 205
Maxwell, James Clerk 74, 75, **75**
Mayall reflector **102**
McCandless, Bruce **227**
McMath solar telescope **104**
measuring 84–93
Merbold, Ulf 228, **233**
mercury 14, 30
Mercury 122, 128, 129, **129**, 208, **209**
    orbit of **127**
    structure of **129**
    surface of **129**
mercury thermometer **41**
Messier, Charles 185
metallic bonds 34, **37**
metals, atomic arrangement of 34, **34**
meteor 126, 146, 199
meteorite 126, 134, 144, 146, 150
    composition of 151
meteoroid 126, 150
meteorology 198, 201
Meteosat 201
methane 38, 144, 145
metric measures 85
microbalance 87
micrometeoroids 224
micrometer **85**
microscopes 94–99
microscopy, fluorescence 96
microtome 96
microwaves 75
Milky Way 167, 174, **174**, 182, 183, 184, 190
    dimensions of **183**
Mimas 158
Mir space station 220, 226, 228, 234–235, **234**, **235**
Mira 172
Mira Ceti 172
mirages **65**
Miranda 206, **207**
mirrors 65, 66, 67
    curved 66–67
    concave 66, **66**, 101, 103
    convex 66, **66**
mission specialists 232
mixtures 31
models, computer 36, **37**
    space-filling 36
molecular cloud 177
molecules 20–21, 28, 38
    benzene **37**
    forces between 13, 14
    movement of 10, 11, **11**, 13, 16, 18, 40, 41, 43
    sodium chloride (salt) **37**
    water 36, **37**
Molniya satellites 201
momentum 79
Monoceros 166
Moon 124, 153–155, 204, 208, 210, 211
moon buggy (*see* lunar roving vehicle)
moons 152–159
morning star 130
motion, laws of 76, 78
motors 62–63
Mount Wilson Observatory 184
moving-coil meter **61**, 92, **92**
multiple docking adaptor (MDA) 230, 231
Multispectral Scanner 202
muon 24, **25**

N

NASA 205
nebula, Crab **180**
    Helix **176**
    Orion **177**
    Ring **179**
nebulae 173, 174, **175**, **176**, **177**, **179**, 180
    dark 173, 174
    emission 173
    planetary 176, **179**
    reflection 173
Nelson, George **218**, **224**
neon 53
Neptune 126, **127**, 136, 145, **145**, 204, 205, **206**
Nereid 159
neutral point **58**
neutrinos 24, 196
neutrons 20, 22, **22**, 26
    Big Bang and 194
    dying stars and 180
Newton, Isaac 68, 76 , 78, 82, 101, 199
newton 80
Newtonian reflector **103**
NGC 2997 **182**
NGC 4319 **192**
NGC 5128 186
NOAA satellite 201
Noctis Labyrinthus **133**
node 45
North Pole 57
nuclear energy 28–29
nuclear magnetic resonance 60
nucleus 20, 21, 22, **22**, 26, 28

O

Oberon 158
observatories, astronomical 104–105
Ocean of Storms 155
Ockels, Wubbo **223**
Oersted, Hans Christian 56, 59
ohm 51
Ohm, Georg Simon 51
Ohm's law 51
Olympus Mons 128, **132**, 134
Orbita 201
orbital maneuvering system (OMS) 217
orbitals 22, **23**, 38
Orbiting Astronomical Observatories 108
Orion **160**, 165
Orion's Belt **165**
oscilloscope 93, **93**

P

Pallas 148
Panavia Tornado **76**
parallax 89, 161
Paramecium bursaria **96**
particle accelerator 24, 60, 120
particle accumulator **121**
particle detectors 207
particle tracks **25**
payload specialists 232
penumbra 124
Penzias, Arno 195
periodic comets 149
Periodic Table 32–33, **32–33**
period-luminosity law 172
Perkins, William H. 111
Perseus 169
petrol, manufacture of **39**
Phobos 147, 156, **156**
Phoebe 158
phosphorus 54
photons 64, 69, 72, 194
photosphere 124, **125**
photosynthesis 38
pion 24, **24**
Pioneer 10 205
Pioneer 11 205, **205**
planets 126–145
    far 136–145
    near 128–135
    orbits of **127**
    sizes of **126–127**
    statistics of 127
Plaskett's star 166
plasma 17
Plough 165
Pluto 126, **127**, 136, **145**, 159
    discovery of 145
    orbit of **127**
Pointers **165**
poles, magnetic 57
Pole star 162
pond skater 14
pores 124
positron **47**
potential difference 50
potentiometer 93
power stations, hydroelectric **14–15**
nuclear 28
pressure, air 17
principal focus 66
prism 42, **64**
prominences **125**
proper motion 169
protons 20, 124
    Big Bang and 194
    electrical charge of 22, **22**, 47
Proxima Centauri 160
pulsars 100, 180
    Crab 180
pumice 10
Pup (*see* Sirius B)
pure research 111

Q

quarks 22, **22**, 24, 121
quartz 12
quasars **82**, 161, 188–189, 192, 195
    double **189**

R

radar aerial **75**
radial motion 169
radiation 74–75, 124, 138, 194, 195
    alpha 26, **26**
    beta 26, **26**
    gamma 26, **26**
    heat 42
    fireball 195
radiation detectors 207
radiative zone **125**
radio, CB **75**
radioactivity 26–27
radiocarbon dating 26, **27**
radiography 114–115
radioisotopes 115
radio telescope 106–107, **106**, **107**
    Very Large Array **106**
radio waves 74, 106, 182, 186
railway tracks, expansion of **13**
rainbow **71**
rangefinders 89
rays, cosmic **24**
reaction control system (RCS) 217
reactions, chemical 38–39, **38**, **39**
real images 66, 67
Reber, Grote 107
red giants 170, 171, 176, **178**, 179, **179**
Red Planet (*see* Mars)
red shift 89, 169, 193
reflection 65, **65**
    law of 65
    and rainbows **71**
    of sound 45
reflector 101, 102–103, **102**, **103**
refraction 65, **65**
    and rainbows **71**
    of sound 45
refractor 101, 102–103, **102**, **103**
research 111
resistance 51, 52
resistance thermometer 93
retrorockets 213, 215
Rhea 158
Rho Ophiuchi **173**
rifle, Royal Enfield **88–89**
rilles 152
ringlets 142
rings 142
Romanenko, Yuri 228
Rontgen, Wilhelm 114
Roque de los Muchachos Observatory 103, 104, **104**
RTGs 205
Russell, Henry 170
Rutherford, Ernest 21

S

Sagittarius **167**
Sails (*see* Vela)
salt **35**, **37**, 53
Salyut 228, 229, **229**
sand-glass **85**
satellite moons 152–159
satellites 198, 199, 218
    communications 200, 218
    Earth-survey 202, **202**, 203
    weather 201, **201**
Saturn 126, **127**, 136, **136**, 140–143, **141**, **143**, 205, 206, 207, **207**
    moons of 158
Saturn V rocket 208, **210**, 211
scales **85**, 86
Schiaparelli, Giovanni 128

238

Schmidt reflector **103**
Schmitt, Harrison 155
Sea of Clouds 155
Sea of Crises 155
Sea of Moscow 155
Sea of Showers 155
Sea of Tranquility **152**, 210
seismic surveying 117
semiconductors 54–55
  n-type 54
  p-type 54
separation methods 112
service module 211
sextant **85**
Seyfert, Carl 186
Shepard, Alan 209
shepherd moons 142, 143, 145, 206
shooting stars (see meteors)
shuttle suit 224, **225**
sidereal day 162
sidereal time 162
siderites **151**
siderolites **151**
Siding Spring Observatory 104, **104**
silica 217
Sirius 164, 168, 206
Sirius B 176
silicon 54
silicon chips 54, **54**, **55**, 105
silicon wafers **54**, **55**
Skylab 123, 226, 228, 230–231
  health aboard 231
  solar panels on 230
Slayton, Donald 212
Slipher, Vesto 193
Small Magellanic Cloud 182, 184, **185**, 190
smoke 16
Socorro 106
sodium chloride (see salt)
sodium hydroxide 53
Solar Max satellite 218, **218**, 226
Solar System 122–159
  debris of 146–151
  moons of 152–159
  planets of 126–145
  spots and flares in 124–125
  Sun in our 123
solenoid 59, **59**
solids 10–11, **11**, 12–13
  change of state in **11**, 13, 18, 19, **19**
  heating of 43
Solovyov, Anatoli 226
solubility 15
solvents 15
sonar 117
sound 40, 44–45
sound-to-light 70
sound waves 44, 45
Southern Cross **174**
South Pole 57
Soyuz 208, 212–213, **213**
Soyuz 11 229
space adaptation syndrome 223, 232
Spacelab 218, 232–233
space medicine 222, 231
space probes 130, 132, 133, 134, 138, 142, 156, 198, 204
Space Race 213
Space Shuttle **78**, 160, 208, 214–215, 216–217, 232
space stations 228–235
  eating aboard 222
  exercise aboard 223
  shower on **228**
  sleeping aboard **223**
spacesuits 224–225, **224**, **225**
space telescopes 108–109, **109**
spacewalking 226–227, **226**, **227**
  first **226**
spectral class 170
spectroscope 105, 167
spectroscopy, absorption 113
  emission 113
  mass 113
  nuclear magnetic resonance (NMR) 113
  X-ray diffraction 13
spectrum **64**
  electromagnetic 74–75
  stellar 167, **167**, 188
Spica 167
spiracle **94**
spokes 142
SPOT satellite 202, **202**
Sputnik 1 200, **200**
Sputnik 2 220
Stafford, Thomas 212
stars 167

birth of 176–177
brightness of 164, 168
chemicals between 174
death of 176, 178–179, 180–181
life cycle of **178–179**
mapping of **163**
movement of **162**, 169
neutron 10, **170**, 180
size of 168–169
speed of 169
temperature of 167
variable 172
stimulated emission 73
stony meteorites 150, 151
stony-iron meteorites 150, 151
Sturgeon, William 60
sublimation **19**
sulphuric acid 130
Sun **17**, 122, **122**, 160, 170, 171, 179
  orbit of 182
  size of 123, **123**
  structure of **125**
sundial **85**, 90
sunspots 124, **125**
superconducting coils 60
supergiants 170, 171, **178**, 179, 180
supernova star 1987A **109**, **178**, 179, 180, **180**
surface tension 14
Sword Handle (see Perseus)
symbols, chemical 38
synchrotrons 120, **120**, 121, **121**
Syrtis Major 133
Systeme International (SI) 85

T
tau 24
Taurus-Littrow Valley **155**
tektites 151, **151**
telescopes 100–109
  early 101
  optical 100
  radio 106–107, **106**, **107**
  space 108–109, **109**
  types of 103
temperature 41
terminal velocity 83
Tethys **141**, 158
Tevatron 120, **120**, 121
Tharsis Ridge 134, **134**
Thematic Mapper 202
therapy, gamma-ray **74**
thermal images **43**
thermocouple thermometer **41**
thermograph, infrared **75**
thermometers 41, 93
Thermos flask **42**
Thomson, J.J. 21, **21**
Thomson, Sir William 41
Thornton, William **232**
thyroid gland 115
tides 153
time 90–91
  solar 162
  star 162
Titan 158, **158**
Titania 158
titanium alloy 33
Titov, Gherman 221, **221**
Titov, Vladimir 220
Tombaugh, Clyde 145
totality 124
transformer **61**
triangulation 88
triple alpha reaction 179
Triton 159, **159**, 206
Tunguska River 150
tungsten carbide 33

U
ultrasound 40, 116, **116**, 117, **117**
ultraviolet light **74**
umbilical suit 224
umbra 124
Umbriel 158
Unicorn (see Monoceros)
Universe, birth of 194
  closed 196, **197**
  expansion of 192, 193, **193**
  open 196, **197**
  place of Solar System within **161**
  size of 160, 161
uranium **22**, 28, **28**
uranium–235 28
Uranus 126, **127**, 136, **144**, 206, **207**
Ursa Major **164**, 165
UV Ceti 172

V
vacuum flask **42**
Van Allen belts 124, 140
Van de Graaf generator 121
vaporization, latent heat of 19
variables, eruptive 172
Vehicle Assembly Building 211, 214
Vela **166**
Venera 130
Venus 122, **126**, 128, 130–131, 204
Vesta 148
Viking 132, 134, 135
Viking probes 205, **205**
Virgo cluster 190, **190**
virtual images 66, 67, 96
virus **98**
viscosity 14, 15
volcano **11**, 131, 132, 134, 152
Volta, Alessandro 50, 93
voltmeter 92–93
volts 50
Voshkod 1 209
Voshkod 2 226
Vostok 208, **209**
Vostok 1 209
Voyager 1 138, 159, 206
Voyager 2 136, 141, 142, 143, 144, 145, 198, 205, 206–207, **207**

W
watches 90
  digital 91, **91**
water, generating electricity with **15**
  liquid and solid 10, **10**
  surface tension of 14
water clock **85**, 90
Watt, James 88
wave of darkening 133
waves, sound 44, 45
weather 201
weight 86–87
weightlessness 220, 221, 222, **223**
white dwarfs 170, 171, 176, 179, **179**
Whitworth, Joseph 88
Wilson, Charles 24
Wilson, Robert 195
wind 17, 141
work 80–81

X
xenon 94
X-rays **74**, 75, 108, 114, 115, **115**, 181
X-ray tomography 115
X-ray tube **115**

Y
Yerkes Observatory 102
Young, John 214
Young, Thomas 68
yttrium-aluminium-garnet crystals 73

Z
Zelenchukskaya 100, 102
zeolite **39**
zones 137, 138, 140

Picture Credits

b = bottom, t = top, c = centre, l = left, r = right.
AAT Anglo Australian Telescope Board. APL Adams Picture Library, London. CD Chemical Designs, Oxford. ESA European Space Agency. FSP Frank Spooner Pictures, London. GSF Geoscience Features, Ashford, Kent. NHPA Natural History Photographic Agency, Ardingly, Sussex. OSF Oxford Scientific Films, Long Hanborough, Oxford. RHPL Robert Harding Picture Library, London. RF Rex Features, London. ROE Royal Observatory, Edinburgh. SC Spacecharts. SCL Spectrum Colour Library, London. SPL Science Photo Library, London.

10 Images Colour Library, Leeds. 11 SC. 12l Imitor. 12tr Paul Brierly. 12br GSF. 13 Hutchison Library. 14bl Frank Lane Agency. 14tr APL. 14–15 J L Knill. 15 RHPL/J Green. 16 SC. 17l Zefa. 17br SC. 17tr R Kerrod. 20 SPL/Patrice Loiez, CERN. 21 University of Cambridge, Cavendish Laboratory. 22 SPL/Dr Mitsuo Chtsuki. 23 (all pics) CD. 24t SPL/Powell, Fowler & Perkins. 24b University of Cambridge, Cavendish Laboratory. 25 SPL/Lawrence Berkeley Laboratory. 27t FSP/Gamma. 27bl Susan Griggs Agency. 27br British Museum. 28 RHPL/Alan Carr. 29 SPL/U.S. Army. 30 Bridgeman Art Library. 31l Equinox Archive/Institute of Archaeology. 31r GSF. 32t SCL. 32b Popperfoto. 32–33 British Coal. 33tl RHPL/Ian Griffiths. 33tr Zefa/Croxford. 34l SPL/Manfred Kage. 34r SPL/Prof. Edwin Mueller. 35 SPL/Dr J Burgess. 36, 37 CD. 38 Zefa. 39l Topham Picture Source. 39tr Brian & Sally Shuel. 39cr CD. 40 SCL. 42 SC/R Kerrod. 43t SPL/Dr R Clark & M R Goff. 43b APL/Gerard Fritz. 45 SCL/J Bradbury. 46 Vautier de Nanxe. 47 SPL/Lawrence Berkeley Laboratory. 48l SPL/John Howard. 48r Minolta. 49 Zefa/T Ives. 50 SPL/David Taylor. 51 ESA. 52b Tim Woodcock. 52t Susan Griggs Agency. 53 RHPL. 54l Art Directors. 54r OSF/Manfred Kage. 55 SPL/David Parker. 56 SPL/U.S. Department of Energy. 57 SPL/Vaughan Fleming. 60l SPL/Alex Bartel. 60r SPL/Pacific Press Service. 61 Zefa. 63 Hutchison Library/R Aberman. 64 SPL/David Parker. 65 R Kerrod. 66 Perkin Elmer. 67l John Watney. 67r R Kerrod. 68l Paul Brierly. 68r SPL/Martin Dohrn. 69t ACE Photo Agency/Jerome Yeats. 69b SPL/Jeremy Burgess. 70 Zefa/K L Benser. 71 SPL/Phil Jude. 72 SPL/Hank Moran. 73l Hughes Research Lab, Malibu. 73r SPL/Alex Tsiaras. 74t SPl/N M Tweedie. 74c SPL. 74bl SPL/Martin Dohrn. 74br SCL. 75t Institute of Electrical Engineering. 75ct SPL/David Parker. 75cr SCL. 75bl SPL. 75br SPL/Martin Dohrn. 76 Jerry Young. 77 RHPL/A Carr. 78t FSP. 78b SCL. 80 SPL/Andrew Clarke. 81t Zefa. 81c Hutchison Library/R House. 81bl Bridgeman Art Library. 81bc David Redfern/Stephen Morley. 81br Zefa/F Damm. 82 SC/NASA. 83t ACE Photo Agency. 83b National Maritime Museum. 84 The Ridgeway Archive. 86t RF. 86c Sinclair Stammers. 86b Mark Fiennes/Science Museum. 87 Brian and Cherry Alexander. 88–89 Royal Ordnance Factory. 88bl Clyde Surveys Ltd. 88br SPL/D Luria. 90 The Ridgeway Archive. 92 Thorn EMI. 93 SPL/D Parker. 94 K Wheeler. 95c, 95b Ann Ronan Picture Library. 95r Michael Holford. 96l SPL/Sinclair Stammers. 96r OSF/D J Patterson. 97l, 97r Paul Brierly. 98 SPL/CNRI. 99bl SPL. 99br SPL/M Serraillier. 100 R Kerrod. 101tl Biblioteca Nazionale Centrale/G Sansoni. 101b The Royal Society. 101r Instituo e Museo di Storia della Scienza. 102l R Kerrod. 102r, 103 SC. 104t SC/Royal Greenwich Observatory. 104b R Kerrod. 105l Anglo Australian Telescope Board. 105r SC/Royal Greenwich Observatory. 106l Art Directors. 106r SPL/Dr R J Allen et al. 107 Bell Labs/Photo Files. 108l SC. 108r Iain Nicholson. 109 SC. 110 SPL/P Loiez, CERN. 111l Gesellschaft Liebig Museum. 111r SPL/Geco UK. 112 AEI/John Price Studios. 113tl, 113cl Pye-Unicam Ltd. 113tc SPL/Geco UK. 113cr New Methods Research Inc. 113b Zefa/Stockmarket. 114 AERE Harwell. 114–115, 115 Phillips Industrial X-Ray, Hamburg. 116t, 116 inset CEGB. 117 SPL/NASA. 118t SPL/J Walsh. 119b SPL/Lawrence Livermore Lab. 120–121b Fermilab. 121tl Michael Freeman. 121tr SPL/H Schneebeli. 121cl SPL/J Collumbet. 121b SPL. 122 SPL. 123l SPL/NASA. 123r High Altitude Observatory/NCAR/National Science Foundation, USA. 124 SPL/J Finch. 128 US Geological Survey, Flagstaff/A McEwan. 129, 130t NASA. 130b Sovfoto. 132, 132–133 NASA. 133c SPL/NASA. 134, 135t NASA. 135c, 135b Washington University/NASA/Mary Dale Bannister. 136, 138 NASA. 139 SPL/NASA. 141t, 141b, 142l, 142r, 143t, 143c, 143b NASA. 144l SC. 144r R Kerrod. 145t SC. 145b Patrick Moore/CCD Hawaii. 146 James H Baker. 147t NASA. 147b R Kerrod. 149l SPL/Dr Klingelsmith/NASA. 149r SC. 150–151 SPL/Jerry Mason. 150b Sovfoto. 151cr, 151bc, 151bl Paul Brierley. 152 SC/NASA. 154 SPL/Dr F Espenak. 155l, 155r, 156tl, 156bl NASA. 157tr, 157bl, 157br, 158t, 158b NASA. 159t SC/NASA. 159b SC. 160 ROE. 162 SPL/Robin Scagell. 164 SC. 165tl, 165tc Hatfield Polytechnic Observatory. 165b Mansell Collection. 166 ROE. 167l Kitt Peak National Observatory. 167r Hansen Planetarium. 169l Smithsonian Institution. 169r Planetarium, Co. Armagh/AAT. 169c RAS/Hale Observatory. 172 Harvard College Observatory. 173 ROE. 174 SPL/Dr E I Robson. 174–175 ROE. 176 Planetarium, Co. Armagh/AAT. 177bl University of Massachusetts. 177cr, 177br SPL. 179l Planetarium, Co. Armagh/AAT. 179r SPL/U.S. Naval Observatory. 180c, 180bl AAT. 180br California Institute of Technology and Carnegie Institute of Washington. 182 Planetarium, Co. Armagh/AAT. 183t SPL/Mt Palomar Observatory. 184 Henry E Huntington Library. 185t SPL/Dr Jean Lorre. 185b ROE. 186 SPL/Dr Steve Gull. 187t Planetarium, Co. Armagh/AAT. 187b SPL/B Cooper & D Parker. 189t SPL/Dr A Stockton. 189bl, 189br SPL. 190l SPL/Allen. 190r ROE. 192 Planetarium/AAT. 198 SC. 200–201t FSP. 200–201b SC. 202l New Scientist/SPOT. 202r Sinclair Stammers. 203 EISCAT. 204 SPL/Max Planck Institut für Aeronomie/D Parker. 204–205 NASA. 206l, 206b SC/NASA. 207cr, 207b SC. 208, 209 SC/NASA. 210, 211 SC. 212, 213t SC/NASA. 213b NASA. 214 SC/NASA. 215 SC. 216 SC/NASA. 217 SC. 218t SC/NASA. 218b, 219 SC. 220 SC/NASA. 221l, 221r SC. 222, 223t SC/NASA. 223bl, 223br, 224 SC. 225 Image/Anthony Wolff. 226l SC. 226r NASA. 227t SC/NASA. 227b SPL/NASA. 228 SC. 230l SPL/NASA. 230r, 231 SC. 232–233 ESA. 232b, 233bl SC/NASA. 233br, 235 SC.